Complex Population Dynamics

Complex Population Dynamics: Theory and Data brings together over two decades of research and reflection on experimental nonlinear population dynamics. The broad theme of the book is the interface of population data and mathematical models. The authors establish a cornerstone example of a low-dimensional mathematical population model for which quantitative predictions, often unexpected, were borne out in controlled and replicated experiments. Central messages include the importance of low-dimensional, mechanistic models that serve as testable hypotheses; the importance of model validation on independent data; the prediction of novel outcomes in response to parameter manipulation; the interaction of nonlinearity and stochasticity—and how stochasticity illuminates, rather than obscures, deterministic forces; abrupt transitions in dynamic regimes in response to interventions; and how chaotic dynamics are expressed in discrete-state, noisy population systems. The book explores nonlinear phenomena in experimental data, including equilibria, cycles, bifurcations, invariant loops, multiple attractors, resonance and attenuance, saddles, stable and unstable manifolds, basins of attraction, basin boundaries, and chaos, and shows how these dynamics manifest in real data in both time series and state space plots.

The book offers an invaluable resource to professional ecologists and applied mathematicians. Although primarily a reference text, it is written to be accessible and engaging to a student audience and could be used as supplementary reading in an advanced ecological modeling class.

Features

- Written in an engaging style with minimal mathematical prerequisites, making it accessible to an audience of general ecologists as well as a more specialized mathematical audience
- A complete accompanying data set from all experiments is freely available on Dryad
- Readers can conveniently explore the model dynamics under discussion with an online LPA simulator.

Chapman & Hall/CRC Mathematical Biology Series

Series Editors
Ruth Baker, Mark Broom, Adam Kleczkowski, Doron Levy, Sergei Petrovskiy

About the Series
This series aims to capture new developments in mathematical biology, as well as high quality work summarizing or contributing to more established topics. Publishing a broad range of textbooks, reference works, and handbooks, the series is designed to appeal to students, researchers, and professionals in mathematical biology.

We will consider proposals on all topics and applications within the field, including but not limited to stochastic modelling, differential equation modelling, dynamical systems, game theory, machine learning, data science, evolutionary biology, cell biology, oncology, epidemiology, ecology, and more.

An Introduction to Physical Oncology
How Mechanistic Mathematical Modeling Can Improve Cancer Therapy Outcomes
Vittorio Cristini, Eugene Koay, Zhihui Wang

Introduction to Mathematical Oncology
Yang Kuang, John D. Nagy, Steffen E. Eikenberry

Stochastic Dynamics for Systems Biology
Christian Mazza, Michel Benaim

Systems Biology
Mathematical Modeling and Model Analysis
Andreas Kremling

Cellular Potts Models
Multiscale Extensions and Biological Applications
Marco Scianna, Luigi Preziosi

Quantitative Biology
From Molecular to Cellular Systems
Edited By Michael E. Wall

Game-Theoretical Models in Biology, Second Edition
Mark Broom, Jan Rychtář

Mathematical Modeling in Biology
A Research Methods Approach
Shandelle M. Henson, James L. Hayward

Complex Population Dynamics
Theory and Data
Shandelle M. Henson, Robert A. Desharnais, J. M. Cushing, Brian Dennis, R. F. Costantino

For more information about the series, visit: https://www.routledge.com/Chapman--HallCRC-Mathematical-Biology-Series/book-series/CRCMBS

Complex Population Dynamics
Theory and Data

Shandelle M. Henson, Robert A. Desharnais, J. M. Cushing, Brian Dennis, and R. F. Costantino

CRC Press
Taylor & Francis Group
Boca Raton London New York

CRC Press is an imprint of the
Taylor & Francis Group, an **informa** business
A CHAPMAN & HALL BOOK

Designed cover image: Shutterstock, with additional design elements by Robert A. Desharnais

First edition published 2026
by CRC Press
2385 NW Executive Center Drive, Suite 320, Boca Raton FL 33431

and by CRC Press
4 Park Square, Milton Park, Abingdon, Oxon, OX14 4RN

CRC Press is an imprint of Taylor & Francis Group, LLC

ISBN: 978-1-032-73659-4 (hbk)
ISBN: 978-1-032-73987-8 (pbk)
ISBN: 978-1-003-46712-0 (ebk)

DOI: 10.1201/9781003467120

Typeset in CMR10 font
by KnowledgeWorks Global Ltd.

Contents

Preface

Population numbers fluctuate, often erratically. Prior to the seminal work of Lord Robert M. May in the 1970s, ecologists viewed such fluctuations as random effects due to environmental noise and/or measurement error. According to this view, population numbers would either equilibrate or settle into regular cycles in the absence of noise. May broke the news to ecologists that simple deterministic processes can cause erratic fluctuations. He popularized within ecology the now familiar dynamic of *chaos*, an extraordinarily complicated phenomenon that can arise from simple one-dimensional dynamic rules. Recognizing that chaos can be difficult to distinguish from noise, May warned that documenting deterministic chaos in populations would not be easy. Indeed, in the intervening years many ecologists concluded that chaos is rare or nonexistent. While statistical detection methods applied to time series have suggested that chaos may be more widespread than previously thought, experimental evidence of chaotic dynamics is difficult to obtain for natural systems. The more general problem of connecting nonlinear theory to observed population fluctuations remains a challenge.

Although mathematical models are quantitatively descriptive and predictive in other disciplines, they have met with less success in ecology. The ubiquitous presence of noise, a lack of fundamental "laws," the paucity of long time series, difficulties of replication and manipulation, and inadequate statistical methodologies have made "hard science" approaches difficult in ecology. Because of this, ecologists have been skeptical, and rightly so, of work concluding the existence of specific nonlinear phenomena in population fluctuations—especially complicated phenomena such as chaos.

These difficulties point to the need for controlled laboratory experiments designed to test the predictions of nonlinear population theory. This book's authors (also known as the "Beetle Team") began our collaboration in the 1980s, representing a spectrum of disciplines, from experimental ecology to applied mathematics to applied and theoretical statistics. We used flour beetles (genus *Tribolium*) as the experimental animal. The *Tribolium* system features accurate census counts, long data sets (over many generations), the minimization of noise, and a life cycle with identifiable stages and interesting nonlinear feedbacks. We focused on biological mechanisms and embraced stochastic structure. Our goal was to establish a cornerstone example of a low-dimensional mathematical population model for which quantitative predictions were borne out in manipulative experiments. We were able to identify and predict, convincingly we think, many types of nonlinear dynamics in experimental population data, including equilibria, cycles, bifurcations, resonance, aperiodicity, saddles, stable and unstable manifolds, and multiple attractors. The documentation of chaotic dynamics in a biological population became a central highlight of the project.

While the data used in this book come from experiments with flour beetles, from a broader and deeper point of view, the book is not about flour beetles per se, but about the interface of population data and mathematical models. It is important to distinguish between parameterizing (calibrating) a model and validating a model [1]. The goal is to obtain confidence not only in the model's ability to fit data, but also in the model's ability to predict data. Prediction is, of course, at the heart of the scientific method. In his book on methods of ecological research, E. D. Ford writes [2]:

> Model assessment against data provides a test of a model's effectiveness. Three levels of assessment can be made for complete models: fitting, predicting, and revealing different results. Fitting is not a strong assessment criteria—yet it can be difficult to achieve and when it is achieved there has to be understanding of how that was done. ... Prediction is often considered as more valuable than fitting and is widely used in both statistical modeling and systems simulation as validation. ... Revealing results of a different kind is the strongest assessment, i.e., that the model predicts something not expected and when searched for through more research, is found to occur.

In this book the reader will find numerous projects, concerning a variety of dynamic phenomena, that accomplish the three criteria described in this quote: model parameterization, model prediction, and the revelation and prediction of new, sometimes unexpected, results (validated by controlled and replicated experiments). This includes, but is not limited to, the highlight phenomenon of chaotic dynamics. As Ford goes on to say, "If chaos is found in ecological systems then it will be an example of the third level of model assessment, where a model explains an unexpected result." However, incorporating the chaotic behavior predicted by models into ecological theories "is likely only if fitting and/or predicting have already been attained."

This book pulls together more than two decades of our experimental and analytical studies of nonlinear population dynamics. The original papers were published in numerous journals across several disciplines. We felt that the entire body of work, moderated by a subsequent decade of our discussion and reflection, should be presented to a new generation of ecologists in one package.

We intend this book for ecologists in general, and not just theoreticians. Although we showcase rigorous research, we also begin each chapter (as well as the entire book) with intuitive motivations and explanations. Likewise, each chapter ends (and the entire book ends) with an intuitive recapitulation of the material, along with a discussion of why it matters in ecology. In this way, we have endeavored to speak to the entire community of ecologists in an accessible way.

A comprehensive collection of data sets for the studies presented in this book is freely available on Dryad at https://doi.org/10.5061/dryad.qjq2bvqmp. Furthermore, an online LPA model simulator is available at https://lpasim.org/ which can be used to examine interactively the dynamics of the LPA model. URLs to preloaded examples are available at the end of several sections in the book. These features aid accessibility and further exploration by the reader. Although this book is a research text rather than a textbook, it could be used as supplemental reading in a course on ecological modeling.

The book is organized into five parts. Part I contains three chapters of introductory material. Readers familiar with these topics can skip some or all of these chapters and go directly to Part II. Chapter 1 is a brief intuitive introduction to the major themes. Chapter 2 provides a philosophical framework for our approach to modeling and stochasticity. Chapter 3 gives an introduction to the mathematics of dynamical systems used in this book. Readers thoroughly familiar with discrete-time models, linearization, stability, saddles, aperiodicity, and bifurcations likely will wish to skip or skim this chapter.

Part II contains two chapters specific to our animal model, the flour beetle *Tribolium castaneum*. These chapters present information important for understanding the rest of the book, and include the life history of flour beetles, laboratory protocols, and the construction of our mechanistic model.

Part III, in some ways the most important section of the book, focuses on mathematical models as testable scientific hypotheses in the context of complex dynamics. *Here we present the experimental confirmation of transitions (bifurcations) between attractors, and we document experimentally a model-predicted route to chaos.* In addition, we consider lattice effects and the anatomy of chaotic attractors, and show how these are manifest in chaotic experimental data. Also in Part III, we confirm empirically the inverse square root rule for demographic stochasticity in complex dynamics—that is, we show how increased population size tightens the data around a chaotic attractor—and we present an experimental documentation of controlling chaos.

Part IV presents three chapters on recognizing nonlinear phenomena in population data. We show how the deterministic architecture of state space—that is, the deterministic structure imparted by stable and unstable invariant sets, including saddles with their stable and unstable manifolds, and multiple attractors with their basins of attractions and basin boundaries—is illuminated by stochasticity and can be recognized and documented in experimental data.

Part V contains further exploration of the mathematical-experimental system, including experiments designed to investigate competition, spectral analyses, synchrony in metapopulations, natural selection, and spatial dynamics. We also present a continuous-time model that invites further exploration by readers.

Throughout the book, we use the term "state variable" for the dependent variable, and thus we use "state space" rather than "phase space." We reserve the word "phase" for discussing phases of cycles. We use the term "trajectory" instead of "orbit" for the graphs of solutions in state space. Other terms and notations are defined within each chapter to enhance readability.

We acknowledge our collaborators on the work discussed in this book: Aaron King, Jeffrey Edmunds, Suzanne Robertson, Dan Reuman, Joel Cohen, Serj Danielian, Lev Ginzburg, Hal Caswell, Charlotte de Vries, and Warren Cheung. We also acknowledge Rosalyn Rael and Tom Vincent, our collaborators on projects related to the material in this book but not directly included herein. Finally, we are grateful for the support of the U. S. National Science Foundation. Award numbers can be found in the original publications cited.

We hope this body of work, presented here for the community of ecologists, will prove to be an accessible, instructive, and convincing demonstration of the value of nonlinear concepts and low-dimensional, mechanistic, mathematical models in ecology.

Bibliography

[1] Aber JD (1997) Why don't we believe the models? Bulletin of the Ecological Society of America 78:232–233. http://www.jstor.org/stable/20168170

[2] Ford ED (2000) Scientific Method for Ecological Research. Cambridge, Cambridge University Press. pp. 390–391.

About the Authors

Shandelle M. Henson is an emeritus Professor of Mathematics and Professor of Ecology at Andrews University. She uses dynamical systems theory and bifurcation theory to study nonlinear population dynamics. She also studies the effects of climate change on the behavior of marine organisms.

Robert A. Desharnais is an emeritus Professor of Biological Sciences at the California State University, Los Angeles. He conducts both theoretical and experimental research focused on nonlinear models in ecology and population genetics. He has also collaborated with colleagues at Cal State LA to develop and test spatially-explicit models of marine mussel beds. In addition, he develops innovative web-based applications for science education that are widely used in high schools, colleges, and universities.

J. M. Cushing is an emeritus Professor at the University of Arizona, affiliated with the Department of Mathematics and the Interdisciplinary Program in Applied Mathematics. He uses dynamical systems theory and bifurcation theory to study nonlinear population, ecological, and evolutionary dynamics. He has authored and co-authored over 180 journal articles and 7 books. He is a Fellow of the American Mathematical Society and winner of the 2021 Bernd Aubach Prize for contributions to discrete dynamical systems and their applications.

Brian Dennis is an emeritus Professor of Wildlife and Statistics at the University of Idaho, Moscow, ID. He received his master's degree in statistics and PhD in ecology at Pennsylvania State University (G Evelyn Hutchinson is his professional grandfather). His research and teaching career has been devoted to improving ways in which ecological models could be connected to ecological data, and to studying how stochastic modeling could enhance ecological theory. His scientific papers have been cited over 10,000 times; recently, two of his papers were selected by a panel of ecological editors for inclusion in volume 2 of the Foundations of Ecology series.

R. F. Costantino is an emeritus Professor of Zoology from the University of Rhode Island. His research interests are experimental and theoretical ecology. He designs and conducts laboratory tests of nonlinear dynamics theory. He also teaches honors courses in nonlinear dynamics at the University of Arizona.

I

Introduction

CHAPTER 1

Sources of Patterns in Population Dynamics

In this chapter, which is presented from an intuitive point of view, we ask whether fluctuations in population data contain patterns and discuss the sources of such patterns. We address nonlinear biological mechanisms, the number of state variables that should be tracked, the idea of signal vs. noise, and the structure of noise. We consider the purpose and meaning of mathematical models, and the role of laboratory studies in identifying patterns.

1.1 NONLINEAR MECHANISMS CREATE PATTERNS IN DATA

Systems are sets of interconnected objects. *Dynamical systems* are those that change through time. All real systems, including population systems, change on some time scale and are hence dynamical systems. Whether population fluctuations exhibit dynamic patterns is clearly of great interest, given that we would like to understand the mechanisms behind any such patterns and also gain the ability to predict into the future. *State variables* (dependent variables) are those aspects of the system that we measure and follow through time in order to look for patterns. Common state variables in ecology include population abundance, population density, and the number of animals in a particular life cycle stage.

In Figure 1.1 we see two examples of patterns in population data. Figure 1.1A shows numbers of flour beetle (*Tribolium castaneum*) larvae in a laboratory population exhibiting an up-down cyclic pattern when cultured in a constant habitat [6]. In Figure 1.1B we see a smooth growth curve, perhaps sigmoidal, for the number of active bald eagle (*Haliaeetus leucocephalus*) territories in Washington State, USA, as their populations have recovered following the banning of Dichlorodiphenyltrichloroethane (DDT). We also see a concomitant initial rise and then decline in the number of glaucous-winged gull (*Larus glaucescens*) territories at a Washington seabird colony [4]. The patterns in Figure 1.1 are obvious.

By contrast, patterns are not as readily apparent in the time series shown in Figure 1.2, which, like Figure 1.1A, shows the number of *Tribolium* larvae cultured in a constant habitat but in another experiment [2, 3]. The peculiar "flat" episode marked with the arrow might be particularly puzzling [5].

What are we to make of these examples? One challenge in identifying patterns is that we are wired to "see" them even in sequences of random events; hence it is necessary to determine whether an observed pattern is likely to occur by chance alone. Another challenge is that of mechanism. If a significant pattern does exist, can we provide a mechanistic explanation for it? And how are mechanisms quantified and tested?

In Figure 1.1, the mechanisms are, in fact, explicable. Cycles of period two in *Tribolium* time series such as those in Figure 1.1A are explained rigorously by the mechanism of

DOI: 10.1201/9781003467120-1

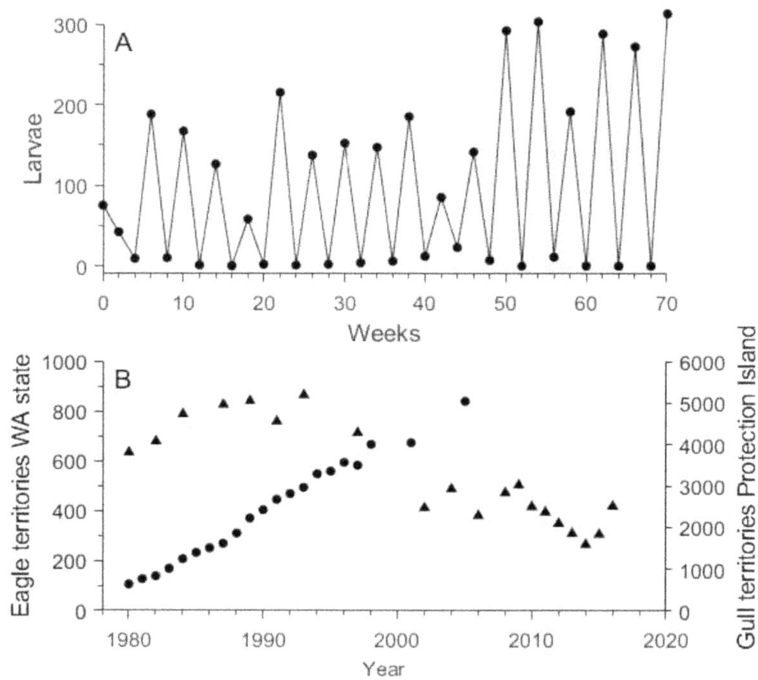

Figure 1.1 Patterns in time-series data. A. Experimental *Tribolium* larval data; control replicate 13 from [6]. B. Numbers of bald eagle territories (solid circles) in Washington State, USA, and numbers of gull territories (solid triangles) at a Washington seabird colony [4].

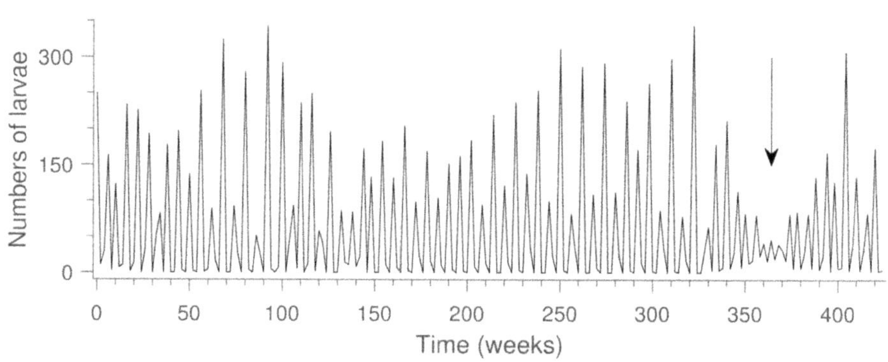

Figure 1.2 Time-series data for *Tribolium* larvae cultured in a constant habitat; replicate 13 from [2, 3]. Note the peculiar "flat" episode (arrow).

cannibalism of eggs by larvae and adults [1]. We will address that particular mechanism in detail in this book. The patterns in Figure 1.1B were explained as a predator-prey interaction [4]. Figure 1.2 appears to have either no pattern or perhaps a mixture of episodic pieces of patterns; or are these only imagined? This book explains the methods by which complex patterns such as those in Figure 1.2 can be understood.

Perhaps a larger question is, should we even expect population data to show patterns? Certainly, field data will be expected to show correlations with environmental drivers. But are there endogenous patterns in population dynamics, arising merely from the biological mechanisms of intraspecific processes, that would persist under controlled circumstances in which environmental perturbations are minimized?

If indeed there are a few major mechanisms involving a small number of state variables that largely drive the dynamics of a population in a controlled environment, then we should be able to detect their signature in data from uncontrolled circumstances as well, although the dynamic patterns may be quite noisy. If there are no major endogenous mechanisms, then beyond the study of environmental correlates, the quest to understand population numbers, and perhaps ecology, seems quixotic.

But surely there are dominant mechanisms and patterns in nature that can be detected if one has an idea what to look for in data. This brings us to the idea of dimension.

1.2 DIMENSIONALITY

As ecologists well know, many mechanisms are at work in populations. *The goal, however, is to discover a few major mechanisms that drive most of the variability in the system.* The intrinsic dynamic quantities involved in these mechanisms are the state variables that should be tracked and measured. The concept of *dimensionality* refers to the number of state variables required to explain most of the variability in the system. A *low-dimensional system* is one that can be described with a small number of state variables.

Some modelers use high-dimensional models in order to bring more realism to simulations. One problem with assuming high dimension is: A system too complicated to understand is being replaced by a model too complicated to understand. In the same way that a road map shows only features of major importance for navigation, so a model should incorporate only a few major mechanisms responsible for the generation of dynamic patterns.

Another problem with high-dimensional models is the large amount of data required. Models typically have constants, called *parameters*, or model coefficients, with unknown values that must be estimated. High-dimensional models have large numbers of parameters, and parameters are gluttonous consumers of data. In addition, each parameter estimated contributes its uncertainty of estimation to the overall uncertainty of system prediction. Too many parameters risks a grossly uncertain model of limited usefulness.

The scientific art of modeling is a careful balancing act: What minimum amount of complexity in the model is needed to capture and explain most of the variability in the system? Virtually everyone would agree that each individual deer in a population has its own unique annual survival rate, but is it necessary that we know the survival rate of every individual for useful predictions? Can we get away with grouping the animals into categories, perhaps by ages which have similar survival probabilities, for our prediction goals?

1.3 SIGNAL AND NOISE

Any model, mathematical or otherwise, that simplifies a real system will deviate from observations of that system. This can be due in part to measurement error (count error in the context of populations), but let us assume for the moment that there is no measurement error. In that case, we can view the departures of observations from model predictions from two different points of view. We can view them as model errors, arising because the model, although it captures the core behavior of the system, does not completely capture reality due to leaving out details; or we can view them as process errors, in which the real system itself suffers stochastic perturbations that cause it to deviate from predictions.

In either case, we can view the core behavior of the system as the deterministic *signal* and these errors as *noise*. The goal is to model the deterministic signal by capturing the mechanisms that generate it and to model the noise with appropriate stochastic structures.

Many systems in classical physics do not have much noise in comparison to the signal. In fact, the mathematical equations used to model these systems often are called "laws" because of their accuracy. In contrast, some systems in the social sciences may be dominated by noise. Ecology falls somewhere between these extremes. *Both signal and noise are important in population dynamics.* Furthermore, and most notably, *signal and noise can interact with each other in important ways*—a central theme in this book.

It is worth mentioning two more points about stochasticity. First, it has structure: There are types and colors of noise, and they interact with the signal in different ways. Second, whether one views noise as high-dimensional deterministic chatter that does not systematically affect the main signal, or as variability arising from random perturbations, testable hypotheses should underlie the way noise is incorporated into models. For example, the noise may be due primarily to environmental perturbations, or it may be due largely to demographic differences among individuals in the population. Each of these types of noise must be treated differently, as we will explain in this book.

1.4 MATHEMATICAL MODELS

Mathematical models as envisioned in this book are equations or systems of equations involving a small number of measurable state variables, mechanistic interactions between state variables, and a limited number of parameters. That is, we are talking about low-dimensional, mechanistic mathematical models, in which the main mechanisms hypothesized to generate the observed signal are encoded into the precise language of mathematical expressions.

There are two complementary ways to approach mathematical modeling. The first is with simple proof-of-concept models, in which the outcomes of one or two mechanisms of interest are analyzed theoretically in isolation, without attempting to achieve biological realism. The second approach is to build a model that is tied rigorously to data and captures the main dynamics of the observed system. The first approach is supportive of the second, and both types of models are discussed in this book.

Mathematical models that are meant to be realistic should serve as testable hypotheses that can be confronted with data. Furthermore, one can pose *alternative models* as alternative hypotheses in order to select the best model.

Model *validation* is important. The selected model should be validated on independent data sets to which it was not fitted, preferably data from novel situations. All of these topics will be addressed in detail in this book.

1.5 A MARRIAGE OF THE THEORETICAL AND EXPERIMENTAL FOR THE SAKE OF THE APPLIED

Theoretical, empirical, and applied ecologists—do they talk? In this book, we endeavor to integrate theory and experiment in a way that speaks to applied ecologists in useful ways.

Some ecologists have had a fraught relationship with the laboratory. How could a system so complicated be studied in such a controlled environment? Is not the very complexity of so many interactions on so many scales the essence of ecology?

Likewise, many are puzzled over the role of mathematical models. Do they ever really tell us anything about actual systems? Has a theoretical concept such as equilibrium or limit cycle ever explained and predicted the behavior of a real population? Are not mathematical models a step further from reality than even the laboratory? Isn't a mathematical model merely a tiny laboratory of the mind in which we play games and prove theorems unrelated to the real world?

We take the point of view that the introduction of systematic controlled perturbations is the primary way to infer causal mechanisms responsible for population fluctuations. This is exactly the purpose of the experimental laboratory when feasible. In the words of E. O. Wilson [7],

> When observation and theory collide, scientists turn to carefully designed experiments for resolution. Their motivation is especially high in the case of biological systems, which are typically far too complex to be grasped by observation and theory alone. The best procedure, as in the rest of science, is first to simplify the system, then to hold it more or less constant while varying the important parameters one or two at a time to see what happens.

Furthermore, causal mechanisms are best described mathematically, requiring a clear understanding of the state variables and interactions involved. In this book, we show how mathematical models can serve as scientific hypotheses and be connected rigorously to data. We demonstrate that models can describe, explain, and predict dynamics, not only within a circumscribed set of conditions in which the model is attached to the data, but also in novel settings.

1.6 CONCLUSIONS

Our work, presented in this book, demonstrates how theory and experiments can be integrated. Although we are not applied ecologists, we suggest that the nonlinear phenomena described in this book—predicted by mathematical models and observed in the laboratory—are useful concepts for all ecologists and constitute real aspects of free-living populations.

Bibliography

[1] Costantino RF, Cushing JM, Dennis B, Desharnais RA (1995) Experimentally induced transitions in the dynamic behavior of insect populations. Nature 375:227–230. https://doi.org/10.1038/375227a0

[2] Costantino RF, Desharnais RA, Cushing JM, Dennis B (1997) Chaotic dynamics in an insect population. Science 275:389–391. https://doi.org/10.1126/science.275.5298.389

[3] Dennis B, Desharnais RA, Cushing JM, Henson SM, Costantino RF (2001) Estimating chaos and complex dynamics in an insect population. Ecological Monographs 71:277–303. https://doi.org/10.1890/0012-9615(2001)071[0277:ECACDI]2.0.CO;2

[4] Henson SM, Desharnais RA, Funasaki ET, Galusha JG, Watson JW, Hayward JL (2019) Predator-prey dynamics of bald eagles and glaucous-winged gulls at Protection Island, Washington, USA. Ecology and Evolution 9:3850–3867. https://doi.org/10.1002/ece3.5011

[5] King AA, Costantino RF, Cushing JM, Henson SM, Desharnais RA, Dennis B (2004) Anatomy of a chaotic attractor: Subtle model-predicted patterns revealed in population data. Proceedings of the National Academy of Sciences 101:408–413. https://doi.org/10.1073/pnas.2237266100

[6] Jillson DA (1980) Insect populations respond to fluctuating environments. Nature 288:699–700. https://doi.org/10.1038/288699a0

[7] Wilson EO (2002) The Future of Life. New York, Alfred Knopf.

CHAPTER 2

Connecting Models with Data

In our approach to nonlinear population dynamics, why did we use low-dimensional mechanistic models, given the other choices available? Why is modeling stochasticity important, and how do we identify and infer dynamic behavior? In this chapter, we focus on these issues. We show how prediction for us was not only about forecasting a time series (short and medium term) but was also about predicting *responses to interventions*. This viewpoint is based on the quantitative traditions of theoretical population and community ecology as they arose in the 1900s. Our discussion is descriptive, attempting to explain the "why" of our approach without equations flying about (of which there are no lack elsewhere in this book).

2.1 A BRIEF HISTORY OF MATHEMATICAL MODELING IN ECOLOGY

Since its inception in the early 20th-century Pearl-Lotka-Volterra era, theoretical ecology has hypothesized that temporal and spatial patterns in population abundances might be understood with mathematical models of dynamical systems. Although the exponential and logistic models of population growth go back several centuries (for instance, [4, 33, 52]), it was Raymond Pearl, a biologist who studied biostatistics under Karl Pearson, who forcefully advocated the logistic model as a serious hypothesis and made attempts to fit the model statistically to population data [40, 41]. In 1925, Alfred Lotka, a theoretical chemist who worked as a postdoc in Pearl's laboratory, explained and applied dynamical system theory to biology and social sciences in a famous tome [31]. Vito Volterra, a prominent mathematician in Italy, had a biologist son-in-law who brought quantitative population growth scenarios (predator-prey, competition, rarity) to his attention for mathematical analysis [27]. Mathematical models became the subject of early laboratory population studies [9, 19, 37] and inspired decades of ecological research.

Concurrently, scientists in the closely related field of epidemiology began exploring the usefulness of mathematical models. In 1911, Ross [45] proposed a system of two logistic-style differential equations representing populations of susceptible humans and mosquitoes in a model of malaria dynamics. The susceptible-infected-recovered/removed (SIR) system of three differential equations championed in 1927 by Kermack and McKendrick [26] still yields useful principles of epidemic management.

DOI: 10.1201/9781003467120-2

2.1.1 Density Dependence

Population regulation was at the forefront of ecological thought. That a population cannot grow exponentially forever was obvious, but what is the regulatory process? Are the factors that cull a population largely independent of population abundance, like rainfall and temperature, or are they more dependent on population abundance, such as food and disease?

Early on, ecologists put the logistic model of Verhulst to work, as championed by Pearl [40, 52]. The logistic is a minimalist model of population regulation that formulates the per-unit-abundance growth rate as a simple linear decreasing function of population abundance. In the resulting differential equation, the population growth rate is a quadratic function of population abundance with roots at zero and at a positive stable equilibrium value. The sigmoidal trajectory of population growth, leveling off at a "carrying capacity," appears in virtually every ecology textbook. The model is somewhat problematic in that it lacks biological process, but it can be regarded as an approximation of more detailed models. Chapman [9] interpreted the linear decreasing term in the model as "environmental resistance"; and, indeed, the model can be obtained as a chemostat model in which the resource nutrient is transformed into biomass [11].

The prospect of a lower critical abundance below which a population would be at heightened risk of extinction caught the attention of ecologists. In 1931, Allee [1] qualitatively described biological ways that the per unit abundance growth rate might increase instead of decrease with population abundance, implying a reduced average fitness if population members were scarce. If deaths were higher than births at low population sizes, an unstable equilibrium size would exist below the stable equilibrium for higher population sizes. The phenomenon of an increasing per-unit-abundance growth rate is now termed an *Allee effect*. In 1938, Volterra [55] proposed and analyzed the population model having a quadratic rather than linear per unit abundance growth rate, increasing at first, then decreasing, in response to a biological puzzle posed by his son-in-law. The result is a differential equation with a cubic function of population size as growth rate. In the current era of conservation biology, the cubic model with Allee effect is frequently rediscovered and reanalyzed.

2.1.2 State Space Portrayal

Poincaré's ideas of state space portrayal [42]—that is, the representation of the trajectories of n state variables as a single trajectory in n-dimensional space—first appear in ecology through the work of Lotka. In Lotka's 1920 paper, his namesake predator-prey model is a system of two coupled differential equations as might be used to describe a chemical reaction between two substances in solution [30]. This paper, which is mathematical and without graphs, demonstrated that the model trajectories are sustained oscillations, which, had he graphed them in state space, would have been closed loops in the predator-prey plane. In his 1925 book, Lotka utilized many state space graphs, illustrating stable equilibria (critically damped or damped oscillatory approach), unstable equilibria (direct or oscillatory repulse), neutral cycles (as for his earlier predator-prey model), and limit cycles [31]. Volterra, working independently of Lotka, displayed state spaces and closed-loop trajectories of his namesake predator-prey model in a 1926 paper [53]. Volterra later kindly acknowledged Lotka's priority for the predator-prey model, and the pair exchanged enthusiastic correspondence for some years.

In 1928, Volterra took the lead in modeling two populations in competition for resources [54]. Lotka's 1932 competition paper [32] came later and acknowledged Volterra's work, but contributed new analytical results and presented a single state space graph with isoclines and trajectories for the case where one species always wins. The famous state space diagrams of the four possible linear isocline configurations for competition did not appear until 1935 in a paper by Gause and Witt [21].

2.1.3 Population Biology in the Laboratory

The possibility of testing the predicted dynamic behaviors in laboratory populations, or of at least measuring model quantities, fascinated biologists in the 1920s and 1930s. Yeast, *Paramecium*, fruit flies, and flour beetles were popular model systems. Carrying capacity and the logistic model were special targets. To this day, the graphs from that era of population abundance data with superimposed logistic curves grace our ecology texts, and the model systems are used for hands-on ecology experiments in schools and colleges.

Flour beetles (*Tribolium* spp.) received particular attention due to their ease of husbandry and counting (see Chapter 4). Their population dynamics proved challenging and interesting. In the 1930s, Park and Ford [17, 38, 39] gave nice reviews capturing the pressing questions of the times. Chapman [9] was responsible for launching general interest in *Tribolium* as a population model for studying regulation. Gause [18] analyzed Chapman's data, obtaining one of the now iconic logistic graphs.

By far the gold standard of laboratory population ecology for the time was set by the series of experiments by Gause; the publications were gathered together and described in Gause's famous 1934 book [20]. Here we read of the logistic model in relation to yeast and *Paramecium* cultures, of the Lotka-Volterra predator-prey models in relation to predation of *Paramecium* on yeast, and of predation of *Didinium* on *Paramecium*. Data and model trajectories swirling in state space! Moreover, Gause subjected the Lotka-Volterra model of competition between two species to experimental test: Two species of *Paramecium* grown separately and together produced such elegant pictures of Lotka-Volterra trajectories that ecologists wrote about "Gause's principle of competitive exclusion" for decades. Moreover, the models were carefully discussed from a scientific standpoint, and even the failings of the models often gave new insights into the biological workings in the laboratory microcosms.

In Gause's 1934 book [20], we have physics-style mathematical models phrased as hypotheses, critical experimental tests of the models with multiple species, careful laboratory protocols explained for future replication and development, and (spectacularly) all of the raw data printed in the back of the book. In some ways, Gause's book eventually would be the ideal on which the Beetle Team would strive to model their work.

2.1.4 Theoretical Ecology

After Gause's work, in which experiments were designed around testing the predictions of specific population models, theory and applications in ecology drifted apart. Until the 1980s, quantitative population ecology consisted mostly of the analysis of deterministic models, unconnected to data [27]. These models reflected the physical science and mathematics traditions of their practitioners and took the form of differential equations (as in Lotka's and Volterra's work discussed above), difference equations (such as in the work of Leslie [28] and Ricker [44]), or partial differential equations (e.g., early spatial models of Skellam [46]).

A notable exception occurred in the late 1970s when a (deterministic and low-dimensional) model, parameterized from data, was used to predict what outcomes would occur if various microorganisms were placed together in (exploitative) competition for a single nutrient. These predictions were successfully documented in experiments using laboratory chemostats [25, 47].

Despite the general lack of connection of models to data during this era, the models did provide ecologists with the concepts and vocabulary needed to think more clearly about the pressing problems of animal and plant abundance. For example, the influential book by Robert May [34], a physicist who became interested in ecological questions after meeting ecologist Robert MacArthur, brought clarity to the issues of diversity and stability of ecological communities.

Nevertheless, mathematical ecology became controversial as an entire literature sprang up in which theoretical ecologists proposed models and proved theorems about dynamic behavior. What were field ecologists to make of these mathematical population models? Were they intended as real hypotheses describing the workings of real, particular systems, or were they toys for teaching concepts? One leading skeptic described mathematical ecology as "prim," portraying the subject as one of unearned formalism, brittle to criticism from outside [48].

One problem was the absence of "laws" in ecology that theoreticians could use to build their models. Physical sciences have reliable, well-tested quantitative building blocks as a foundation for deriving predictions and making reliable progress. However, in ecology, the Lotka-Volterra formulations for interacting species, even after the modest laboratory successes, could never be more than a tentative assumption when applied to a new population system.

Furthermore, the theories of ecology cried out for stochastic ingredients which the theorists (however well trained in mathematics) were ill-equipped to supply. What would deterministic dynamical system concepts, such as stable equilibrium, cycles, and multiple attractors look like in the presence of stochastic forces? In fact, the troubling chasm between models and data could not be bridged without stochastic techniques.

2.2 THE QUEST FOR REALISM

After the 1980s, ecology saw a resurgence of studies connecting models and data. These studies, mediated by the massive increase in computational power, have not in general looked like the early low-dimensional approaches discussed above that followed in the tradition of the hard sciences. In this section, we consider briefly some of the methodological and philosophical issues involved in contemporary approaches before discussing our own approach within this context.

At its best, a dynamic mathematical model is a hypothesis about some aspect of a real dynamic system that is vulnerable to empirical testing. The quest for more realism in modeling usually means dividing the system into finer and more numerous state variables. Such division is intended to reflect the heterogeneity in the system. In Lotka's [31] and Leslie's [28] modeling work, populations were divided into age classes to account for heterogeneity in demographic rates. The division into age or stage classes is a major component of theoretical ecology [8, 10].

2.2.1 Agent-based Models

The quest for realism also led to *individual-based models* [23], with the term *agent-based models* being used more often recently, especially in other scientific fields. Agent-based models take advantage of modern-day computing power. For instance, modern weather forecasting models divide the earth into a fine grid and use various physics laws (prominently the Navier-Stokes equation) to calculate the atmospheric changes from each point on the grid to its neighbors [36]. Agent-based models in ecology attempt to incorporate an important source of variability into mathematical accounts of population dynamics: the heterogeneity of demographic rates among individual population members. Agent-based models designate individual demographic rates for each population member, along with interactions between individual population members. Each population member becomes a state variable.

Agent-based models face two difficulties. First, although there are a handful of defined ecological systems in which every individual has been monitored (such as in the Grants' studies of Darwin's Galapagos finches [22]), such data for individuals are not usually available in most ecological systems. The modelers then need to quantify the heterogeneity with assumed probability distributions that might not be well-estimated. Second, the agent-based models, as in the weather forecasting example, need well-tested physics-style "laws" by which the individuals affect each other. Population ecology has a dearth of "laws" other than underlying chemistry and physics. As a result of these two problems, agent-based population models typically serve more in the role of untested (and sometimes untestable) hypotheses.

2.2.2 Hierarchical Models

Hierarchical models are large statistical or machine learning approaches with many parameters that organize systems into nested and layered relationships. Hierarchical models incorporate temporal or spatial variability in model quantities as stochastic variation. The idea is that the variation in, for instance, demographic rates can be handled as "missing data" (latent variables), for which a statistical distribution model can be substituted. Spatial variation in birth rate, for instance, might be addressed by assuming that the organisms at different locales have birth rates drawn from a gamma distribution, with prior distributions assumed for the two gamma parameters. This approach can be used to account for sampling variability in a state variable or for imputing missing data on a state variable, even when the variable is entirely absent from the observations.

Such hierarchical models in the 1970s and 1980s could not be handled by existing Bayesian or frequentist methods. For non-normal hierarchical models, the likelihood function itself involved a multi-dimensional integral. Computing and maximizing the integral were insurmountable problems.[1]

Then, in the 1990's, increased computing power enabled the discovery that observations from the posterior distribution in a Bayesian framework could be simulated using Markov chain Monte Carlo simulations, which bypassed the need to calculate a multi-dimensional integral. Hierarchical models offered a huge jump in the degree of realism that one could include in a model to be fitted to data. Many hierarchical levels in the model such as random effects or unobserved variables could be included with hardly any extra difficulty. While data might be limited, there seemed to be no limit to the complexity of models that could be fitted to data. Practically every ecological journal, starting around the late 1990s, featured expository articles on "Bayesian hierarchical models in ecology." This revolution within ecology, however, contained little reflection on inferential philosophy.

2.2.3 Probability and Inferential Philosophy

Stochasticity in the macroscopic world is in reality the work of deterministic forces that are so high dimensional and so sensitive to initial conditions that they are near-impossible to analyze with deterministic system models. Often, however, long-term frequency patterns in such systems can be captured with stochastic (or probabilistic) models. To incorporate probabilistic ingredients into mathematical modeling, one must be clear about the meaning of probability.

The three probability axioms of Kolmogorov are the foundation of mathematical probability [50]. The first axiom gives the mathematical definition of probability as just a

[1]For random effects in the normal linear models family, the integral could be evaluated analytically, yielding a multivariate normal distribution for the observations (reflecting the dependencies in observations that arise due to the random effects), giving the now-standard normal mixed models (for example, PROC MIXED in SAS).

non-negative real number attached (like a sticky note) to a set. The second axiom envisions an uber-set called the sample space consisting of a population of sets, each set with its own probability attached, with the probability of the entire sample space defined to be one. The third axiom provides for obtaining the probability of a set consisting of any union of disjoint sets as the sum of their individual probabilities. From the three axioms can be derived the familiar properties of probability, such as: All probabilities are bounded above by one, the probability of the complement of a set (all of the sample space not in the set) is one minus the probability of the set, the probability of a subset of a set is less than or equal to the probability of the set.

The three axioms of Kolmogorov governing probability give no guidance for how scientists are supposed to use probability to study nature. However, the properties of probability that emerge mathematically from the axioms follow the intuitive/empirical properties of games of chance that had been studied for hundreds of years before Kolmogorov. The properties provide a framework that allows us to study the frequencies of events in the macroscopic world. The possible distinct outcomes of a random process are the sets, and their sticky notes tell how frequently they occur. The sample space is the collection of possible outcomes. We can call the outcomes and their unions events, such as the event that the next card to be dealt is either heart or spade.

Probability is natural for modeling systems in which small changes are multiplied into large changes. The quintessential random system of flipping a coin is such a multiplier. If one could know the exact initial magnitudes of the coin's upward and rotational velocities, one could calculate the heads or tails outcome with near certainty using deterministic mechanics. A tiny change in those initial magnitudes, though, potentially produces a large change in the system outcome. A colored two-dimensional map of the initial conditions of the two velocities that produce heads or tails would be speckled, and flipping a coin casually would be like throwing a dart at the map while blindfolded. Another such system is a pinball machine. Round bumpers magnify any tiny changes in the initial velocity of the ball. Players differ in their skills at timing the flippers and at subtle below-cancel-threshold "tilting."One can envision that a player's pinball scores after many repetitions would form a histogram of scores characterizing the player-machine system. The framework for using probability to model such systems is that repetitions of these dynamic systems have outcomes that occur with predictable and stable frequencies.

This notion of probability, used in the sense of the proportion of repetitions that we expect a particular outcome to occur, as in the coin and pinball examples above, is *frequentist probability*. It can be made more mathematically rigorous by associating the probability of an event in a hypothetically repeatable system with the mathematical result derived from Kolmogorov's axioms that the proportion of event occurrences when repeating the random process approaches a limit quantity. This *Law of Large Numbers* guaranteeing a stable limiting proportion of outcomes actually has several mathematical forms, some strong and some weak. Our macroscopic world runs on this law. Insurance, casinos, health practices, manufacturing, and finance are examples of social institutions that rely on stable long-run event frequencies.

Bayesian probability has a different definition of probability, which involves the concept of a *personal probability*, also known as *subjective probability*. A typical way to understand personal probability is through betting. A parameter in a model is a quantity that needs to be estimated with data. Before the data are to be gathered or studied, an investigator could envision the odds he/she would give in a bet that the parameter's true value is less than, say, 10. If the investigator thinks that such a low value is unlikely, then perhaps odds of 1 to 5 might be accepted (the investigator would wager to lose 1/5 dollar against winning 1 dollar if the parameter value is less than 10). Now, repeat the betting thought except for setting the parameter threshold at 20 instead of 10. The investigator might rate a parameter value of less than 20 a little more likely than a value of less than 10, so the investigator could extend greater odds, maybe 1 to 3 (lose 1/3 dollar against winning 1 dollar). Keep increasing the threshold for the parameter and eliciting odds. The quantity lose/(1 + lose), where "lose" is the amount of dollars to be lost against winning 1 dollar, increases from zero toward one as a function of the parameter betting threshold and is therefore a cumulative distribution function for some probability distribution on the parameter's range of possible values. The distribution describes that investigator's beliefs about the value of the parameter.

The betting concept of personal probability is somewhat limited in that an investigator would likely bet differently if the process were phrased in thousands of dollars instead of dollars, given that the utility of money changes depending on how much money one has. Nonetheless, subjective probabilists consider the Bayesian definition to be a coherent way of defining the probability of inherently unrepeatable, one-time events such as the outcome of a sporting event. From the point of view of frequentist probability, the unrepeatability lies in the fact that the probability of an outcome varies from repetition to repetition due to changes in initial or prevailing conditions, and that frequencies in do-overs of the experiment would indeed converge, not to a probability of one trial but to the mean of the probabilities of the ensemble of do-overs.

Although ecological modelers by and large have not been much concerned about the philosophical issues behind the choice of inferential approach, in discussions one can get the sense that modelers using Bayesian methods often have a somewhat frequentist view of the meaning of the posterior distribution, regarding the posterior percentiles as being like confidence intervals. And in this, they are not far from the truth if the data are of high quality. According to Walker's Theorem [56], as the sample size increases, the posterior distribution approaches in the limit the multivariate normal distribution that is the asymptotic distribution of maximum likelihood (ML) parameter estimates from frequentist statistics. With large sample size, the Bayesian framework becomes an algorithm for calculating ML parameter estimates; thus, the effect of the prior distribution on the inferences disappears provided the data are abundant. Machine learning models, for instance, which are almost all "trained" (i.e., parameters estimated) using Bayesian approaches, may have thousands of parameters, and the training often requires millions of observations.

It is important to note that some ecologists have suggested (e.g., [6]) that the model selection methods based on information theory, for example the Akaike Information Criterion (AIC) and its variants, constitute another inferential class alongside with Bayesian and frequentist methods. The AIC essentially measures the Kullback-Leibler divergence of a model from the actual data-generating process. The AIC methodology for choosing the "best" model from a list of alternatives consists of computing the AIC value for each model, subtracting the lowest AIC value of the group from each AIC, and then ranking the resulting differences, denoted ΔAIC. The model with $\Delta\text{AIC} = 0$ (that is, the model with the lowest AIC), is inferred to be the "best" model. It turns out that ΔAIC is a statistically consistent estimate, and thus AIC-type model selection is similar to obtaining a sample mean for data from a normal distribution in a frequentist fashion [13, 14, 49].

Agent-based models, as well as large hierarchical models and machine learning models, are built mainly for projecting the past and present states of a system to predict its future states. In the case of machine learning, the past and present states form training data for the model. The hypothesis is that the displayed patterns of system change will continue and can be mimicked by flexible mathematical structures to predict the recurrence of the patterns in the future. The mathematical model itself need not contain much about the causal workings in the system. For example, a popular structure used for machine learning is a web of connected logistic regression models (regressions for predicting binary outcomes) called a (statistical) neural net. Such projective prediction models have a potentially serious flaw. If an intervention or perturbation occurs in the system that was not a part of the patterns in the training data, the predicted system response might go badly wrong. The machine learning models of the present day require millions of observations to train, and can suffer "hallucinations" if the conditions under which they handle complex tasks change. Because of this, they are not suitable for predicting the outcomes of novel interventions.

In the case of our studies, we did not need agent-based models, hierarchical models, or model selection criteria. Individuals in the population cultures were counted carefully. While we cannot claim that counting errors were not made, we do claim they are so small and infrequent as to be negligible to the conclusions. In experiments consisting of "replicate" populations (usually in incubated bottles, each containing 20 g of flour media), we did not observe heterogeneity of parameters among replicates and occasionally did statistical testing for the presence of heterogeneity (likelihood ratio chi-square tests). Given that we were working with a single model, along with hypothetical interventions in the model, we had no need for model selection from among a collection of models. The model's ability to fit the data and *predict the outcomes of experimental interventions* served as our selection criteria. Parameter estimation, confidence intervals, and hypothesis tests were possible through the ordinary frequentist methods, although they were just beginning to be applied to nonlinear time series problems [51].

2.3 OUR APPROACH

It was electrifying. Robert May's 1974 [35] descriptions of complex, random-seeming dynamical behavior in simple low-dimensional deterministic population models represented a gauntlet-like challenge to ecology: Could such fluctuations in some real population systems be caused, *to an important degree*, by low-dimensional deterministic dynamics? Here was something potentially testable. After all, hadn't ecologists seen sigmoid population curves, competition victories of one species over another, and sustained coupled oscillations of prey and predator, in experimental settings? Could the quantitative tradition from Pearl-Lotka-Volterra of low-dimensional models really work and be useful in population ecology?

We decided to find out. The ideas of exploring chaos and other nonlinear behaviors in population systems were exciting to everyone in our team. There were, however, lots of challenges to tackle.

The experimental model system of flour beetles was clear enough. R. F. Costantino and R. A. Desharnais were specialists in laboratory *Tribolium* culture, having carried on the pioneering approach of Thomas Park to studying population ecology with beetle population experiments. Moreover, they had some substantial preliminary data that showed periodic cycles caused by life-stage interactions. A key property of the data was that the populations were censused every two weeks, rather than the traditional monthly census of earlier work. Also, they had preserved the time series from individual population replicates (as in the appendix of Gause's book!) rather than averaging them together as Park did in his publications. The preliminary data would turn out to be instrumental toward building model hypotheses, statistical estimation of model parameters, and ultimately producing results intriguing enough to procure funding to conduct more extensive experiments.

An early problem we faced was how much detail to include in modeling. The data had three state variables: feeding larvae (L), pupae plus large non-feeding larvae and callow adults (P), and the mature adults (A). We knew the major nonlinearity was cannibalism. Our earliest modeling attempts ambitiously focused on two state variables, juveniles ($J = L + P$) and adults (A). We spent almost a year and a half considering two-dimensional approaches, but nothing seemed to work well for existing data. We reluctantly turned to a three-dimensional approach, L, P, and A, and swiftly found it to work remarkably well. The three state variables favorably exploited a coincidental time interval in the beetle life cycle: Both the larval and the pupal stages lasted about two weeks under the prevailing experimental conditions, corresponding to the census interval.

The three-dimensional model gave superb descriptions of the existing data, but it did something else in an intriguing way. It turned out to have a vast repertoire of nonlinear dynamic behaviors when parameter values were changed. Stable equilibria, stable cycles, multiple stable attractors, aperiodic fluctuations on invariant loops, chaos: The model could do it all. The bifurcation diagrams, depicting how attractors change as a parameter changes, were crazy complex. Furthermore, the parameter changes involved could be manipulated experimentally, with population treatments as simple as adding or removing life stage members from a population to change a survival rate. The model not only could predict how trajectories should proceed, also how dynamical behavior should change in response to interventions.

But how would we obtain the parameter values (things like survival rates and cannibalism rates) for the populations in the extant, historical studies? Some parameters, such as cannibalism rates, could conceivably be measured in separate assays by putting known amounts of adults and eggs in cultures and counting the results in a given time interval. However, the assay conditions likely would be different from the population-culture conditions despite all attempts to control the differences, and the assay parameters likely would be different from the parameters that best describe population growth trajectories. (We did undertake some judicious assays to test the functional forms of model ingredients. See Chapter 4 for examples.)

Instead of assays, we looked to the time series data of population sizes for parameter estimation. Finding parameter values that produce model trajectories that best describe the data can be done in a variety of ways. Of course, none of the ways produce model trajectories that fit the data perfectly well. Population data can depart from model trajectories for a variety of reasons, and to obtain quality estimates of parameters, it is important to model those reasons adequately. One must model the noise as well as the signal.

2.3.1 Stochasticity

Ecological stochasticity has received a great deal of theoretical attention. One of the main concepts is that stochasticity in population growth data (as defined by departures of data from an otherwise adequate model) can be separated into two broad categories. First, there can be sampling error; ecological population abundances in most situations must be estimated, and the estimates typically have random errors. Even Gause's 1934 data [19] were obtained by sampling small well-mixed volumes of the culture media. Data from counting individuals in flour beetle cultures, done carefully, are about as close as ecologically possible to being free from sampling error. Second, ecological populations fluctuate intrinsically and naturally from a variety of random forces collectively known as process noise. Process noise can be as simple as the random outcome of n birds each independently facing an annual death risk of 10%. The deterministic model would simply decrease the population by exactly 10% every year. The stochastic model would roll the "dice of life" every year for each animal; the outcome would likely not be exactly 90% remaining alive each year, and, if a hypothetical do-over for a particular year could be envisioned, the outcome likely would be different from the first outcome. Process noise, importantly, exists even when the population is not being observed or sampled.

Various broad types of process noise affect populations in different ways, and, unlike observational sampling error, have an effect on the future states of the population. The above mortality model becomes essentially deterministic in a large population: If there are 10,000 birds, then the proportion remaining alive next year will be close, relatively, to 9000 under the model, while if there are only 10 birds, then there is a 26% chance that there will be 8 or less birds next year. This is an example of what ecologists call *demographic noise*, and the determinism at large population sizes is a consequence of the law of large numbers. Another type of process noise occurs when demographic parameters vary randomly from time period to time period. For instance, the mortality rate for the bird population could fluctuate from, say, as little as 5% to over 30%, depending on how favorable conditions were for survival. This type of process noise, termed *environmental noise* by ecologists, would cause noteworthy random fluctuations (departures from the deterministic model) in large as well as small populations. Either type of process noise, or both, could affect a population system, and the types of process noise are not limited to demographic and environmental (*catastrophic noise* being an important third type).

The point is, *for estimating model parameters from time series data, one must model the noise adequately.* Otherwise, serious biases can be introduced into the resulting estimates. Various follow-up statistical analyses include diagnostic checks of the probability model used for the noise, similar to the analyses of model residuals in ordinary regression methods.

We started out using a parameter estimation approach that assumed environmental noise was the main driver of fluctuations from deterministic predictions. The approach analyzes populations on a logarithmic scale. The population sizes in these first studies were in intermediate ranges (high tens to hundreds) where demographic and environmental noises both have similar effects and are hard to tell apart statistically. However, as our experiments chased new nonlinear phenomena, it became clear that the noise was acting more like demographic noise. Our later publications use parameter estimation under demographic noise. A critical experiment, performed precisely on this question and involving populations with tens of thousands of flour beetles, bolstered the conclusion that demographic noise was the principal source of fluctuations (see Chapter 10).

2.3.2 Inferring Dynamic Behavior

Nonlinear models have dynamic behaviors that can be categorized into "types": behaviors such as locally stable equilibrium points, limit cycles, aperiodic cycles on invariant loops, and the ultimate attention-getter of chaos. These behaviors typically arise as attracting sets in state space to which all trajectories within a neighborhood region of state space are drawn asymptotically. Multiple attractors are separated by saddles, which are attracting on some lower-dimensional surface but repelling elsewhere, and allow the system to change behaviors upon perturbation.

How can one infer that a type of behavior is being displayed by an empirical time series? How can one argue that here is an example of a stable three-cycle, or that here is an example of chaos?

A number of nonparametric or semiparametric statistical approaches exist. Historically, time series data were modeled with autoregressive (and possibly integrated) moving average statistical models [7]. These models are built from linear difference equations and have difficulty diagnosing nonlinear behaviors. They are best suited to short-term forecasting; forecasts decay rapidly to the time series average, making them unsuitable for nonlinear model predictions.

Another time series approach is frequency domain analysis, characterized by the estimation of the spectral density [43]. Frequency domain analysis is especially suited for detecting cycles, in particular, for teasing out patterns in systems governed by layers of different cycles such as short term and long term. Cycles appear as sharp peaks in the spectral density. Chaos can be suggested by peaks that are wide, indicating a broad mixture of periodicities as can be produced by a trajectory being influenced by many limit cycles on saddles in the chaotic attractor. Every chaotic attractor has its own anatomy, and there might be some useful model evaluation information to be had by comparing the empirical spectral density with that obtained from model simulations. Good estimation requires many observations. Some of the authors have tried frequency domain analysis on flour beetle data, as described in Chapters 16–17, with interesting results.

Other methods use statistical approaches such as neural net regression or response surface analysis to estimate the types of nonlinear dynamic behavior in a collection of population time series, and have been used to search for chaotic signals [16]. These methods can suggest chaos in stochastic systems with high amount of noise as well as in chaotic systems. However, our goal was to search directly for mathematical chaos as a deterministic phenomenon in a low-dimensional system.

Our approach to inferring dynamic behavior in the experimental system centered on the detection of model-predicted transitions between dynamic states in response to interventions (parameter manipulation). In particular, our argument for observable chaotic dynamics in flour beetle populations would be a "bookend" argument, in which the model predicted a sequence of attractors—from low-period cycles, to chaos, to low-period cycles—as a parameter was manipulated from low to high. If populations experimentally displayed the predicted cycles at the low end and the high end of the parameter values, along with wild and crazy trajectories in between, we would argue that the wild and crazy behavior was the influence of a chaotic attractor. Furthermore, each chaotic attractor has its own anatomy, built of cycles of various periods, and model trajectories episodically visit these cycles. We would discover that an experimental trajectory under the influence of a chaotic attractor will have its own unique, model-predicted, blend of episodic cycle periods.

2.4 CONCLUSIONS

Dynamic mathematical modeling in science is not only about forecasting system behavior; it is about predicting the response of a system to novel situations. This requires models that capture important causal mechanisms in some adequate way. Novel situations may arise from human or natural interventions that perturb system variables or alter parameters, or from generalizing the conditions under which the model is applied.

In our work, we have used low-dimensional, mechanistic mathematical models to predict transitions between dynamic states in response to experimental interventions in the laboratory (parameter and habitat manipulations). The original dynamical behavior of the flour beetle cultures under the specified laboratory conditions was stable two-cycle oscillations, especially in the L and P stages. Our low-dimensional, discrete-time, three-stage model contained cannibalism as the major mechanistic cause of the oscillations and fitted the existing data well. Critically, the model also predicted that various experimental interventions would throw beetle cultures into displaying various different and exotic nonlinear phenomena, including chaos. These experiments were performed, and the predicted dynamical transitions were induced. At every turn, the model was tested, and its predictions became a great source of discovery and surprise.

Generalizing conditions under which a model is applied requires more than manipulating parameters of mechanisms already in the model; it requires generalizing model assumptions. For example, Kepler's Laws [3] are collectively a mathematical model of planetary orbits that magnificently capture patterns in Tycho Brahe's careful 16th-century measurements of planet movements. Newton's gravitational law greatly extends the conditions under which motions of gravitating bodies can be predicted (higher velocities in the two-body problem provide open conic section flybys instead of Kepler's ellipses; three or more bodies, with the help of modern computers and Poincare's original insights, can produce beautifully complex gravitational dances). Einstein's general theory of relativity agreed with Newtonian dynamics for ordinary purposes but greatly expanded the gravitational conditions under which we could make better predictions, giving us a more accurate model of Mercury's orbit, giving us GPS, and predicting the magnificent photos of Einstein rings recorded by the Webb telescope.

Similarly, the mechanistic models in this book work well for a particular set of laboratory conditions. Can the LPA model be generalized to an expanded set of conditions? We welcome contributions in which the durations of the life stages appear as additional parameters, or the time interval is one day [15], or eggs are an additional state variable [24]. Our data are fully archived online, including eight years of data from three replicate populations under conditions predicted to entail chaos. The biologist authors who did the counting are more than happy for someone to play mechanistic Kepler to their Tycho.

Is there practical merit in this work in laboratory population ecology? Ecologists are concerned, and increasingly alarmed, about ecosphere changes. Much study has been devoted to understanding ecosystem responses to environmental perturbations, and the emerging collective wisdom expects many such responses to be nonlinear (a typical example from the modeling literature on coral reefs is found in Blackwood [5]). Our experiments were aimed at trying to envision what nonlinearity would look like in ecological data and at how to analyze the data and draw inferences. We demonstrated that a relatively simple nonlinear model predicted surprising responses of the population system to a wide variety of experimental treatments.

The model was not only helpful for understanding the system; it was necessary.

Bibliography

[1] Allee WC (1931) Animal Aggregations: A Study in General Sociology. Chicago, University of Chicago Press.

[2] Ballentine LE (2016) Propensity, probability, and quantum theory. Foundations of Physics 46:973–1005. https://doi.org/10.1007/s10701-016-9991-0

[3] Bennett JO, Donahue MO, Schneider N, Voit M (2023) The Cosmic Perspective, 10th edition. Hoboken, New Jersey, Pearson.

[4] Bernoulli J (1695) Explicationes, annotationes & additiones ad ea, quae in actis sup. de curva clastica, isochrona paracentrica, & velaria, hinc inde memorata, & paratim controversa legundur; ubi de linea mediarum directionum, alliisque novis. Acta Eruditorum.

[5] Blackwood JC, Okasaki C, Archer A, Matt EW, Sherman E, Montovan K (2018) Modeling alternative stable states in Caribbean coral reefs. Natural Resource Modeling 31:e12157. https://doi.org/10.1111/nrm.12157

[6] Bolker BM (2008) Ecological Models and Data in R. Princeton, Princeton University Press.

[7] Box GEP, Jenkins GM (1970) Time Series Analysis: Forecasting and Control. San Francisco, Holden-Day.

[8] Caswell H (2001) Matrix Population Models: Construction, Analysis and Interpretation. Second edition. Sunderland, Massachusetts, Sinauer Associates.

[9] Chapman RN (1928) The quantitative analysis of environmental factors. Ecology 9:111–122. https://doi.org/10.2307/1929348

[10] Cushing JM (1998) An Introduction to Structured Population Dynamics. CBMS-NSF Regional Conference Series in Applied Mathematics, Vol. 71. Philadelphia, Society for Industrial and Applied Mathematics.

[11] Dennis B (1978) Analytical solution to an open-system model of population growth. Mathematical Biosciences 40:167–169. https://doi.org/10.1016/0025-5564(78)90079-2

[12] Dennis B (2004) Statistics and the scientific method in ecology. In Taper ML, Lele SR (eds.), The Nature of Scientific Evidence, pp. 327–378. Chicago, University of Chicago Press.

[13] Dennis B, Ponciano JM, Taper ML, Lele SR (2019) Errors in statistical inference under model misspecification: Evidence, hypothesis testing, and AIC. Frontiers in Ecology and Evolution 7:372. https://doi.org/10.3389/fevo.2019.00372

[14] Dennis B, Taper ML, Ponciano JM (2024) Evidential analysis: An alternative to hypothesis testing in normal linear models. Entropy 26:964. https://doi.org/10.3390/e26110964

[15] Desharnais RA, Liu L (1987) Stable demographic limit cycles in laboratory populations of *Tribolium castaneum*. Journal of Animal Ecology 56:885–906. https://doi.org/10.2307/4955

[16] Ellner S, Turchin P (1995) Chaos in a noisy world: New methods and evidence from time-series analysis. American Naturalist 145:343–375. https://doi.org/10.1086/285744

[17] Ford J (1937) Research on populations of *Tribulium confusum* and its bearing on ecological theory: A summary. Journal of Animal Ecology 6:1–14. https://doi.org/10.2307/1054

[18] Gause GF (1931) The influence of ecological factors on the size of population. The American Naturalist 65:70–76. https://doi.org/10.1086/280347

[19] Gause GF (1934a) Experimental analysis of Vito Volterra's mathematical theory of the struggle for existence. Science 79:16–17. https://doi.org/10.1126/science.79.2036.16.b

[20] Gause GF (1934b) The Struggle for Existence. Baltimore, Williams and Wilkins.

[21] Gause GF, Witt AA (1935) Behavior of mixed populations and the problem of natural selection. American Naturalist 69:596–609. https://doi.org/10.1086/280628

[22] Grant PR, Grant BR (2014) 40 Years of Evolution: Darwin's Finches on Daphne Major Island. Princeton, Princeton University Press.

[23] Grimm V, Railsback SF (2005) Individual-based Modeling and Ecology. Princeton, Princeton University Press.

[24] Hastings A, Costantino RF (1987) Cannibalistic egg-larva interactions in *Tribolium*: An explanation for oscillations in population numbers. American Naturalist 130:36–52. https://www.jstor.org/stable/2461879

[25] Hansen SR, Hubbell SP (1980) Single nutrient microbial competition: Qualitative agreement between experimental and theoretical forecast outcomes. Science 207:1491-1493. https://www.science.org/doi/abs/10.1126/science.6767274

[26] Kermack WO, McKendrick AG (1927) A contribution to the mathematical theory of epidemics. Proceedings of the Royal Society of London, Series A 115:700–721. https://doi.org/10.1098/rspa.1927.0118

[27] Kingsland SE (1985) Modeling Nature: Episodes in the History Population Ecology. Chicago, University of Chicago Press.

[28] Leslie PH (1945) On the use of matrices in certain population mathematics. Biometrika 33:183–212. https://doi.org/10.1093/biomet/33.3.183

[29] Lasota A, Mackey MC (1985) Probabilistic Properties of Deterministic Systems. Cambridge, Cambridge University Press.

[30] Lotka AJ (1920) Analytical note on certain rhythmic relations in organic systems. Proceedings of the National Academy of Sciences of the United States of America 6:410–415. https://doi.org/10.1073/pnas.6.7.410

[31] Lotka AJ (1925) Elements of Physical Biology. Baltimore, Williams and Wilkins.

[32] Lotka AJ (1932) The growth of mixed populations: Two species competing for a common food supply. Journal of the Washington Academy of Sciences 22:461–469. http://www.jstor.org/stable/24530449

[33] Malthus TR (1826) An Essay on the Principle of Population: A View of its Past and Present Effects on Human Happiness; with an Inquiry into Our Prospects Respecting the Future Removal or Mitigation of the Evils which It Occasions. Sixth Edition. London, John Murray.

[34] May RM (1973) Stability and Complexity in Model Ecosystems. Princeton, Princeton University Press.

[35] May RM (1974) Biological populations with nonoverlapping generations: Stable points, stable cycles, and chaos. Science 186:645–647. https://doi.org/10.1126/science.186.4164.645

[36] Palmer T (2022) The Primacy of Doubt: From Quantum Physics to Climate Change, How the Science of Uncertainty Can Help Us Understand Our Chaotic World. New York, Basic Books.

[37] Park T (1932) Studies in population physiology: The relation of numbers to initial population growth in the flour beetle *Tribolium confusum* Duval. Ecology 13:172–181. https://doi.org/10.2307/1931067

[38] Park T (1933) Studies in population physiology. II. Factors regulating initial growth of *Tribolium confusum* populations. Journal of Experimental Biology 65:17–42. https://doi.org/10.1002/jez.1400650103

[39] Park T (1934) Observations on the general biology of the flour beetle, *Tribolium confusum*. Quarterly Review of Biology 9:36–54. https://doi.org/10.1086/394454

[40] Pearl R (1922) The Biology of Death. Philadelphia, J. B. Lippincott.

[41] Pearl R, Reed LJ (1920) On the rate of growth of the population of the United States since 1790 and its mathematical representation. Proceedings of the National Academy of Sciences of the United States of America 6:275–288. https://doi.org/10.1073/pnas.6.6.275

[42] Poincaré H (1892) Les Méthodes Nouvelles de la Mécanique Céleste. Paris, Gauthier-Villars.

[43] Priestley MB (1981) Spectral Analysis and Time Series. New York, Academic Press.

[44] Ricker WE (1954) Stock and recruitment. Journal of the Fisheries Research Board of Canada 11:559–623. https://doi.org/10.1139/f54-039

[45] Ross R (1911) The Prevention of Malaria. Second Edition. London, John Murray.

[46] Skellam JG (1951) Random dispersal in theoretical populations. Biometrika 38:196–218. https://doi.org/10.1093/biomet/38.1-2.196

[47] Smith HL, Waltman P (1995) The Theory of the Chemostat: Dynamics of Microbial Competition. Volume 13. Cambridge, Cambridge University Press.

[48] Strong DR (1986) Density vagueness: Abiding the variance in the demography of real populations. In Diamond J, Case TJ (eds.), Community Ecology, pp. 257–268. New York, Harper and Row.

[49] Taper ML, Ponciano, JM (2016) Evidential statistics as a statistical modern synthesis to support 21st century science. Population Ecology 58:9–29. https://doi.org/10.1007/s10144-015-0533-y

[50] Thomasian AJ (1969) The Structure of Probability Theory With Applications. New York, McGraw-Hill.

[51] Tong, H (1990) Non-linear Time Series: A Dynamical System Approach. Oxford, Oxford University Press.

[52] Verhulst PF (1838) Notice sur la loi que la population poursuit dans son accroissement. Correspondance Mathématique et Physique 10:113–121. https://doi.org/10.1007/bf02309004

[53] Volterra V (1926) Fluctuations in the abundance of a species considered mathematically. Nature 118:558–560. https://doi.org/10.1038/118558a0

[54] Volterra V (1928) Variations and fluctuations of the number of individuals in animal species living together. International Council for the Exploration of the Sea Journal of Marine Science 3:3–51. https://doi.org/10.1093/icesjms/3.1.3

[55] Volterra V (1938) Population growth, equilibria, and extinction under specified breeding conditions: A development and extension of the theory of the logistic curve. Human Biology 3:1–11. https://doi.org/10.1007/978-3-642-50151-7_3

[56] Walker AM (1969) On the asymptotic behaviour of posterior distributions. Jorurnal of the Royal Statistical Society B 31:80–88. https://doi.org/10.1111/j.2517-6161.1969.tb00767.x

CHAPTER 3

Exploring State Space

In this chapter, we aim to present, in an accessible way, a number of important ideas from dynamical systems theory that underlie this book. The chapter addresses two fundamental concepts: 1) the exponential growth of linear systems, and 2) the approximately exponential growth of nonlinear systems when they are close to an equilibrium (*linearization*). We illustrate these ideas with the famous one-dimensional Ricker model and with two Ricker-type two-dimensional models. We introduce equilibria, stability, cycles, bifurcations, bifurcation diagrams, attractors, saddles, manifolds, basins, and basin boundaries; and we emphasize the conceptual framework of state space.

Readers thoroughly familiar with these topics may wish to skip this chapter.

3.1 DISCRETE-TIME MODELS

All real-world systems are dynamic on some scale, that is, they experience change over time. In some systems, change happens more or less continuously. These systems typically are modeled with equations that involve continuous-time rates of change (i.e., derivatives). Equations that involve derivatives are called *differential equations*. Other systems experience temporal pulses; for example, a population might have a yearly reproductive pulse. For these systems, *discrete-time models* may be more useful, with discrete time step chosen to be appropriate for the application. As we will see in Chapter 4, *Tribolium castaneum* has an approximately two-week feeding-larval stage that greatly affects its population dynamics. A reasonable population model for this species would thus be discrete-time with a time step of two weeks.

This book focuses on discrete-time models, also known as *difference equations*. A few mathematical developments are in order.

3.2 ONE DIMENSION: LINEAR MODELS

A one-dimensional discrete-time model is a recursive equation in one dependent variable (state variable) x, that is, an equation of the form

$$x_{t+1} = f(x_t),$$

where x and $f(x)$ are real numbers. A one-dimensional discrete-time model is also called a *map*. It is *linear* if the state variable appears only in a linear way, that is, to the first power. In one dimension, the linear model of interest in this book is[1]

$$x_{t+1} = bx_t. \tag{3.1}$$

[1] In general, a one-dimensional linear model has the form $x_{t+1} = b(t)x_t + h(t)$. If both coefficient functions b and h are independent of time, then the model is called *autonomous*. If h is identically 0, the model is

DOI: 10.1201/9781003467120-3

What type of assumptions would lead to the linear model (3.1)? Here is one example. Let x_t be the number of individuals in a population at time t. Assume that

(A1) The per capita birth rate, averaged over the population, is a constant number $b > 0$ of offspring per individual per time step. (The population birth rate is therefore bx_t individuals per time step.)

(A2) All individuals die at the end of the time step. (The population death rate is therefore x_t individuals per time step, and generations are non-overlapping).

(A3) The only flows into or out of the population are due to births and deaths.

There are three equivalent ways to think about discrete-time model formulation. The first two are generally inferior to the third, for reasons we will explain after presenting the three methods.

First, in the spirit of "compartmental" modeling from differential equations, we can consider the net change per unit time step, $\Delta x_t = x_{t+1} - x_t$, to be the sum of the inflow rates minus the sum of the outflow rates, which in this case is the birth rate minus the death rate:

$$\begin{aligned} \Delta x_t &= \Sigma \text{ inflow rates} \quad - \quad \Sigma \text{ outflow rates} \\ x_{t+1} - x_t &= bx_t - x_t. \end{aligned}$$

Second, we can think of the new census as equal to the last census plus the new births minus the new deaths:

$$\begin{aligned} \text{new census} &= \text{last census} + \text{births} - \text{deaths} \\ x_{t+1} &= x_t + bx_t - x_t. \end{aligned}$$

Third, and best, we can think of the new census as new recruits plus survivors from the last census:

$$\begin{aligned} \text{new census} &= \text{recruits} + \text{survivors} \\ x_{t+1} &= bx_t + 0. \end{aligned}$$

In each method we obtain equation (3.1). All three methods are mentioned here because many people wonder how discrete-time models relate to the "compartmental" approach in differential equations. However, subtracting deaths (as in the first two methods) instead of adding survivors (as in the third method) can be problematic in discrete-time modeling. If the modeler is not careful to set up proper constraints, subtracting deaths can lead to minus signs that cause solutions to go negative and become nonsensical. Many problems are averted if one thinks in terms of the accounting principle of adding survivors instead of subtracting deaths. We will use the third method in Chapter 5 when we introduce the flour beetle model.

In model (3.1), at each time step the population size is multiplied by b. The solution formula is therefore

$$x_t = x_0 b^t. \tag{3.2}$$

The discrete-time equation (3.1) predicts population levels recursively, conditioned on the previous population size, whereas its solution formula (3.2) predicts the population level at time t based only on the initial population size. Thus, the latter is called a *closed-form solution*. The geometric rate of growth (or decay) b is called the *eigenvalue*.

Note that we also obtain a linear model if assumption A2 above is modified to allow for a constant per capita death rate and overlapping generations:

---

called *homogeneous*. In this book, we typically will mean *linear homogeneous and autonomous* whenever we talk about a linear model. Model (3.1) is linear homogeneous and autonomous.

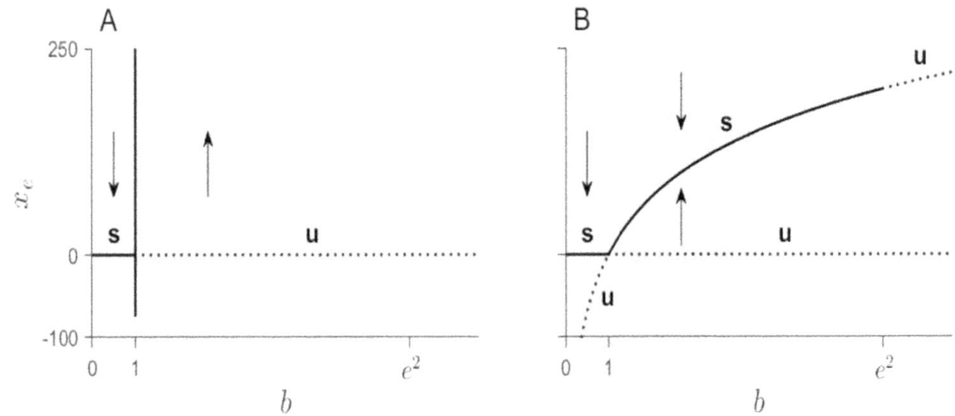

Figure 3.1 Bifurcation diagrams in the b-x_e plane, showing the equilibrium branches as functions of the bifurcation parameter b. Dotted equilibrium branches are unstable (u); solid branches are stable (s). A. The linear model (3.1) has equilibrium branches $x_e = 0$ (solid and dotted horizontal line) and $b = 1$ (vertical solid line). The extinction state $x_e = 0$ is stable for $b < 1$ and unstable for $b > 1$. B. The Ricker model (3.3), which is nonlinear, also has two equilibrium branches. The extinction equilibrium is again stable for $b < 1$ and unstable for $b > 1$. The non-zero branch of equilibria from panel A, however, is now "bent over to the right" by the nonlinearity, and these non-zero equilibria are locally stable for $1 < b < e^2$. Here $c = 0.01$.

(A2*) A fraction $\mu \in [0, 1]$ of the individuals die at the end of each time step. (The average per capita death rate is μ deaths per individual per time step, and the population death rate is therefore μx_t individuals per time step. If $\mu < 1$, then generations are overlapping).

Given assumptions A1, A2*, and A3, we obtain the linear model

$$\begin{aligned} \text{new census} &= \text{recruits} + \text{survivors} \\ x_{t+1} &= bx_t + (1 - \mu)x_t \\ &= rx_t, \end{aligned}$$

where $r = b + 1 - \mu$ now plays the role of b in model (3.1).

3.2.1 Equilibria

For any value of $b > 0$, if the initial condition is $x_0 = 0$, then the solution is a constant $x_t = 0$ for all t. A constant solution is called an *equilibrium* or *fixed point*, and the constant solution $x_t = 0$ is called the *extinction equilibrium*. If $b \neq 1$, then there are no other equilibrium solutions except the extinction equilibrium. If $b = 1$, then every initial condition gives rise to an equilibrium solution $x_t = x_0$ for all t. It is useful to summarize these possibilities by graphing all equilibria as a function of b. The *bifurcation diagram* in Figure 3.1A illustrates the two *branches* of equilibria for equation (3.1), shown in the b-x_e plane. The line $x_e = 0$, which lies on the horizontal axis, illustrates that $x_e = 0$ is an equilibrium for all values of b. The vertical line of equilibria at $b = 1$ illustrates that when $b = 1$, every value of x is an equilibrium.

3.2.2 Stability

Does the linear model (3.1) predict population extinction or persistence? That is to say, what about the behavior of non-equilibrium solutions? Do they approach the extinction equilibrium or move away from it?

If $b > 1$, then all nonzero solutions $x_t = x_0 b^t$, $x_0 > 0$, move away from the equilibrium $x_e = 0$ and grow without bound. In this case, populations do not go extinct. Mathematically, we say that the extinction equilibrium $x_e = 0$ is a *repeller* and refer to it as an *unstable* equilibrium.

On the other hand, if $0 < b \leq 1$, then nonzero solutions do not grow but instead remain less than (or equal to) x_0 for all time. In particular, if $x_0 > 0$ is small, then the solution remains small for all time. This property is called *Lyapunov stability*. Furthermore, if $0 < b < 1$, we see that nonzero solutions $x_t = x_0 b^t$ tend to the extinction equilibrium $x_e = 0$ in the asymptotic limit as t increases without bound. In this case, we say the extinction equilibrium $x_e = 0$ is also attracting or is an *attractor*. We summarize these two properties (Lyapunov stable and attracting) by saying $x_e = 0$ is *asymptotically stable* when $0 < b < 1$. A common practice is to refer to an asymptotically stable equilibrium as simply a "stable" equilibrium. In later chapters of this book, we generally will follow this common convention and use the term stable to mean asymptotically stable.

Note that if $b = 1$, then every x_0 is an equilibrium, and these equilibria are, in a sense, on the border between asymptotic stability and instability. They are called *neutrally stable*.[2] Although we will not discuss neutral stability further in this book, *the point at which stability changes ($b = 1$ in this example) will play a major role throughout the book.* Such critical values are called *bifurcation points*, and the parameter being "tuned" (b in this example) is called a *bifurcation parameter*.

Let us summarize some key points illustrated in this section. First, a linear model arises from the assumption of density-independent per capita birth and death rates, that is, situations in which there are no crowding effects. Second, the linear model has exactly one equilibrium, the extinction equilibrium (if $b \neq 1$), or else a vertical line of equilibria (if $b = 1$). Third, all non-equilibrium solutions grow (or decline) geometrically away from (or toward) extinction. The geometric rate of growth (or decay) is called the eigenvalue. Fourth, if you imagine "tuning" the eigenvalue b from 0 to $+\infty$, as it passes through the bifurcation point $b = 1$, there is a vertical bifurcation of equilibria, and the stability of the extinction equilibrium changes from asymptotically stable to unstable. That is, as the eigenvalue b increases through the critical value $b = 1$, population extinction is replaced by population survival (Figure 3.1A).

3.3 ONE DIMENSION: NONLINEAR MODELS

Clearly, exponential growth cannot be sustained indefinitely with limited resources. What if overcrowding were to decrease the per capita birth rate?

3.3.1 The Ricker Model

Now assume

(A1) The average per capita birth rate is a constant number $b > 0$ of offspring per individual per unit time at low population sizes (i.e., when there is no overcrowding).

(A2) Because of overcrowding, the average per capita birth rate is reduced by the factor e^{-cx_t}, where $c > 0$ quantifies the strength of the crowding effect. (The *Ricker nonlinearity*

[2]The formal definitions of Lyapunov stability, asymptotic stability, neutral stability, and instability involve precise ε-δ representations, which the reader can find in dynamical systems textbooks.

e^{-cx_t}, which satisfies $0 < e^{-cx_t} < 1$, can be interpreted as a survival probability. See the appendix at the end of this chapter for a mechanistic derivation of the Ricker nonlinearity.)

(A3) The average per capita death rate is μ deaths per individual per unit time, where $0 < \mu \leq 1$.

(A4) The only flows into or out of the population are due to births and deaths.

By assumptions A1 and A2, the per capita birth rate in the presence of x_t individuals is $r = be^{-cx_t}$; therefore the population birth rate is $rx_t = be^{-cx_t}x_t$ offspring per unit time. The model can be written

$$\begin{aligned} x_{t+1} &= \text{recruits} + \text{survivors} \\ &= bx_t e^{-cx_t} + (1-\mu)\,x_t. \end{aligned}$$

If generations are non-overlapping ($\mu = 1$), we obtain the famous *Ricker model*

$$x_{t+1} = bx_t e^{-cx_t} \tag{3.3}$$

with parameters $b, c > 0$, historically used in fisheries [6, 7]. The Ricker model is *nonlinear* because the state variable x appears in a nonlinear way.

Equilibria of the Ricker model must satisfy the *fixed-point equation*

$$x_e = bx_e e^{-cx_e}.$$

Note that $x_e = 0$ is an equilibrium. If $x_e \neq 0$, we have $1 = be^{-cx_e}$, which yields a nontrivial equilibrium $x_e = (\ln b)/c$. If we graph the two equilibria $x_e = 0$ and $x_e = (\ln b)/c$ as functions of b, we obtain Figure 3.1B. Note that the nontrivial equilibrium branch $x_e = (\ln b)/c$ is positive if and only if $b > 1$. The value $b = 1$ at which the two equilibrium branches cross is called a *bifurcation point*. The graph of x_e vs. b is called a *bifurcation diagram*, and b is called the *bifurcation parameter*.

Does the population go extinct or does it persist? What are the stabilities of the equilibria of the Ricker model? Do their stabilities depend on the value of b? Clearly we need to find a way to quantify and study stability in nonlinear equations. To this end, we now turn to one of the most important subjects in applied mathematics, the subject of *linearization*.

3.3.2 Linearization

Suppose $f(x)$ is a nonlinear, continuously differentiable function of x. *Continuously differentiable* means that df/dx exists and is continuous. Let's consider the nonlinear equation

$$x_{t+1} = f(x_t) \tag{3.4}$$

with equilibrium x_e. The graph of $f(x)$ as a function of x can be approximated near the point x_e by its tangent line at $x = x_e$; this can be expressed as

$$f(x) \approx f(x_e) + \left[\frac{df}{dx}(x_e)\right](x - x_e) \text{ for } x \approx x_e,$$

where the symbol $\approx$ means "approximately equal to," and the quantity $[df/dx(x_e)]$ is the derivative of f taken with respect to x and then evaluated at x_e. Because x_e is an equilibrium of equation (3.4), we know that $f(x_e) = x_e$. Thus,

$$x_{t+1} = f(x_t) \approx x_e + \left[\frac{df}{dx}(x_e)\right](x_t - x_e) \text{ for } x_t \approx x_e, \tag{3.5}$$

which can be written

$$x_{t+1} - x_e \approx \lambda\,(x_t - x_e) \text{ for } x_t \approx x_e, \tag{3.6}$$

where

$$\lambda \mathrel{\hat{=}} \frac{df}{dx}(x_e)$$

and the symbol $\hat{=}$ means "defined to be." If we write the displacement of the system from its equilibrium as

$$u_t = x_t - x_e,$$

we obtain from equation (3.6) a linear approximation for the displacement of the system from its equilibrium:

$$u_{t+1} \approx \lambda u_t \text{ for } u_t \approx 0.$$

Thus, for small displacements, the displacement at time $t+1$ is approximately λ times the displacement at time t. Stated another way, x_{t+1} is about λ times as far from the equilibrium as x_t was. You can see that if $-1 < \lambda < 1$, this is good news for the stability of equilibrium x_e, because the distance between the system state and the equilibrium is shrinking as time goes on—in fact, it appears that the equilibrium x_e would be asymptotically stable. If $\lambda > 1$ or $\lambda < -1$, however, the distance between the system state and the equilibrium grows whenever the system is near the equilibrium, and so it appears that x_e would be unstable.

The *linearization of* $x_{t+1} = f(x_t)$ *at the equilibrium* x_e is defined to be the linear model

$$u_{t+1} = \lambda u_t,$$

where the *eigenvalue* λ is the number

$$\lambda = \frac{df}{dx}(x_e).$$

Note that the solution of the linearization is

$$u_t = u_0 \lambda^t.$$

Thus, *the displacement of the population from equilibrium grows or decays approximately exponentially when the population size is near its equilibrium value, with exponential rate equal to the eigenvalue* λ. Whether the displacement grows or decays depends on whether $|\lambda| > 1$ or $|\lambda| < 1$.

An equilibrium x_e is *Lyapunov stable* if solutions x_t remain as close to x_e as one wishes provided their initial conditions are selected close enough to x_e. An equilibrium is *unstable* otherwise. An equilibrium x_e is said to be an *attractor* if x_t tends to the equilibrium as t tends to infinity for all initial conditions sufficiently close to the equilibrium. If an equilibrium is both Lyapunov stable and an attractor, then it is called *(locally) asymptotically stable.* It is common practice to refer to a (locally) asymptotically stable equilibrium as simply a stable equilibrium, a custom we will follow in this book.

The *Linearization Theorem* says that

$|\lambda| < 1$ implies x_e is (locally) asymptotically stable.

$|\lambda| > 1$ implies x_e is unstable.

If $|\lambda| = 1$, we say the equilibrium is *nonhyperbolic*, and the Linearization Theorem does not apply. In this situation, linearization does not tell us anything about the stability of the original nonlinear system and we would need to examine higher-order effects.

Linearization is one of the most important tools in applied mathematics.

3.3.3 Linearizing the Ricker Model

Consider again the Ricker model (3.3), for which $b, c > 0$. For this model, $f(x) = bxe^{-cx}$. We already have found that the equilibria are $x_e = 0$ and $x_e = (\ln b)/c$. Let's linearize the Ricker model about each of its equilibria. In both cases, we will need the derivative of f with respect to x:

$$\frac{df}{dx} = be^{-cx}(1 - cx).$$

We first linearize about $x_e = 0$. The eigenvalue is

$$\lambda = \frac{df}{dx}(0) = b,$$

and so the linearization at $x_e = 0$ is

$$u_{t+1} = bu_t.$$

Recalling that $b > 0$, we have, from the Linearization Theorem:

$$\text{If } 0 < b < 1, \text{then } x_e = 0 \text{ is asymptotically stable.}$$
$$\text{If } b > 1, \text{then } x_e = 0 \text{ is unstable.}$$

Now we linearize about the nontrivial equilibrium $x_e = (\ln b)/c$:

$$\lambda = \frac{df}{dx}\left(\frac{\ln b}{c}\right) = 1 - \ln b.$$

The linearization at $x_e = (\ln b)/c$ is therefore

$$u_{t+1} = (1 - \ln b)\,u_t,$$

and so:

$$\begin{aligned}&\text{If } |1 - \ln b| < 1, \text{ then } x_e = (\ln b)/c \text{ is asymptotically stable.}\\&\text{If } |1 - \ln b| > 1, \text{ then } x_e = (\ln b)/c \text{ is unstable.}\end{aligned} \tag{3.7}$$

Recalling that

$$\begin{aligned}&|x| < p \text{ if and only if } -p < x < p\\&|x| > p \text{ if and only if } x < -p \text{ or } x > p,\end{aligned} \tag{3.8}$$

we can obtain from conditions (3.7)–(3.8):

$$\text{If } 0 < b < 1, \text{then } x_e = (\ln b)/c \text{ is unstable.}$$
$$\text{If } 1 < b < e^2, \text{then } x_e = (\ln b)/c \text{ is asymptotically stable.}$$
$$\text{If } b > e^2, \text{then } x_e = (\ln b)/c \text{ is unstable.}$$

The first of these conditions ($0 < b < 1$) does not seem to be of ecological interest given that $x_e < 0$ in that case; however, in the next paragraph we will indicate why it illustrates an important point.

We can now indicate stabilities on the Ricker bifurcation diagram, which we do in Figure 3.1B. When looking at a bifurcation diagram such as Figure 3.1B, one must remember that any particular system has a fixed value of b, for example, $b = 1.5$. Its equilibria are given by the values of the equilibrium branches directly above that specific value of b. Another way to think of it is that any particular system "lives on a vertical line" in Figure 3.1B. Notice how the inclusion of the Ricker nonlinearity has "bent" the vertical branch of equilibria in Figure 3.1A over to the right in Figure 3.1B, creating two equilibria for each value of $b > 1$. Finally, note how the two equilibrium branches experience an *exchange of stability* when they cross at $b = 1$. This exchange is typical and is the reason we pointed out the stability of the negative equilibrium for $0 < b < 1$.

3.3.4 Bifurcation Diagram for the Ricker Model

What is the long-term fate of the Ricker population if $b > e^2$, given that both equilibria are unstable? For these values of b, non-equilibrium solutions cannot equilibrate, because there are no stable equilibria for them to approach. How do solutions of the Ricker model behave for these values of b? Some exploratory computer simulations are in order.

Fix $c = 0.01$ and $x_0 = 75$. Let's set b at various values and inspect the resulting *time series*, that is, the resulting solutions as functions of time. Figure 3.2A–D shows the time series generated for $b = 0.5$, $b = 1.3$, $b = 3.6$, and $b = 5.7$. If $b = 0.5$, solutions equilibrate to zero, whereas if $b = 1.3$, solutions equilibrate to a positive equilibrium $x_e \approx 26.3$. The *attractors* in these cases are the *singleton sets* $\{0\}$ and $\{26.3\}$, respectively. For $b = 3.6$, the equilibrium is larger, about $x_e \approx 128.1$, and for $b = 5.7$, the equilibrium value is even larger, about $x_e \approx 174.0$. Interestingly, solutions oscillate as they approach this latter equilibrium. The attractors are the sets $\{128.1\}$ and $\{174.0\}$, respectively.

Figure 3.2E–H shows time series for $b = 8$, $b = 14$, $b = 14.6$, and $b = 17$. We know from our previous linearization calculations that for $b > e^2 \approx 7.389$, solutions no longer equilibrate. The computer simulation for $b = 8$ in Figure 3.2E suggests solutions approach a periodic oscillation with period two, called a *two-cycle*; that is, they begin to oscillate between the two values $x \approx 138.6$ and $x \approx 277.3$. The attractor in this case is the set of *two* points $\{138.6, 277.3\}$. Further numerical explorations near $b = e^2$ suggests that as b is increased through the value e^2, the point attractor (for $b < e^2$) bifurcates into two-point attractors (for $b > e^2$, but near e^2). The value $b = e^2$ is called a *bifurcation value* or *bifurcation point*.

For $b = 14$, solutions approach a *four-cycle* and the attractor is a set of four points, and for $b = 14.6$, solutions approach an *eight-cycle* and the attractor is a set of eight points. As b is increased, the periods of these cycles continue to double, until at a critical value of b there is a transition to an aperiodic, complicated dynamic called *chaos*. The attractor is then an infinite set of points. This sequence of bifurcations is called a *period-doubling cascade to chaos*.

Two chaotic time series are shown in Figure 3.2H for $b = 17$, one with initial condition $x_0 = 75$ (solid curve) and another with initial condition $x_0 = 75.1$ (dashed curve). The two chaotic time series are close together for several time steps, but soon diverge and look nothing like each other. This is due to *sensitivity to initial conditions*: A small change or error in initial conditions is quickly magnified.

If the attractor, also called the *final state*, of each time series is plotted against the parameter b, we can continue the Ricker bifurcation diagram in Figure 3.1B beyond $b = e^2$, with the result shown in Figure 3.3. Note that unstable equilibria and cycles are not shown in Figure 3.3, since computer iterations only identify the stable ones.

Chaos has a complicated mathematical definition that we will not address here. For our purposes here we simply note three important characteristics of chaos. First, *chaos is deterministic*. Note that the two chaotic time series shown in Figure 3.2H are completely determined by the simple iterative rule $x_{t+1} = 17x_t \exp(-0.01x_t)$ and the initial condition ($x_0 = 75$ for the solid curve; $x_0 = 75.1$ for the dashed curve). Second, chaotic dynamics, although deterministic, appear random and unpredictable. Third, chaotic dynamics exhibit sensitivity to initial conditions.

Let's summarize some key points illustrated in this section. First, the inclusion of density dependence renders a model nonlinear. Second, nonlinear models can have more than one equilibrium. In this example, the inclusion of the Ricker nonlinearity "bent" the vertical branch of equilibria in Figure 3.1A over to the right (Figure 3.1B), creating two equilibria for each value of $b > 1$. Third, a non-equilibrium solution that begins sufficiently close to an equilibrium moves away from (or moves toward) the equilibrium at an approximately

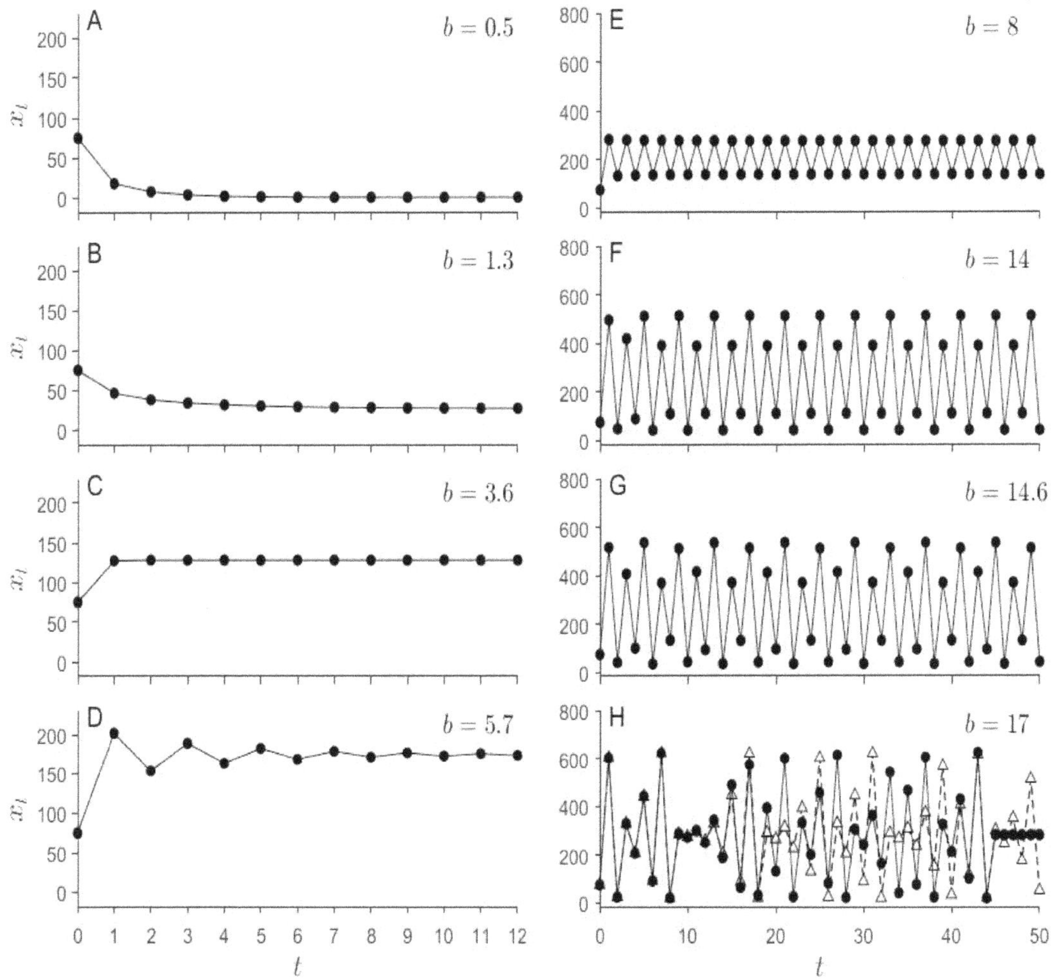

Figure 3.2 Time series for Ricker model (3.3) with $c = 0.01$ and $x_0 = 75$, illustrating the period-doubling cascade to chaos. A. Extinction for $b = 0.5$. B. Positive equilibrium for $b = 1.3$. C. Positive equilibrium for $b = 3.6$. D. Positive equilibrium for $b = 5.7$. E. Two-cycle for $b = 8$. F. Four-cycle for $b = 14$. G. Eight-cycle for $b = 14.6$. H. Two chaotic time series for $b = 17$, with initial condition $x_0 = 75$ (solid curve) and initial condition $x_0 = 75.1$ (dashed curve). The two solutions soon diverge, showing sensitivity to initial conditions.

geometric rate. The geometric rate is called the eigenvalue. Fourth, if you increase the eigenvalue b for the extinction equilibrium through $b = 1$, there is, in this Ricker example, a bifurcation of stable positive equilibria, and the stability of the extinction equilibrium changes from stable to unstable. Fifth, the positive equilibria on the bifurcating branch can destabilize and further bifurcations can occur that create nonequilibrium attractors. In the Ricker model example, the positive equilibrium branch destabilizes as b increases through $b = e^2$, and a branch of stable two-cycles is born.

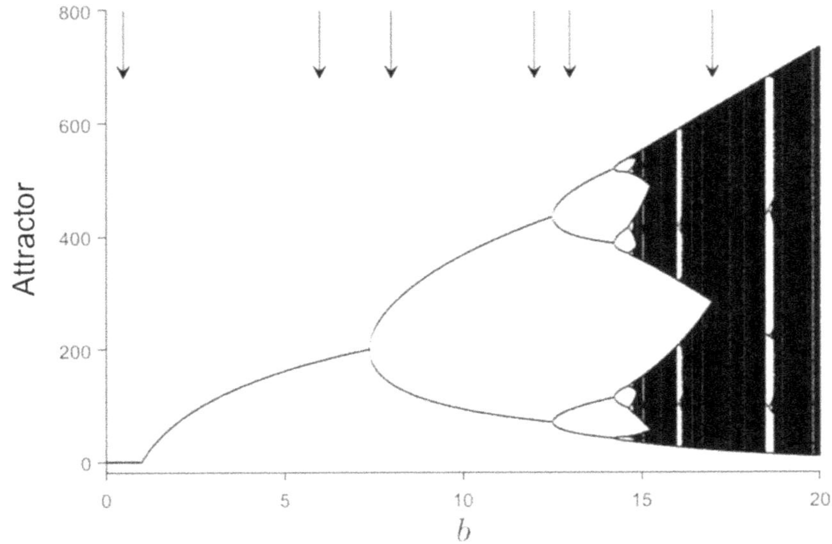

Figure 3.3 Bifurcation diagram for the Ricker model (3.3) with $c = 0.01$. This is also the bifurcation diagram for the Ricker model with sensescence (3.18). The arrows placed at $b = 0.5, 6, 8, 12, 13, 17$, which correspond to the panels in Figure 3.4, are associated, respectively, with extinction, positive equilibrium, two-cycle, two-cycle with larger amplitude, four-cycle, and chaotic attractors.

3.4 HIGHER DIMENSIONS: LINEAR SYSTEMS

Higher-dimensional systems have more than one state variable. A linear system is one in which all state variables appear to the first power. Consider, for example, the linear two-dimensional system[3]

$$\begin{aligned} x_{t+1} &= ax_t + by_t \\ y_{t+1} &= cx_t + dy_t. \end{aligned} \tag{3.9}$$

Our current goal is to understand the behavior of this linear system by finding formulas for its solutions. Although we illustrate the solution process in two dimensions, the techniques are the same for dimensions greater than two.

In one dimension, we found that solutions of linear models were exponential, of the form b^t. Let's propose an exponential Ansatz (trial solution) of system (3.9) of the form

$$\begin{aligned} x_t &= v_1\lambda^t \\ y_t &= v_2\lambda^t \end{aligned} \tag{3.10}$$

and see if it works. Plugging the Ansatz into system (3.9), we obtain

$$\begin{aligned} v_1\lambda^{t+1} &= av_1\lambda^t + bv_2\lambda^t \\ v_2\lambda^{t+1} &= cv_1\lambda^t + dv_2\lambda^t. \end{aligned}$$

[3]Again, we are using the term linear to mean linear homogeneous and autonomous.

Assuming $\lambda \neq 0$, we can algebraically rearrange this as

$$\begin{aligned} (a-\lambda)\, v_1 + b v_2 &= 0 \\ c v_1 + (d-\lambda)\, v_2 &= 0. \end{aligned} \tag{3.11}$$

In the (v_1, v_2)-plane, these equations describe two lines passing through the origin; therefore, this system of equations has a nontrivial solution if and only if the lines are coincident. A fact from linear algebra asserts this is true if and only if the determinant[4] of their coefficients is zero:

$$\begin{vmatrix} a-\lambda & b \\ c & d-\lambda \end{vmatrix} = 0.$$

The two λ-solutions (real or complex) of the resulting *characteristic equation*

$$\lambda^2 - (a+d)\,\lambda + ad - bc = 0 \tag{3.12}$$

are the *eigenvalues*.

For example, consider the linear system

$$\begin{aligned} x_{t+1} &= 3x_t + 2y_t \\ y_{t+1} &= \frac{1}{2}x_t + 3y_t. \end{aligned}$$

The characteristic equation is $\lambda^2 - 6\lambda + 8 = 0$, and so the eigenvalues are $\lambda = 2, 4$.

If $\lambda = 2$, then from equation (3.11), we see that v_1 and v_2 must satisfy $v_1 + 2v_2 = 0$. One such pair is $v_1 = -2$ and $v_2 = 1$. The vector $[v_1, v_2] = [-2, 1]$ is called an *eigenvector*[5] associated with eigenvalue $\lambda = 2$.

If $\lambda = 4$, the v_1 and v_2 must satisfy $-v_1 + 2v_2 = 0$, and so an associated eigenvector is $[2, 1]$. This provides two solutions in the form of the Ansatz (3.10), written in vector form as

$$\begin{pmatrix} x_t \\ y_t \end{pmatrix} = \begin{pmatrix} -2 \\ 1 \end{pmatrix} 2^t \text{ and } \begin{pmatrix} x_t \\ y_t \end{pmatrix} = \begin{pmatrix} 2 \\ 1 \end{pmatrix} 4^t. \tag{3.13}$$

These two solutions are *independent* because they are not multiples of each other.[6]

There are other choices for eigenvectors associated with $\lambda = 2$ and $\lambda = 4$. For example, for $\lambda = 2$ the pair $v_1 = 1$ and $v_2 = -1/2$ also satisfies the equation $v_1 + 2v_2 = 0$ which gives the eigenvector $[v_1, v_2] = [1, -1/2]$ and yet another solution

$$\begin{pmatrix} x_t \\ y_t \end{pmatrix} = \begin{pmatrix} 1 \\ -1/2 \end{pmatrix} 2^t.$$

Clearly there infinitely many solution pairs of the equation $v_1 + 2v_2 = 0$ and hence infinitely many eigenvectors associated with $\lambda = 2$ (and with $\lambda = 4$ as well), each of which produces a solution of the linear system. However, we need not consider all these other eigenvectors and solutions for the following reason: A fundamental fact from linear algebra is that *two independent solutions span the whole space of solutions.* That is to say, every solution $[x_t, y_t]$

[4]The determinant of matrix $\begin{pmatrix} \alpha & \beta \\ \gamma & \delta \end{pmatrix}$ is denoted $\begin{vmatrix} \alpha & \beta \\ \gamma & \delta \end{vmatrix}$ and is equal to $\alpha\delta - \beta\gamma$.

[5]We use square-bracket notation throughout this book to indicate transpose. That is, $[v_1, v_2]$ means the column vector $(v_1, v_2)^T$, which is the transpose of the row vector (v_1, v_2).

[6]In general, a set of vectors is *linearly independent* if you cannot form any nontrivial linear combination of the vectors that equals the zero vector.

must be a linear combination of any pair of two independent solutions. Thus, the linear combination of two independent solutions is called the *general solution* of the equation.

In the example above, the two solutions (3.13) yield the general solution

$$\begin{pmatrix} x_t \\ y_t \end{pmatrix} = c_1 \begin{pmatrix} -2 \\ 1 \end{pmatrix} 2^t + c_2 \begin{pmatrix} 2 \\ 1 \end{pmatrix} 4^t,$$

where c_1 and c_2 are real numbers. The general solution also can be written as

$$\begin{aligned} x_t &= -2c_1 (2)^t + 2c_2 (4)^t \\ y_t &= c_1 (2)^t + c_2 (4)^t . \end{aligned}$$

Specification of an initial condition leads to particular values of c_1, c_2, which determine the *particular solution* for the initial value problem.

What if the eigenvalues are complex numbers? In that case, the associated eigenvectors, obtained from equations (3.11) as before, are complex. The key to understanding the effects of complex eigenvalues on solutions is to know that any complex number $\lambda = \alpha + \beta i$ can be expressed, in so-called polar coordinates, as $\lambda = re^{i\varphi} = r(\cos\varphi + i\sin\varphi)$, where $r = |\lambda| = \sqrt{\alpha^2 + \beta^2}$ is the modulus of the complex number and φ is the angle whose tangent is β/α.[7]

Thus, the Ansatz formulas (3.10) with $\lambda^t = r^t e^{i\varphi t} = (r^t \cos\varphi t) + i(r^t \sin\varphi t)$ yield a complex-valued solution. A lemma from linear algebra says that the real and imaginary parts of a complex-valued solution are two independent real-valued solutions. Therefore, the linear combination of the real and imaginary parts of the complex solution is the general solution of the equation. Since these two independent solutions involve $\sin\varphi t$ and $\cos\varphi t$, it turns out that all solution trajectories plotted in the (x, y)-plane form spirals around the origin. In addition, since both independent solutions have a factor of r^t, we see that all solutions have a factor of r^t. This implies solutions spiral in toward the origin if $r = |\lambda| < 1$ or spiral outward (unbounded) if $r = |\lambda| > 1$. If $r = |\lambda| = 1$, then solutions are cycles of period n if $\varphi = 2\pi/n$ for some integer n; and they are closed loops (*invariant loops*) if $\varphi \neq 2\pi/n$ for any integer n.

3.5 HIGHER DIMENSIONS: NONLINEAR SYSTEMS

Let's consider the following nonlinear system. Suppose a population has a juvenile (non-reproductive) stage x and an adult (reproductive stage) y and can be modeled with a two-dimensional Ricker-type model of the form

$$\begin{aligned} x_{t+1} &= by_t e^{-cy_t} \\ y_{t+1} &= x_t + sy_t. \end{aligned} \tag{3.14}$$

Here, the time unit is the juvenile maturation period, so that no juvenile from time t remains in the juvenile class at time $t+1$, and the only juveniles at time $t+1$ are those produced by adults present at time t. The coefficient $0 < s < 1$ is the fraction of adults that survive each time period.

[7]The real part of a complex number $\lambda = \alpha + \beta i$, denoted $\mathrm{Re}(\lambda)$, is given by $\mathrm{Re}(\lambda) = \alpha$ and the imaginary part, denoted $\mathrm{Im}(\lambda)$, is $\mathrm{Im}(\lambda) = \beta$. One can graph complex numbers as points (α, β) in the complex plane, in which $\mathrm{Im}(\lambda)$ is plotted against $\mathrm{Re}(\lambda)$. The modulus $|\lambda| = \sqrt{\alpha^2 + \beta^2}$ is the length of the line segment between the origin and the point. The angle φ subtended by this line has tangent β/α, i.e. $\tan\varphi = \beta/\alpha$. The set of points for which $|\lambda| = 1$ is the unit circle in the complex plane. When $|\lambda| < 1$, we say λ is *inside the unit circle in the complex plane.*

First, we find the equilibria from the fixed-point equations

$$x = bye^{-cy}$$
$$y = x + sy.$$

Clearly $[0, 0]$ is an equilibrium. From the second equation, we obtain $x = (1 - s)y$, which, from the first equation, yields $1 - s = be^{-cy}$, which can be solved for y. In this way we obtain two equilibrium branches as functions of b:

$$\begin{pmatrix} 0 \\ 0 \end{pmatrix} \quad \text{and} \quad \begin{pmatrix} (1-s)\left(\ln \frac{b}{1-s}\right)/c \\ \left(\ln \frac{b}{1-s}\right)/c \end{pmatrix}. \tag{3.15}$$

Does the population go extinct or persist? How do we determine the stabilities of these equilibrium branches? In order to answer this question, we turn again to the concept of linearization.

3.5.1 Linearization

Suppose f and g are nonlinear functions of x and y and are continuously differentiable in both variables, and consider the two-dimensional nonlinear system

$$\begin{aligned} x_{t+1} &= f\left(x_t, y_t\right) \\ y_{t+1} &= g\left(x_t, y_t\right). \end{aligned} \tag{3.16}$$

We will write and linearize this system using vector notation in order to highlight the parallels to the one-dimensional situation in Section 3.3.2. For those less familiar with vector notation, the same steps are shown without vector notation in an appendix at the end of this chapter.

In vector form, we can write system (3.16) as

$$\mathbf{x}_{t+1} = F\left(\mathbf{x}_t\right),$$

where

$$\mathbf{x} = \begin{pmatrix} x \\ y \end{pmatrix} \quad \text{and} \quad F = \begin{pmatrix} f \\ g \end{pmatrix}.$$

If $\mathbf{x}_e = [x_e, y_e]$ is an equilibrium, then, in direct parallel to equation (3.5), we can approximate F near $\mathbf{x}_e$ with its linearization at $\mathbf{x}_e$:

$$\mathbf{x}_{t+1} = F\left(\mathbf{x}_t\right) \approx F\left(\mathbf{x}_e\right) + J\left(\mathbf{x}_e\right)\left(\mathbf{x}_t - \mathbf{x}_e\right)$$

for $\mathbf{x}_t \approx \mathbf{x}_e$, where the *Jacobian matrix* J is a kind of generalized derivative whose entries are the partial derivatives[8] of f and g:

$$J \triangleq \begin{pmatrix} \frac{\partial f}{\partial x} & \frac{\partial f}{\partial y} \\ \frac{\partial g}{\partial x} & \frac{\partial g}{\partial y} \end{pmatrix}.$$

Because $\mathbf{x}_e$ is an equilibrium, we have $\mathbf{x}_e = F\left(\mathbf{x}_e\right)$, and thus

$$\mathbf{x}_{t+1} - \mathbf{x}_e \approx J\left(\mathbf{x}_e\right)\left(\mathbf{x}_t - \mathbf{x}_e\right)$$

[8]The partial derivative $\partial f/\partial x$ is simply the derivative of $f(x, y)$ with respect to x while treating y as a constant.

for $\mathbf{x}_t \approx \mathbf{x}_e$, in parallel to equation (3.6). If we define the displacement of the system from its equilibrium as

$$\mathbf{u}_t = \mathbf{x}_t - \mathbf{x}_e,$$

we obtain

$$\mathbf{u}_{t+1} \approx J(\mathbf{x}_e)\,\mathbf{u}_t$$

for $\mathbf{u}_t \approx \mathbf{0}$.

The *linearization of* $\mathbf{x}_{t+1} = F(\mathbf{x}_t)$ *at the equilibrium* $\mathbf{x}_e$ is defined to be the linear system

$$\mathbf{u}_{t+1} = J(\mathbf{x}_e)\,\mathbf{u}_t. \tag{3.17}$$

If the eigenvalues of the Jacobian matrix $J(\mathbf{x}_e)$ are λ_1 and λ_2, then the *Linearization Theorem* guarantees that the following two statements are true:

If $|\lambda_i| < 1$ for all i, then $[x_e, y_e]$ is (locally) asymptotically stable.
If $|\lambda_j| > 1$ for some j, then $[x_e, y_e]$ is unstable.

3.5.2 Ricker Model with Senescence

Before linearizing the juvenile-adult Ricker model (3.14), let's turn briefly to a different two-dimensional extension of the Ricker model (3.3), in which individuals go through a second, senescent (post-reproductive), life stage after one time step. Suppose that all individuals survive until senescence, and that each senescent individual survives each subsequent model time step with probability s. If the number of senescent individuals at time t is denoted y_t, then we have the two-dimensional model

$$\begin{aligned} x_{t+1} &= b x_t e^{-c x_t} \\ y_{t+1} &= x_t + s y_t, \end{aligned} \tag{3.18}$$

where $b, c > 0$ and $0 < s < 1$.

Model (3.18) is a trivial extension of the Ricker model, because the first equation does not involve y, and hence the dynamics for x are identical to those of the Ricker model (3.3). We pose this two-dimensional model in order to visualize the nonlinear Ricker dynamics in state space.

The fixed-point equations for model (3.18) are

$$\begin{aligned} x &= b x e^{-cx} \\ y &= x + s y, \end{aligned}$$

which yield, as a function of b, two equilibrium branches

$$\begin{pmatrix} 0 \\ 0 \end{pmatrix} \quad \text{and} \quad \begin{pmatrix} (\ln b)/c \\ (\ln b)/(c(1-s)) \end{pmatrix}, \tag{3.19}$$

the second of which is positive when $b > 1$.

The Jacobian is

$$J(x,y) = \begin{pmatrix} b e^{-cx}(1-cx) & 0 \\ 1 & s \end{pmatrix}.$$

When evaluated at the trivial equilibrium, the Jacobian is

$$J(0,0) = \begin{pmatrix} b & 0 \\ 1 & s \end{pmatrix},$$

which has characteristic equation $(b-\lambda)(s-\lambda)=0$ and eigenvalues $\lambda=b$ and $\lambda=s<1$. Thus, as for the Ricker map, the trivial equilibrium is asymptotically stable for $b<1$ and unstable for $b>1$.

Evaluated at the nontrivial equilibrium, the Jacobian is

$$J=\begin{pmatrix} 1-\ln b & 0 \\ 1 & s \end{pmatrix},$$

with eigenvalues $\lambda=1-\ln b$ and $\lambda=s<1$. Thus, when positive (which it is for $b>1$), the nontrivial equilibrium is asymptotically stable for $1<b<e^2$ and unstable for $b>e^2$, as it was for the Ricker map. Indeed, the bifurcation diagram for x vs. b for the Ricker model with senescence (3.18) is identical to that of the Ricker model (3.3), shown in Figure 3.3. The two-dimensionality of model (3.18), however, will allow us to visualize the sequence of attractors as they appear in state space.

3.6 STATE SPACE

State space (also called phase space) is the space spanned by the dependent variables (state variables), that is, the joint region of all possible values of the state variables. For model (3.18), state space is the x-y plane. One of the most helpful conceptual tools in dynamical systems theory is the ability to visualize attractors and other invariant sets in state space. Invariant sets in state space are like fixed furniture in a room about which model trajectories move through time as governed by the properties of these invariant sets. In this section, we will consider examples of equilibria, cycles, invariant loops, and chaotic attractors in state space.

Consider the bifurcation diagram for model (3.18) in Figure 3.3. Six arrows along the period-doubling cascade mark the values of $b=0.5$, $b=6$, $b=8$, $b=12$, $b=13$, and $b=17$. The attractors for these values of b are, respectively, the extinction equilibrium, a positive equilibrium, a two-cycle, a two-cycle with larger amplitude, a four-cycle, and a chaotic attractor. Figure 3.4 shows these six attractors, taken from the period-doubling cascade, in state space; and to the right of each state space plot is an associated time series.

3.6.1 Saddles

What happens to the positive equilibrium of model (3.18) when it gives up its stability to the new-born two-cycle at $b=e^2$ (Figure 3.3)? It still exists, now as an unstable equilibrium, but does it continue to influence the dynamics? This is an important question, not only for this example, but in general, and the answer is a resounding "Yes." As we shall see throughout this book, *unstable invariant sets continue to influence the trajectories of populations in state space.*

We noted previously that the positive equilibrium, given in (3.19), of model (3.18) has eigenvalues $\lambda=s$ and $\lambda=1-\ln b$. As b increases through the critical value $b=e^2$, one eigenvalue ($\lambda=s$) remains between 0 and 1, with eigenvector $[0,1]$, whereas the other ($\lambda=1-\ln b$) decreases through the value -1, with eigenvector $[1-s-\ln b, 1]$. An equilibrium with one eigenvalue satisfying $|\lambda|<1$ and another satisfying $|\lambda|>1$ is called a *saddle*. Locally, the line in state space through the saddle equilibrium that is determined by the eigenvector $[0,1]$, which is a vertical line in this example, is tangent to the *stable manifold*, along which trajectories in state space approach the saddle equilibrium. The line through the saddle that is determined by the vector $[1-s-\ln b, 1]$, which is a line with slope $1/(1-s-\ln b)$, is tangent to the *unstable manifold*, along which trajectories move away from the saddle equilibrium (in an oscillatory fashion in this example) and approach the two-cycle attractor. Trajectories that land close to the stable manifold move toward the

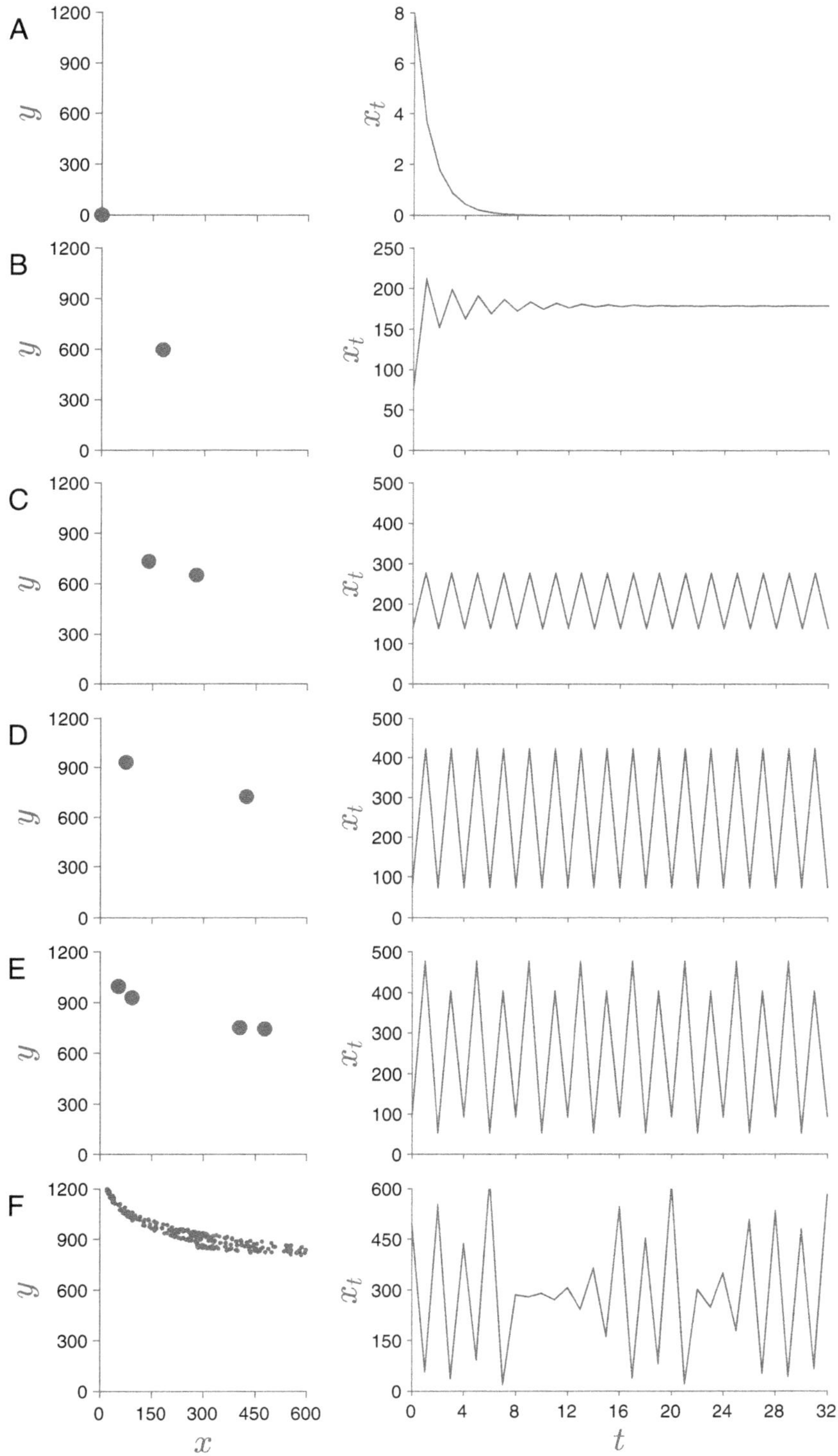

Figure 3.4 Period-doubling cascade for model (3.18) with $c = 0.01$ and $s = 0.7$. Attractors are shown in state space along with corresponding time series. A. $b = 0.5$, extinction equilibrium. B. $b = 6$, positive equilibrium. C. $b = 8$, two-cycle. D. $b = 12$, larger amplitude two-cycle. E. $b = 13$, four-cycle. F. $b = 17$, chaotic attractor. These attractors correspond to the arrows in Figure 3.3.

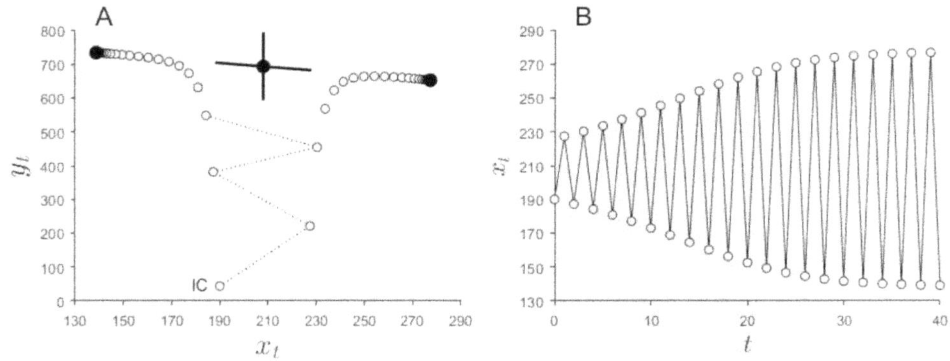

Figure 3.5 Saddle flyby for model (3.18) with $b = 8$, $c = 0.01$, $s = 0.7$, and initial condition (IC) $[190, 44]$. A. Shown in state space, the trajectory approaches the saddle (intersection of cross) along the stable manifold (vertical line), and exits along the unstable manifold as it moves toward the two-cycle attractor (solid circles not on cross). The first few steps are connected with dotted lines. B. Corresponding time series for x.

equilibrium before moving away in the direction of the unstable manifold; such a trajectory is called a *saddle flyby*.

Figure (3.5) shows a saddle flyby, both in state space and as a time series, for model (3.18) with $b = 8$, $c = 0.01$, $s = 0.7$, and initial condition IC $= [190, 44]$. The saddle, shown with a cross, has coordinates $[207.94, 693.15]$. The eigenvalues are $\lambda = 0.7$ with eigenvector $[0, 1]$ and $\lambda = -1.08$ with eigenvector $[-1.78, 1]$. Close to the saddle, the unstable manifold is tangent to a line with slope $-1/1.78 = -0.561\,80$. The trajectory, which begins close to the stable manifold, first approaches the saddle in the direction of the (vertical) stable manifold and then departs, oscillating back and forth in the direction of the unstable manifold toward the stable two-cycle. The distance between the points on the trajectory indicate the speed. Locally (close to the saddle), along the stable manifold, the trajectory's displacement from the saddle decays like $(0.7)^t$, slowing down as it approaches the saddle. Locally, along the unstable manifold, the displacement grows like $(-1.80)^t$, speeding up as it leaves the saddle (but slowing again as it approaches the two-cycle).

3.6.2 Invariant Loops

Now let's return to the juvenile-adult Ricker model (3.14) with equilibria (3.15). The Jacobian is

$$J(x, y) = \begin{pmatrix} 0 & be^{-cy}(1 - cy) \\ 1 & s \end{pmatrix}.$$

For the $[0, 0]$ branch of equilibria, we have

$$J(0, 0) = \begin{pmatrix} 0 & b \\ 1 & s \end{pmatrix},$$

with characteristic equation

$$\lambda^2 - s\lambda - b = 0.$$

The eigenvalues are

$$\lambda = \frac{1}{2}s \pm \frac{1}{2}\sqrt{s^2 + 4b},$$

both of which are real since the discriminant $s^2 + 4b > 0$. The larger eigenvalue (with the "+") is positive, and the other is negative since $\sqrt{s^2 + 4b} > s$. The positive eigenvalue is

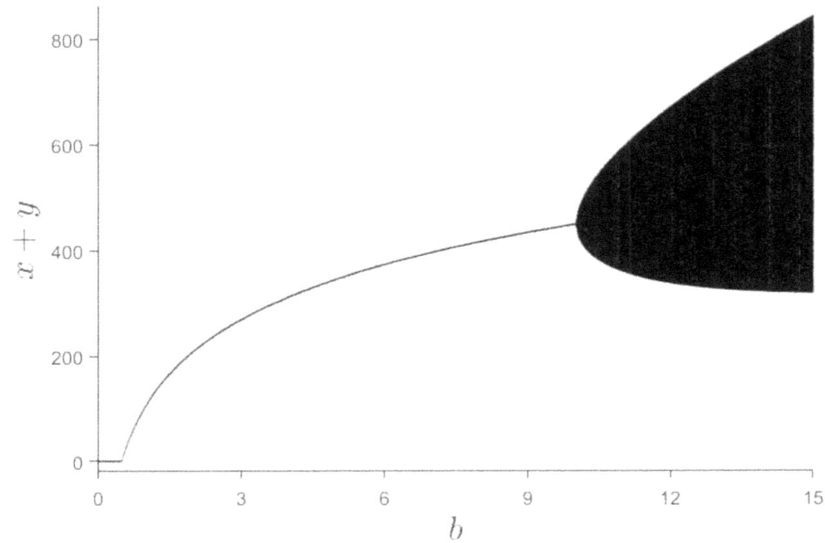

Figure 3.6 Bifurcation diagram for system (3.14) with $c = 0.01$ and $s = 0.5$. The primary bifurcation occurs at $b = 1 - s = 0.5$, and the invariant loop bifurcation occurs at $b = 10.043$, as given by formula (3.20).

also *dominant*, in the sense that it is the largest in absolute value. If $b > 0$ is small, the eigenvalues are close to 0 and s, and so both eigenvalues satisfy $|\lambda| < 1$. Thus, the extinction equilibrium is asymptotically stable for small b. Furthermore, the dominant eigenvalue is increasing in b, and increases through the value $+1$ as b increases through the value $b = 1 - s$. At this point a branch of nontrivial equilibria bifurcates to the right (Figure 3.6). The other eigenvalue at this primary bifurcation point is $-(1-s)$.

For the positive branch of equilibria, given in (3.15), the Jacobian is

$$J = \begin{pmatrix} 0 & (1-s)\left(1 - \ln \frac{b}{1-s}\right) \\ 1 & s \end{pmatrix},$$

with eigenvalues $\lambda = \frac{1}{2}s \pm \frac{1}{2}\sqrt{\Delta}$, where this time the discriminant is

$$\Delta = s^2 + 4(1-s)\left(1 - \ln \frac{b}{1-s}\right).$$

Note that the derivative of Δ with respect to b is $\partial\Delta/\partial b = -4(1-s)/b < 0$ and thus Δ is decreasing in b.

At the primary bifurcation point (that is, when $b = 1 - s$), we have $\Delta = (2-s)^2 > 0$ and $\lambda = 1, s - 1$, so as b increases through $b = 1 - s$, both eigenvalues remain real with $|\lambda| < 1$ until Δ decreases through $\Delta = 0$, at which point the eigenvalues become the complex conjugate pair $\lambda = \frac{1}{2}s \pm \frac{1}{2}i\sqrt{|\Delta|}$, still inside the unit circle in the complex plane (that is, still satisfying $|\lambda| < 1$).

As b continues to increase, an *invariant loop bifurcation* occurs as the two conjugate eigenvalues exit the unit circle in the complex plane (that is, as they simultaneously pass through the value $|\lambda| = 1$). An invariant loop bifurcation is also called a *Neimark-Sacker bifurcation* or a discrete Hopf bifurcation. The invariant loop bifurcation in this model occurs when

$$\left| \frac{1}{2}s \pm \frac{1}{2}i\sqrt{-s^2 + 4(1-s)\left(\ln \frac{b}{1-s} - 1\right)} \right| = 1,$$

that is, when

$$\left(\frac{1}{2}s\right)^2 + \left(\frac{1}{2}\sqrt{-s^2 + 4(1-s)\left(\ln\frac{b}{1-s} - 1\right)}\right)^2 = 1,$$

which occurs when

$$b = (1-s)\exp\frac{2-s}{1-s}. \tag{3.20}$$

Figure 3.6 shows the bifurcation diagram for system (3.14) for $c = 0.01$ and $s = 0.5$. The primary bifurcation occurs at $b = 1 - s = 0.5$, and the invariant loop bifurcation occurs at approximately $b = 10.043$, as calculated by equation (3.20).

The birth of the invariant loop is illustrated in Figure 3.7A with a sequence of state space plots for $b = 9.8$, $b = 10.2$, and $b = 11$. Figure 3.7B shows the time series on the invariant loop for $b = 11$.

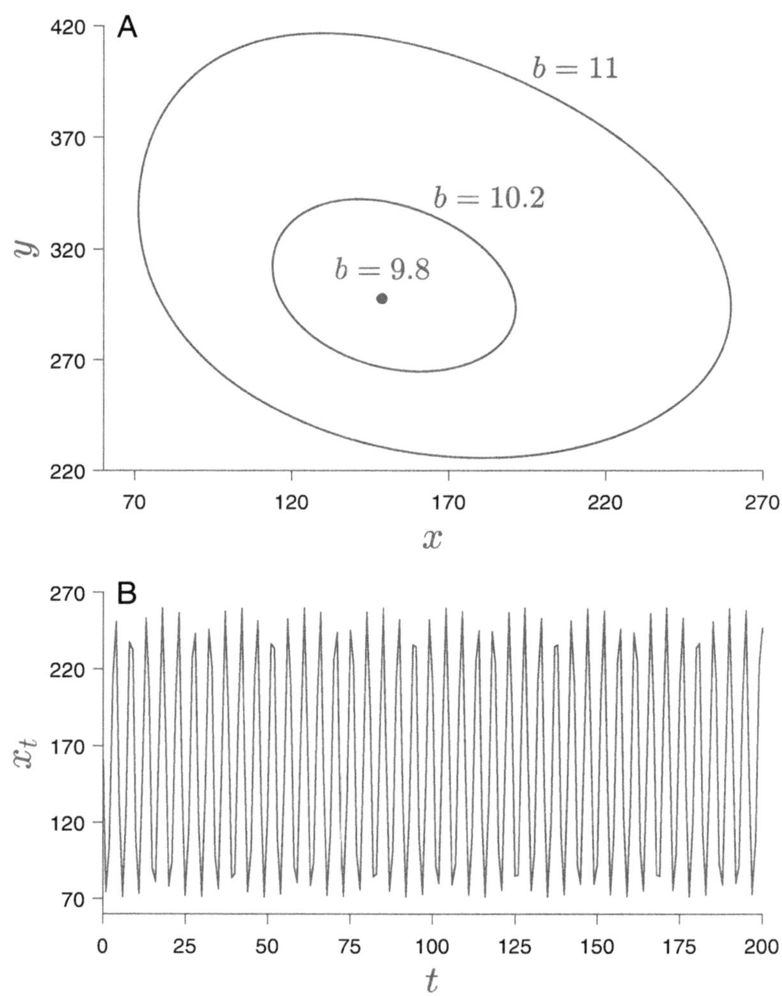

Figure 3.7 Dynamics of system (3.14) with $c = 0.01$ and $s = 0.5$, showing the birth of an invariant loop and a time series on an invariant loop. A. State space for $b = 9.8$ (equilibrium), $b = 10.2$ (invariant loop), and $b = 11$ (larger invariant loop). B. Time series with $b = 11$.

3.7 CONCLUSIONS

Populations without density dependence grow exponentially or decline exponentially toward extinction. Populations with density dependence also behave this way when close to extinction. Mechanisms of density dependence, however, also can lead to a plethora of other types of dynamic phenomena, including positive equilibria, saddles, bifurcations, period-doubling cascades to chaos, invariant loops, and many more. *The thesis of this book is that nonlinear mechanisms in populations can be identified, incorporated into mathematical models, analyzed mathematically, and used to explain and predict the dynamics of real populations.*

3.7.1 Thinking in State Space

State space is like a room in which model trajectories move through time. For a fixed set of parameter values, the room has fixed "furniture," that is, structure, including attracting and repelling invariant sets, saddles that attract in some directions and repel in others, basins of attraction, and basin boundaries. This structure results from the nonlinearities in the model and governs the movement of the model trajectories through state space.

Empirical population data also move about in state space through time. The structure of empirical state space comes from the dominant mechanistic forces that undergird the empirical population system. If these mechanistic forces are well-captured in a mathematical model, then we can conceptualize empirical trajectories as moving within the state space determined by the mathematical model.

Empirical trajectories, however, are stochastic. As we shall discover and emphasize in this book, *stochasticity can allow data trajectories to visit and revisit all of the entities in state space, illuminating the deterministic structure of the "room," and providing a mix of dynamic patterns in time series.*

3.7.2 New Paradigms Arise from Chaos

Determinism generally has been equated with *predictability.* If a system is deterministic, then the output is completely determined by the input, and is thus predictable, given the input. Similarly, *stochasticity* has been identified with *unpredictability.* Chaos, however, significantly alters this classical paradigm. Chaos is deterministic, but has sensitivity to initial conditions. Therefore, if the initial conditions are not known exactly, then the future state of the system is, for all practical purposes, unpredictable. So in a practical sense, there is such a thing as "deterministic unpredictability."

Another classical paradigm is that simple causes give rise to simple results, and equivalently, complex results require complex causes. Chaos shows why this is not necessarily the case. The Ricker model is a simple deterministic recursion formula, and yet its dynamics are so complicated that one can spend a lifetime exploring them. In fact, complex results do *not* necessarily imply complex causes. *Very simple rules can generate extraordinarily complicated results.*

In 1976, Robert May wrote [5]:

> Quite apart from their intrinsic mathematical interest, the above results raise very awkward biological questions. They show that simple and fully deterministic models, in which all biological parameters are exactly known, can nonetheless (if the nonlinearities are sufficiently severe) lead to population dynamics which are in effect indistinguishable from the sample function of a random process. Apparently chaotic population fluctuations need not necessarily be due to random environmental fluctuations, or sampling errors, but may reflect the workings of some deterministic, but strongly density dependent, population model.

May's hypothesis is the idea that apparently random fluctuations of population abundances might be explainable largely by low-dimensional, nonlinear, deterministic forces.

In 1997, chaos was documented in the dynamics of a biological population for the first time when it was experimentally induced in laboratory populations of insects [1, 2, 4]. We will discuss these experiments in this book.

3.8 APPENDIX: THE RICKER NONLINEARITY

The Ricker nonlinearity e^{-cx} arises from random contacts between individuals in a population. The contacts could be deleterious or advantageous. For example, in fisheries or insect models, the Ricker function can result from random cannibalistic encounters between individuals within a population.

Suppose, for example, that we are modeling an insect population with J juveniles and A adults using a discrete-time model. Suppose the adults cannibalize juveniles through random encounters. Let Δt be a small fraction of one model time step, say one m^{th} of the time step, so that $1 = m\Delta t$. Assume that the probability that one adult encounters (and hence cannibalizes) one juvenile during Δt units of time is proportional to the time elapsed Δt, that is, $c\Delta t$. The coefficient $c > 0$ can be thought of as the continuous-time encounter rate or cannibalism rate. The probability that one juvenile survives cannibalism by one adult in Δt units of time is therefore $1 - c\Delta t$. Assuming the encounters are independent events, the probability that a juvenile survives all of the A adults in Δt units of time is the product

$$(1 - c\Delta t)^A .$$

Now, each model time step is composed of $m = 1/\Delta t$ successive times of duration Δt. The probability that a juvenile survives all of the adults during this time is the product

$$(1 - c\Delta t)^{Am} = (1 - c\Delta t)^{A/\Delta t} .$$

If we let $\Delta t \to 0$, we find

$$\lim_{\Delta t \to 0} (1 - c\Delta t)^{A/\Delta t} = e^{-cA}.$$

So, the probability that a juvenile survives cannibalism during one model time step in the presence of A adults is e^{-cA}, which is the Ricker nonlinearity.

Interestingly, the Ricker nonlinearity also can be derived for encounters that arise from some non-random spatial or temporal processes. The idea is that the number of encounters, say x, can be modeled as a stochastic "birth process," that is, a discrete nondecreasing process on the nonnegative integers in which encounters are one-by-one "arrivals" in space or time. Random encounters are in fact such a birth process with a constant arrival rate known as a Poisson process. However, the birth process model can be extended instead to have an arrival rate that is a function of x, the current number of encounters: An increasing function represents aggregated arrivals (finding some means a tendency to find more) and a decreasing function represents disaggregation (arrivals overly spread out). Remarkably, if the background intensity of encounters is proportional to A (the density of the items being encountered), then the "zero encounters" probability for all such birth process models is of the form e^{-cA} [3].

3.9 APPENDIX: LINEARIZATION

In this appendix we will obtain the linearization and Jacobian with scalar equations. Suppose f and g are nonlinear functions of x and y and are continuously differentiable in both variables. Consider the nonlinear system

$$\begin{aligned} x_{t+1} &= f(x_t, y_t) \\ y_{t+1} &= g(x_t, y_t). \end{aligned} \tag{3.21}$$

An equilibrium $[x_e, y_e]$ of system (3.21) satisfies the fixed point equations $x_e = f(x_e, y_e)$ and $y_e = g(x_e, y_e)$. The surface $z = f(x, y)$ can be approximated near an equilibrium (x_e, y_e) by its *tangent plane*:

$$f(x, y) \approx f(x_e, y_e) + \frac{\partial f}{\partial x}(x_e, y_e)(x - x_e) + \frac{\partial f}{\partial y}(x_e, y_e)(y - y_e)$$

for $x \approx x_e$ and $y \approx y_e$. Thus, near the equilibrium, we have

$$\begin{aligned} x_{t+1} &\approx x_e + \frac{\partial f}{\partial x}(x_e, y_e)(x_t - x_e) + \frac{\partial f}{\partial y}(x_e, y_e)(y_t - y_e) \\ y_{t+1} &\approx y_e + \frac{\partial g}{\partial x}(x_e, y_e)(x_t - x_e) + \frac{\partial g}{\partial y}(x_e, y_e)(y_t - y_e). \end{aligned}$$

If we define the displacement from equilibrium by the changes of variables $u_t = x_t - x_e$ and $v_t = y_t - y_e$, and change the $\approx$ to $=$, we obtain the *linearization*:

$$\begin{aligned} u_{t+1} &= \frac{\partial f}{\partial x}(x_e, y_e) u_t + \frac{\partial f}{\partial y}(x_e, y_e) v_t \\ v_{t+1} &= \frac{\partial g}{\partial x}(x_e, y_e) u_t + \frac{\partial g}{\partial y}(x_e, y_e) v_t, \end{aligned}$$

which can be written in vector form as

$$\mathbf{u}_{t+1} = J(\mathbf{x}_e)\,\mathbf{u}_t,$$

where

$$\mathbf{u} = \begin{pmatrix} u \\ v \end{pmatrix}; \quad J = \begin{pmatrix} \frac{\partial f}{\partial x} & \frac{\partial f}{\partial y} \\ \frac{\partial g}{\partial x} & \frac{\partial g}{\partial y} \end{pmatrix}.$$

Bibliography

[1] Costantino RF, Desharnais RA, Cushing JM, Dennis B (1997) Chaotic dynamics in an insect population. Science 275:389–391. https://doi.org/10.1126/science.275.5298.389

[2] Cushing JM, Costantino RF, Dennis B, Desharnais RA, Henson SM (2003) Chaos in Ecology: Experimental Nonlinear Dynamics. Theoretical Ecology Series. Volume 1. New York, Academic Press.

[3] Dennis B (1989) Allee effects: Population growth, critical density, and the chance of extinction. Natural Resource Modeling 3:481–538. https://doi.org/10.1111/j.1939-7445.1989.tb00119.x

[4] Dennis B, Desharnais RA, Cushing JM, Henson SM, Costantino RF (2001) Estimating chaos and complex dynamics in an insect population. Ecological Monographs 71:277-303. https://doi.org/10.1890/0012-9615(2001)071[0277:ECACDI]2.0.CO;2

[5] May RM (1976) Theoretical Ecology: Principles and Applications. Philadelphia, WB Saunders.

[6] Ricker WE (1954) Stock and recruitment. Journal of the Fisheries Research Board of Canada 11:559–623. https://doi.org/10.1139/f54-039

[7] Ricker WE (1975) Computation and interpretation of biological statistics of fish populations. Bulletin of the Fisheries Research Board of Canada 191. Information Canada, Ottawa. https://waves-vagues.dfo-mpo.gc.ca/Library/1485.pdf

II

Population Data and Models

CHAPTER 4

The *Tribolium* Laboratory Model

Flour beetles of the genus *Tribolium* are small insects that complete their entire life cycle in stored grain products. There are 26 species in the genus [22] and, for several species, many mutants and genetic strains have been identified. They have been widely used as an experimental organism in a variety of research fields [43]. The flour beetle laboratory system often is referred to as the "*Tribolium* model." In this chapter, we provide a brief overview of the *Tribolium* model in ecology, including its historical contributions, biological attributes, and laboratory protocols. We also discuss the approach used for applying the *Tribolium* model to the study of nonlinear population dynamics.

4.1 THE *TRIBOLIUM* MODEL IN ECOLOGY

Flour beetles of the genus *Tribolium* have had a role in the study of ecology for nearly a century. These insects are easily cultured in the laboratory and their life stages can be sifted from the medium and counted without estimation error. They complete their life cycle in about a month, so multi-generational population experiments are feasible. These species have been used to study a variety of ecological phenomena, including population growth, species competition, population stochasticity, and nonlinear population dynamics. This section highlights some milestone studies in these topics. See the reviews by Mertz [33], Costantino and Desharnais [5], Cushing et al. [8], and Desharnais et al. [20] for a more complete picture.

4.1.1 Early Research

The *Tribolium* flour beetle was introduced as an experimental organism for the study of populations in 1928, with the publication of a paper titled "The quantitative analysis of environmental factors" by Royal Norton Chapman at the University of Minnesota [3]. Although not a mathematician, Chapman emphasized the critical role of mathematics in science and suggested that ecology could benefit from a more quantitative approach. He wrote:

> Those of us who are engaged in the study of animal ecology may do well to pause and consider the future of our branch of science in the light of the work of Fourier, who more than a century ago, began to apply mathematical methods to the study of the transmission of heat through solids. In this day when it is possible for physicists and engineers to calculate so exactly the amount of heat required to warm a given space surrounded by walls with given coefficients of

DOI: 10.1201/9781003467120-4

> conduction and subject to known external conditions, it is hard to realize that there was once a day when it seemed to be impossible to apply quantitative calculations to the transmission of heat. All of the objections which now stand in the way of the use of these methods in ecology then stood in the way of the use of these methods in connection with thermodynamics.

Chapman conducted an experiment in which the size of the experimental habitat was varied according to a geometric series with 4, 8, 16, 32, and 64 g of medium. Each population was initiated with one pair of adults for each 4 g of flour. The populations were counted at regular intervals over the course of 156 days. Chapman found that the density of the total number of insects approached a similar equilibrium in each habitat. Borrowing from Ohm's Law, where the current of an electrical circuit is equal to the voltage divided by the resistance, he wrote

$$C = \frac{B}{R}, \tag{4.1}$$

where C is the equilibrium concentration of insects, B is the "biotic potential," and R is the "resistance." Chapman's equation is problematic because he defined the biotic potential B as the product of the equilibrium number of insects C and the per capita fecundity f, that is $B = fC$, in which case $f = R$ and $C = C$. Nevertheless, a similar formulation is obtained from the often-used logistic population model, $dN/dt = N\,(r - aN)$, which has an equilibrium population density of $N^* = r/a$, mirroring Chapman's equation (4.1), where, in the logistic model, r is the reproductive rate and a is the density sensitivity parameter.

In addition to recommending that ecology embrace a quantitative approach, Chapman also defended the use of laboratory microcosms to study ecology:

> That objections will be raised to the use of laboratory results for field conditions, may be predicted on the basis of past experience. In answer to such objections it is sufficient to say that there is no surer way to progress than by the comparison of the results from careful determinations and calculations under known conditions with those which are to be found in the field. If the theoretical predictions account for nearly 100 per cent of what is found in the field, we may consider the problem as solved for all practical purposes. If there is a wider difference, which is probable, our search for the unknown may be guided by a logical process of elimination and more minute analysis. The history of the physical sciences offers abundant support for such procedure. By following this precedent, ecology may become one of the more exact of the biological sciences.

Despite a long and productive history, the use of laboratory studies is still a topic that has merited discussion in the ecological literature [2, 13].

It would be tempting to believe that the application of quantitative approaches to the study of population dynamics using *Tribolium* involved a gradual increase in complexity. The work of John Stanley, a student of Chapman, contradicts this assumption. In a series of nine papers [45, 46, 47, 48, 49, 50, 51, 52, 53], published from 1931 to 1966, he presented amazingly detailed mathematical formulations for the growth of laboratory populations of flour beetles. His models involved hundreds of equations dealing with the minutiae of *T. confusum*: numbers of eggs, six larval instars, pupae, callow and reproductive adults. He specified every possible cannibalistic pathway among the various life stages. He treated cannibalism as a random collision process that considers beetle movements, the size of the life stages, and the volume and condition of the medium. He also considered beetle tunneling in an effort to relate this behavior to egg-laying and egg-eating. This work was initiated before the advent of digital computing, so there was little opportunity to test his models with experimental data. Stanley sacrificed generality and tractability in favor of realism, which limited the application of his work to other ecological systems.

4.1.2 Species Competition

The *Tribolium* model is well known among ecologists for its application to the study of ecological competition. This is due largely to Thomas Park and his colleagues, who conducted a 20-year research program focused on competition between two closely related species, *T. castaneum* and *T. confusum*, in mixed species laboratory cultures [25, 35, 36, 37, 38, 39, 41]. In an attempt to obtain stable coexistence, several experimental factors were varied: the genetic strain of each species [38], the environmental conditions [36], initial numbers of beetles [37, 25], and the degree of genetic variability among the founding individuals [34]. While coexistence was never achieved, the surviving species varied among the experimental replicates under a variety of experimental conditions, a phenomenon they termed *competitive indeterminacy*. Two explanations for competitive indeterminacy were proposed: genetic founder effects and demographic stochasticity. After much debate and further experimentation, the latter explanation seemed more parsimonious with the empirical data [34]. See Chapter 15 for more details and an alternative view of these competition studies.

Thomas Park's research program often involved a close collaboration between experimentalists and theorists. Two influential colleagues were the biomathematicians Jerzy Neyman and Patrick Holt Leslie. For example, Leslie analyzed a simplified discrete-time stochastic model for the adults of two competing species [24]:

$$\begin{aligned} N_{1,t+1} &= \frac{b_1 N_{1,t}}{1 + c_{11} N_{1,t} + c_{12} N_{2,t}} + \varepsilon_{1,t}, \\ N_{2,t+1} &= \frac{b_2 N_{2,t}}{1 + c_{21} N_{1,t} + c_{22} N_{2,t}} + \varepsilon_{2,t}. \end{aligned} \tag{4.2}$$

Here, $N_{1,t}$ and $N_{2,t}$ are the numbers of adults of the two species at time t, b_1 and b_2 are the per capita rates of reproduction, and c_{ij} is the effect of an individual of species j on the population growth of species i. The quantities $\varepsilon_{1,t}$ and $\varepsilon_{2,t}$ are independent normally-distributed random variables with means equal to zero and variances based on approximating the system as a birth-death process. The important point is that, for some parameter values, competitive indeterminacy occurs, where the probabilities of the winning species and the mean times to extinction of the losing species were consistent with experimental outcomes. Leslie's model inspired an empirical study in which the initial numbers of both *T. castaneum* and *T. confusum* were varied in mixed species cultures [25].

4.1.3 Population Stochasticity

Leslie used a single species version of his model (4.2) to compare the predicted stationary distribution of adult numbers to the experimental values. Costantino, Desharnais, and Dennis followed up on this theme, using a stochastic differential equation for the change in adult numbers [4, 9, 16, 19]:

$$\frac{dN(t)}{dt} = N(t)\left(be^{-cN(t)} - \mu + \sigma\varepsilon(t)\right). \tag{4.3}$$

Here, $N(t)$ is the number of adults at time t, $b > 0$ is the reproductive rate, $\mu > 0$ is the adult mortality rate, and $c > 0$ is the rate of cannibalism of adults on immatures. The term $\varepsilon(t)$ is a random variable for Gaussian white noise. Using a diffusion equation approach, they obtained a gamma probability density function (PDF)

$$f(N) = kN^{\alpha-1}e^{-\beta N} \tag{4.4}$$

to describe the stationary probability distribution of adult numbers. The values $\alpha > 0$ and $\beta > 0$ are the shape and scale parameters of the gamma PDF and k is the constant of integration. The shape and scale parameters are functions of the parameters b, μ, c, and σ from the stochastic population model (4.3), allowing one to connect the properties of the stationary distribution to the biological factors controlling population growth.

Costantino and Desharnais applied the probability prediction (4.4) to 30 data sets for the stationary distributions of *T. castaneum* and *T. confusum* [5]. Overall, the gamma PDF (4.4) provided a good fit to these data. Comparisons were made among populations with different environmental conditions, genetic backgrounds, and geographical origins. In most cases, there was a good qualitative agreement between the differences in the fitted parameters of the stationary distributions and known differences in the biological attributes of the populations. These results demonstrated how variation in population numbers could be associated with life history information.

4.1.4 Ecological Genetics

By the time Chapman introduced *Tribolium* into the study of ecology, *Drosophila* had already become established as an animal model for the study of genetics. In 1953, Thomas Park met with A. E. Bell at Purdue University. With the help of Park, Bell established a laboratory facility to study the use of flour beetles in genetical research, with a focus on quantitative genetics and its applications to animal breeding [1].

In the mid-1970s, R. F. Costantino, a student of Bell, launched a research program that used the *Tribolium* model to couple population dynamics with classical population genetics. The focus was on the single-locus corn oil sensitive (*cos*) mutant that inhibits larval growth and viability when corn oil is added to the medium. Populations initiated with the wild type (+) and *cos* alleles reached a stable genetic polymorphism, where both alleles are maintained at equilibrium [31].

In the late-1970s, Costantino and his student, R. A. Desharnais, studied models of ecological genetics that coupled the deterministic population model (4.3), with $\sigma = 0$, to the continuous-time single-locus model of natural selection from population genetics theory [7]. They used reproductive parameters, b_{11}, b_{12}, and b_{22}, for the *cos*/*cos*, +/*cos*, and +/+ genotypes. The reproductive parameter b in model (4.3) is an average of the b_{ij} values that depends on the *cos* allele frequency. They studied a variety of phenomena including polymorphism and stability [14], the maximization of population size [15, 19], and the cost of natural selection [5, 17]. These topics are considered in more detail in Chapter 18.

4.1.5 Nonlinear Dynamics

The Beetle Team came together in the 1980s. We represented several disciplines: experimental ecology, applied mathematics, and statistical theory and applications. The *Tribolium* model offered an opportunity to study nonlinear population dynamics using rigorous integration of mathematical modeling, biological experimentation, and statistical methodology—a rare combination for ecology. Thus began two decades of experimental studies of nonlinear dynamics in populations. We were able to demonstrate many key features of nonlinear dynamics in experimental population data, including attracting cycles, aperiodicity, saddles, bifurcations, phase switching, multiple attractors, and even chaos. Our goal was a cornerstone example of a simple mathematical population model for which detailed, quantitative predictions are borne out in manipulative experiments [20]. The work covers the entire spectrum of interdisciplinary modeling activity, from the formulation of a model and analysis

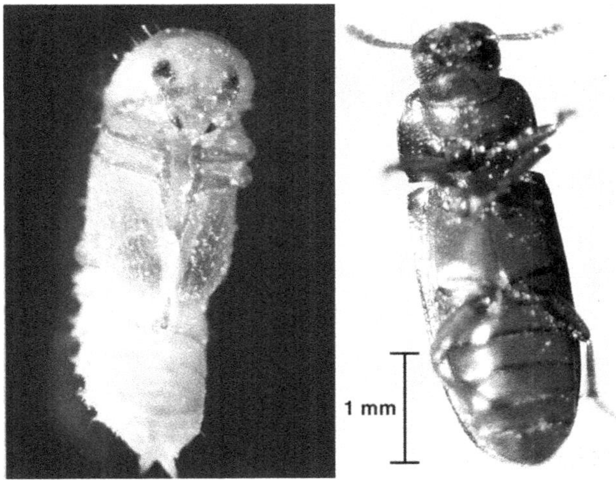

Figure 4.1 Pupa (left) and adult (right) of the flour beetle *Tribolium castaneum.*

of its predictions to the design of experiments and analysis of the resulting data [10]. This book describes the details of this research program.

4.2 THE *TRIBOLIUM* EXPERIMENTAL SYSTEM

Tribolium is a genus of beetles (order Coleoptera) in the family Tenebrionidae (sometimes called darkling beetles). Some common names for *Tribolium* are flour beetles, flour weevils, red weevils, and bran bugs. Their natural habitat is the bark of trees or in rotting logs, but they have been pests of grain, flour, and other stored food products for thousands of years. The species *T. castaneum* ("red flour beetle") and *T. confusum* ("confused flour beetle") are cosmopolitan and have been utilized as a study system in many disciplines [43]. Figure 4.1 shows a pupa and adult of *T. castaneum*. This section discusses their biology and use as an experimental model in ecology.

4.2.1 *Tribolium* Life Cycle

Flour beetles undergo a holometabolous development, with four developmental stages: egg, larva, pupa, and adult (Figure 4.2). The entire life cycle is completed in flour medium. The developmental period is temperature dependent. The time from hatched egg to adult is about four weeks at 32°C [44]. Eggs hatch in three to four days and the larvae begin to tunnel and feed. As larvae grow, they molt; there are usually a total of six larval instars in *T. castaneum* and *T. confusum*. Large larvae become quiescent during their last instar and stop feeding; these are sometimes called prepupae. The duration of the immobile pupal stage is 4–6 days. Sex identification is most easily carried out on pupae. Once adults eclose, it takes another few days for them to become fully scleroticized and pigmented. These callow adults are not yet sexually mature. The mean expectation of life for adults is about 300 days.

In the next chapter, we develop a discrete-time mathematical model (the "LPA model") for studying nonlinear population dynamics using the *Tribolium* model. This model partitions the life cycle into three groups: the L, P, and A stages.[1] The L-stage refers to feeding

[1]The egg stage, which is short in duration, is not included explicitly in the model. The effects of egg

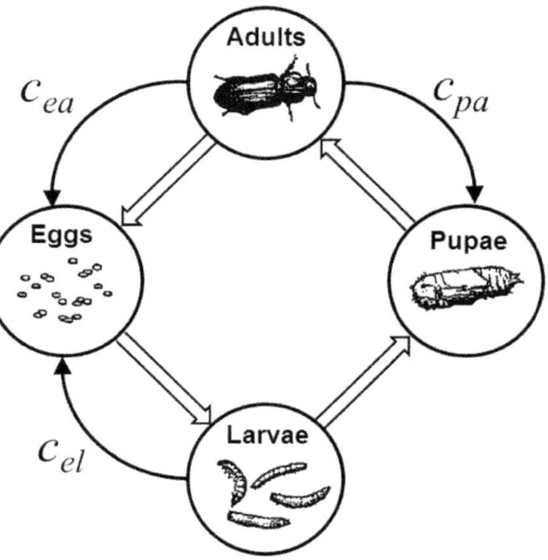

Figure 4.2 Life cycle and life stage interactions in *Tribolium*. The wide arrows represent the developmental cycle and the narrow arrows represent cannibalism.

larvae. The P-stage refers to prepupae, pupae, and callow adults. The A-stage denotes reproductive adults. In the remainder of this book, the terms "larvae," "pupae," and "adults" will be used to denote these three stages. Under the laboratory conditions where experiments designed to test the model predictions were conducted, the durations of the larval and pupal stages are approximately two weeks each. This corresponds to the time step of the model.

4.2.2 Nonlinear Life Stage Interactions

One of the most compelling reasons for using the *Tribolium* model in the study of populations is that it provides a fascinating example of nonlinear demographic dynamics. Laboratory populations of flour beetles maintained under constant environmental conditions exhibit dramatic fluctuations in density and age structure. These fluctuations cannot be characterized as stochastic; they are the result of strong behavioral interactions among the life stages—the most important being cannibalism.

The life stage interactions that drive the dynamics of *Tribolium* populations are summarized in Figure 4.2. Adults eat the immobile eggs and pupae, and mobile larvae eat eggs [40]. The rate at which larvae eat pupae is small and can be ignored. Cannibalism is the biological mechanism that is the basis of nonlinearity in the LPA mathematical model, which is presented in Chapter 5. For example, when eggs hatch, they produce feeding larvae, which are voracious cannibals of eggs. A large cohort of larvae can shut down further recruitment into the larval stage until those larvae become immobile pupae. As a result of the subsequent reduction of egg cannibalism by larvae, many more eggs hatch, and a new cohort of larvae is produced. Thus, in the absence of any experimental manipulations, flour beetle cultures often display large age-class oscillations [10, 11, 12], exemplifying the

eating by adults and larvae (Figure 4.2) are incorporated as density-dependent reductions in the number of larvae produced per adult.

major role cannibalism can play in the nonlinear dynamics of flour beetle populations. Cannibalism does not predominate in all species in the genus *Tribolium*. In the species *T. brevicornis*, for example, the life stage interaction involves the inhibition of metamorphosis, that is, the duration of the larval stage is extended when adult densities are high [23]. See Chapter 20.

4.2.3 Laboratory Protocols

The *Tribolium* model has many characteristics that make it well-suited for ecological studies. Flour beetles spend their entire life cycle in the medium. Cultures can be sifted and an accurate count of all life stages can be obtained (no estimation error). Although they are capable of flight, these insects rarely do so in the laboratory, which contributes to the ease of handling. The insects do not require anaesthetization and they are not affected adversely by manipulation. A population of hundreds of individuals can be maintained on a small quantity of medium, which aids in the replication of experimental treatments. The sexes are easily discriminated in the pupal stage, making it convenient to obtain virgin adults for mating. The four week developmental period allows for experimental population time series that span multiple generations.

The *Tribolium* model experimental protocols are fairly straightforward. Some details differ among laboratories; here we describe the procedures typical to the experiments reported in this book. Populations are contained in small glass vials with 20 g of standard medium (95% whole wheat flour and 5% dried brewer's yeast). The vials are capped with plastic lids containing holes covered with a fine embedded mesh. We maintain cultures in dark incubators at a constant temperature (~33°C) and relative humidity (~56%). To avoid inbreeding, we initiate experimental populations with insects obtained from laboratory stocks of many thousands of individuals. A census of each population is obtained every two weeks by sifting the medium and counting the numbers of each life stage (except eggs). We return all insects (including eggs) to fresh medium after each census. Any required experimental manipulation occurs at the time of census. The length of an experiment may vary, but a typical duration is 80 weeks.

4.2.4 Habitat Size

Altering the volume of flour is a convenient way to introduce constant, periodic, or random changes in habitat size as part of an experimental protocol. This is of particular interest if one can connect habitat size changes to population model predictions.

In the LPA model presented in Chapter 5, parameters are included for the per capita rates of cannibalism for larvae eating eggs (c_{el}), adults eating eggs (c_{ea}), and adults eating pupae (c_{pa}). These cannibalistic interactions are outlined in Figure 4.2. A straightforward assumption is that encounters between cannibals and victims occur at random. In this case, rates of cannibalism should be inversely proportional to habitat size v, measured here as the weight of the medium. As detailed in the next chapter, an exponential function is used to obtain the probability of survival:

$$s_{ij} = \exp\left(-k_0 - \ell_i k / v_j\right), \tag{4.5}$$

where, for example, s_{ij} is the probability that an egg survives cannibalism in one interval of time in the presence of ℓ_i larvae in a habitat with v_j grams of medium.[2] The parameter $k > 0$ is the per capita rate at which larvae cannibalize eggs and $k_0 \geq 0$ represents natural

[2]In our models, we use V to denote habitat size, where $V = 1$ represents 20 g of medium, the standard used in many of the experiments.

TABLE 4.1 Number of eggs surviving cannibalism by larvae.

Number of larvae, ℓ_i	Habitat size, v_j (g of medium) 8	12	16	20	24	28	32
0	87	86	89	90	88	92	95
10	84	85	83	85	86	86	89
20	80	81	83	83	82	87	85
30	83	80	81	79	82	85	87
40	73	70	74	76	75	80	77
50	65	68	73	71	75	77	79

mortality. If there are n eggs and the probability of each egg being eaten is an independent event, then the number of surviving eggs Y_{ij} is a binomial random variable:

$$\Pr\left(Y_{ij} = y_{ij}\right) = \binom{n}{y_{ij}} s_{ij}^{y_{ij}} \left(1 - s_{ij}^{y_{ij}}\right)^{n-y_{ij}}, \tag{4.6}$$

from which we can obtain maximum likelihood parameter estimates for k_0 and k and test the adequacy of the survival model (4.5).

We designed an assay experiment to evaluate the effect of habitat size on the rate of egg cannibalism by larvae. One hundred eggs were placed in each of 42 glass vials. Each vial was assigned a different treatment combination of the amount of medium and the number of larvae in a factorial design. The treatments for various amounts of medium (v_i) were: 8, 12, 16, 20, 24, 28, and 32 g. The numbers of large larvae (ℓ_i) were: 0, 10, 20, 30, 40, and 50. There was one vial for each treatment combination. We placed the vials in a dark incubator at 34°C for 48 h, and counted the numbers of surviving eggs in the vials. The results appear in Table 4.1.

We tested the adequacy of the survival model (4.5). Let y_{ij} represent the number of surviving eggs in a vial with ℓ_i larvae on v_j grams of medium. The binomial distribution (4.6) with $n = 100$ gives the probability of obtaining the observed number y_{ij} of recovered eggs, where s_{ij} is given by equation (4.5). The likelihood function

$$L\left(k_0, k\right) = \prod_i \prod_j \binom{n}{y_{ij}} s_{ij}^{y_{ij}} \left(1 - s_{ij}^{y_{ij}}\right)^{n-y_{ij}} \tag{4.7}$$

gives the joint probability of observing the experimental data in Table 4.1 as a function of the parameters k_0 and k. We obtained the maximum likelihood (ML) estimates $\hat{k}_0 = 0.1235$ and $\hat{k} = 0.05598$ by maximizing $\log L\left(k_0, k\right)$. We conducted a likelihood ratio goodness-of-fit test, with the null hypothesis H_0 being the two-parameter cannibalism survival model (4.5) with the ML estimates, and the alternative hypothesis H_1 being an independent binomial distribution with separate unrestricted survival probabilities $\hat{s}_{ij} = y_{ij}/n$ for each of the vials. If $\hat{L}_0$ and $\hat{L}_1$ are the likelihoods of models H_0 and H_1, evaluated at their ML-estimated survival probabilities, then the likelihood ratio test statistic is given by

$$G^2 = -2 \log \left(\frac{\hat{L}_0}{\hat{L}_1}\right).$$

Under H_0 (cannibalism survival model fits), G^2 has an asymptotic chi-square distribution with degrees of freedom equal to the difference in number of estimated parameters in H_1

TABLE 4.2 Number of pupae surviving cannibalism by adults.

Number of adults, a_i	Habitat size, v_j (g of medium) 8	12	16	20	24	28	32
0	50	48	49	50	49	48	49
50	42	42	43	44	45	47	47
100	41	42	43	42	44	45	46
150	39	40	42	41	43	44	44
200	37	38	38	40	41	43	43
300	31	32	32	34	33	37	39

and in H_0. For the data in Table 4.1, the test fails to reject the null hypothesis, indicating that the proposed survival model fits acceptably well ($G^2 = 30.49$, df $= 40$, $p = 0.86$). Thus, there is no evidence for a lack of fit of the survival model (4.5).

A second assay experiment was designed to evaluate the effect of habitat size on the rate of pupal cannibalism by adults. Fifty pupae ($n = 50$) were placed in each of 42 glass vials. The design was identical to the previously described assay experiment, except that the numbers of adults (a_j) were: 0, 50, 100, 150, 200, and 300. The results appear in Table 4.2. Using the likelihood function (4.7) and the survival model (4.5), where y_{ij} is the number of surviving pupae and a_i replaces ℓ_i, we computed the ML estimates $\hat{k}_0 = 0.3600$ and $\hat{k} = 0.01762$. The likelihood ratio test statistic was $G^2 = 30.95$, indicating, once again, that the proposed survival model (4.5) fits well (df $= 40$, $p = 0.84$).

In summary, the results of the two assay experiments suggest that rates of egg cannibalism by larvae and pupal cannibalism by adults are inversely proportional to the habitat size. No assay experiment is available for egg cannibalism by adults, because the protocols described above would be complicated by the fact that adults lay eggs in addition to eating them. Nevertheless, in the various forms of the LPA model that appear in this book, we feel confident that the rates of cannibalism are inversely proportional to habitat size v and hence the parameters c_{el}, c_{ea}, and c_{pa} can be divided by v when the size of the habitat is altered.

4.3 TESTING NONLINEAR POPULATION THEORY

The research program described in this book was motivated by an uneasiness with the state of mathematical population ecology. In the early 1980s, our Beetle Team began long and introspective discussions, in which, of course, we frequently lamented (as is the duty of theoreticians) the lack of quantitative understanding among our data-oriented colleagues, but also in which we frequently ended up at the same, painful conclusion: Mathematical ecology was simply not delivering the goods. The problem with mathematical ecology, we felt, was that it was long on concepts and short on hypotheses. It was disconnected; it was often not clear, for instance, whether "general insights" arising from a predator-prey model were intended to explain an actual system. Theoreticians would offer up a collection of concepts (stability, equilibrium, limit cycles, etc.) as potentially useful tools. But rarely, if ever, were these concepts pinned as convincing explanations of real systems. The question remained: Has a low-dimensional, nonlinear population model ever been a convincing representation of the workings of a real ecological system? Could such a model ever withstand the excruciating challenges from detailed experiments? Are textbook-style models merely teaching tools, or are they intended as a technology for application to real systems? Even with its implicit distortions, has an ecological model ever captured the essence of its subject? We were hard pressed for examples in which population models constituted reliable knowledge.

If we wanted to address nonlinear population dynamics, we knew we had to try a different approach. We identified three elements that were necessary if we were to closely integrate theory and data. First, we needed a low-dimensional mechanistic model that captured the biology of the flour beetle. Second, we placed an emphasis on examining model predictions and experimental data as both time series and geometric entities in state space. Third, we embraced stochasticity as a necessary feature of our theoretical approach. These three features are discussed below.

4.3.1 Mechanistic Life Cycle Model

Because of the work of Thomas Park and his colleagues, the *Tribolium* model was already well-established as an experimental tool for exploring ecological concepts. Park was strongly influenced by P. H. Leslie's use of the Verhulst and Lotka-Volterra assumptions of density dependence and competition [29, 54, 55]. Leslie brings this argument forward in his work with his colleague, and beetle expert, Park. For example, the two-species competition model (4.2) is a discrete-time analog of the Lotka-Volterra equations. Interestingly, Park held this point of view until 1970 when evidence convinced him to emphasize cannibalism [41].

While cannibalism was long recognized as a biological feature of flour beetles (Figure 4.2), little effort was made to include cannibalism in population models. In their study of stationary probability distributions for population size, Costantino, Desharnais, and Dennis [4, 9, 16, 19] used a single differential equation for adult numbers (4.3) that was based on cannibalism, but it excluded the complex cannibalistic interactions that involved immatures. Desharnais and Liu [18] devised a 300×300 Leslie matrix model that included the important cannibalistic pathways among the flour beetle life stages, but it was too large and unwieldy for making testable model predictions. What was needed was a simple low-dimensional model that captured the important features of the flour beetle life history.

In 1995, the Beetle Team presented their solution to the theoretical challenge: a simple model of three difference equations with six parameters that captured the important features of the *Tribolium* system [10]. The state variables were the numbers of larvae, pupae, and adults. Cannibalism was included: larvae eating eggs, adults eating eggs, and larvae eating pupae. Rigorous methods of parameter estimation and model validation were applied to the demographic data of Desharnais and Costantino [14]. The model succeeded with flying colors; it passed statistical scrutiny and did remarkably well in predicting the nonlinear life stage oscillations present in the data. More importantly, these analyses allowed us to reliably locate the experimental populations in model parameter space.

The flood gates were opened. Armed with an effective low-dimensional mechanistic difference equation model of an easily-manipulated experimental system, the Beetle Team was in a good position to examine rigorously general predictions arising from nonlinear population dynamics.

4.3.2 Attractors in State Space

Visual representations have played an important role in dynamical systems theory. For example, in 1890, Henri Poincaré discovered chaotic dynamics within the three-body problem in an essay on celestial mechanics [42]. He devised a novel topological approach to the motion of an asteroid under the influence of the planet Jupiter and the Sun. Poincaré contemplated the movement of the asteroid in state space by considering a "recurrence map" of when the asteroid would intersect a two-dimensional surface that was transverse to the path of the asteroid. He demonstrated that one possibility was an extremely complex aperiodic motion

[21]. Similarly, in 1963 Edward Lorenz used deterministic nonlinear differential equations to represent hydrodynamic flow [28]. He looked at bounded solutions of these equations in three dimensional state space. This gave rise to the now famous "Lorenz attractor," an iconic visual representation of chaotic dynamics. The term "chaos" was coined in a seminal 1975 paper by Tien-Yien Li and James A. Yorke [26] in which they established that extremely complex dynamics can arise from even simple one-dimensional difference equations (also see [27]). This fact was also popularized at that time by Robert May [30].

In our experimental and theoretical studies with the *Tribolium* model, we investigated the dynamics in terms of three aggregated life stage variables: larvae (L), pupae (P), and adults (A). The deterministic nonlinear LPA model predicts bounded trajectories that approach an attractor. These attractors reside in three-dimensional LPA state space. In our analyses of the model predictions and experimental data, it was often helpful to supplement time series analysis with visualizations in state space, to see if the trajectories approach and fluctuate around the predicted attractors. Indeed, throughout this book, there are many examples of three-dimensional LPA state-space plots that are an integral part of the analysis of "complex population dynamics."

4.3.3 Incorporating Stochasticity

An essential part of our research program has been the key role that stochasticity plays in population dynamics. We believe that stochastic forces form one of the crucial aspects of population fluctuations usually omitted from simple textbook models. The word "stochastic" derives from the Greek word "stochos" (guess). Stochastic forces are high dimensional, unexplained, unpredictable forces influencing population change. Stochastic forces can often be represented effectively with probability models, as described in Chapter 2.

A crucial function of stochastic models is that they serve as a bridge between the deterministic dynamics and the experimental data. All statistical methods have stochastic models at their foundation. By integrating stochasticity into the mechanistic LPA model, our data analyses are tied directly to the beetle biology and the system's predicted dynamics. This allows us to conduct seamlessly many of the tasks necessary for rigorous scientific investigation: parameter estimation, model validation, and inferential tests of hypotheses.

Perhaps more importantly, adding stochasticity to our models allowed us to explore a wider range of theoretical outcomes. Often, stochasticity is viewed as troublesome noise that obscures the deterministic signal coming from a time series. But stochasticity is much more than that. By moving the system around in state space, a population can fall under the influence of a variety of nonlinear beasts, including, for example, multiple attractors, saddle manifolds, lattice cycles, and Arnol'd tongues. These phenomena may be revealed episodically within a single experimental time series. This is why a modeling effort that seeks to explain population data should not stop with a deterministic skeleton. A crucial next step is to add stochasticity as the flesh on the bones.

4.4 CONCLUSIONS

The *Tribolium* model refers to the flour beetle laboratory system in all its manifestations. It has a long history of use in the study of ecology. Flour beetles have been used to study a variety of population phenomena, including density regulation, species competition, population stochasticity, ecological genetics, and nonlinear demographic dynamics. Examples of these are found in the pages of this book.

The *Tribolium* model has many features that make it well-suited as an experimental system for population ecology. The beetles are easily cultured and manipulated in the laboratory. The relatively short developmental period makes feasible multigenerational time series. Several species and genetic strains have been characterized. A wealth of data is available about the life history of these insects. Finally, the cannibalistic life stage interactions that occur in population cultures make them an ideal model system for the study of nonlinear dynamics.

In the early 1980s, the Beetle Team embarked on an ambitious research program to employ the *Tribolium* model for the study of nonlinear population dynamics. They identified three key elements that were crucial to their success: (1) a low-dimensional mechanistic model was an effective predictor of the demographic dynamics in flour beetle populations, (2) a focus on population trajectories in state space, and (3) the use of stochastic versions of the deterministic models for statistical analyses and model predictions. Demonstrations of these elements fill the pages of this book.

Bibliography

[1] Bell AE (1982) The *Tribolium* Model in Animal Breeding Research. Second World Congress on Genetics Applied to Livestock Production 5:26–42.

[2] Bonsall MB, Hassell MP (2005) Understanding ecological concepts: The role of laboratory systems. Advances in Ecological Research 37:1–36. https://doi.org/10.1016/S0065-2504(04)37001-7

[3] Chapman, RN (1928) The quantitative analysis of environmental factors. Ecology 9:111–122. https://doi.org/10.2307/1929348

[4] Costantino RF, Desharnais RA (1981) Gamma distributions of adult numbers for *Tribolium* populations in the region of their steady states. Journal of Animal Ecology 50:667–681. https://doi.org/10.2307/4128

[5] Costantino RF, Desharnais RA (1991) Population Dynamics and the *Tribolium* Model: Genetics and Demography. Monographs on Theoretical and Applied Genetics 13. New York, Springer-Verlag.

[6] Costantino RF, Desharnais RA, Cushing JM, Dennis B, Henson SM, King AA (2005) Nonlinear stochastic population dynamics: The flour beetle *Tribolium* as an effective tool of discovery. In Desharnais RA (ed.), Advances in Ecological Research, pp. 101–141. New York, Academic Press.

[7] Crow JF, Kimura M (1970) An Introduction to Population Genetics Theory. New York, Harper and Row.

[8] Cushing JM, Costantino RF, Dennis B, Desharnais RA, Henson SM (2002) Chaos in Ecology: Experimental Nonlinear Dynamics. San Diego, Academic Press.

[9] Dennis B, Costantino RF (1988) Analysis of steady-state populations with the gamma abundance model: Application to *Tribolium*. Ecology 69:1200–1213. https://doi.org/10.2307/1941275

[10] Dennis B, Desharnais RA, Cushing JM, Costantino RF (1995) Nonlinear demographic dynamics: Mathematical models, statistical methods, and biological experiments. Ecological Monographs 65:261–282. https://doi.org/10.2307/2937060

[11] Dennis B, Desharnais RA, Cushing JM, Costantino RF (1997) Transitions in population dynamics: Equilibria to periodic cycles to aperiodic cycles. Journal of Animal Ecology 66:704–729. https://doi.org/10.2307/5923

[12] Dennis B, Desharnais RA, Cushing JM, Henson SM, Costantino RF (2001) Estimating chaos and complex dynamics in an insect population. Ecological Monographs 71:277–303. https://doi.org/10.1890/0012-9615(2001)071[0277:ECACDI]2.0.CO;2

[13] Desharnais R (2005) Population Dynamics and Laboratory Ecology. Vol. 37. San Diego, Elsevier Academic Press.

[14] Desharnais RA, Costantino RF (1980) Genetic analysis of a population of *Tribolium*: VII. Stability: Response to genetic and demographic perturbations. Canadian Journal of Genetics and Cytology 22:577–589. https://doi.org/10.1139/g80-063

[15] Desharnais RA, Costantino RF (1983) Natural selection and density-dependent population growth. Genetics 105:1029–1040. https://doi.org/10.1093/genetics/105.4.1029

[16] Desharnais RA, Costantino RF (1985) Genetic analysis of a population of *Tribolium*. VIII. The stationary stochastic dynamics of adult numbers. Canadian Journal of Genetics and Cytology 27:341–350. https://doi.org/10.1139/g85-051

[17] Desharnais RA, Costantino RF (1982) Natural selection and fitness entropy in a density-regulated population. Genetics 101:317–329. https://doi.org/10.1093/genetics/101.2.317

[18] Desharnais RA, Liu L (1987) Stable demographic limit cycles in laboratory populations of *Tribolium castaneum*. Journal of Animal Ecology 56:885–906. https://doi.org/10.2307/4955

[19] Desharnais RA, Dennis B, Costantino RF (1990) Genetic analysis of a population of *Tribolium*. IX. Maximization of population size and the concept of a stochastic equilibrium. Genome 33:571–580. https://doi.org/10.1139/g90-084

[20] Desharnais RA, Henson SM, Costantino RF, Dennis B (2023) Capturing chaos: A multidisciplinary approach to nonlinear population dynamics. Journal of Difference Equations and Applications 30:1–38. https://doi.org/10.1080/10236198.2023.2260013

[21] Galison P (2004) Einstein's Clocks and Poincaré's Maps: Empires of Time. Illustrated edition. New York, W. W. Norton & Company.

[22] Hinton HE (1948) A synopsis of the genus *Tribolium* Macleay, with some remarks on the evolution of its species-groups (Coleoptera, Tenebrionidae). Bulletin of Entomological Research 39:13–55. https://doi.org/10.1017/S0007485300024287

[23] Jillson DA, Costantino RF (1980) Growth, distribution, and competition of *Tribolium castaneum* and *Tribolium brevicornis* in fine-grained habitats. American Naturalist 116:206–219. https://doi.org/10.1086/283623

[24] Leslie PH (1962) A stochastic model for two competing species of *Tribolium* and its application to some experimental data. Biometrika 49:1–25. https://doi.org/10.2307/2333464

[25] Leslie PH, Park T, Mertz DB (1968) The effect of varying the initial numbers on the outcome of competition between two *Tribolium* species. Journal of Animal Ecology 37:9–23. https://doi.org/10.2307/2708

[26] Li TY, Yorke JA (1975) Period three implies chaos. The American Mathematical Monthly 82:985–992. https://doi.org/10.1080/00029890.1975.11994008

[27] Li, TY, Yorke JA (2004). Period three implies chaos. In Hunt BR, Li TY, Kennedy JA, Nusse HE (eds.), The Theory of Chaotic Attractors. New York, Springer. https://doi.org/10.1007/978-0-387-21830-4_6

[28] Lorenz EN (1963) Deterministic nonperiodic flow. Journal of Atmospheric Sciences 20:130–141. https://doi.org/10.1177/0309133308091948

[29] Lotka AJ (1932) The growth of mixed populations: Two species competing for a common food supply. Journal of the Washington Academy of Sciences 22:461–469. https://www.jstor.org/stable/24530449

[30] May R (1976) Simple mathematical models with very complicated dynamics. Nature 261:459–467. https://doi.org/10.1038/261459a0

[31] Moffa AM, Costantino RF (1977) Genetic analysis of a population of *Tribolium*. VI. Polymorphism and demographic equilibrium. Genetics 87:785–805. https://doi.org/10.1093/genetics/87.4.785

[32] Mertz, DB (1969) Age-distribution and abundance in populations of flour beetles. I. Experimental studies. Ecological Monographs 39:1–31. https://doi.org/10.2307/1948563

[33] Mertz, DB (1972) The *Tribolium* model and the mathematics of population growth. Annual Review of Ecology and Systematics 3:51–78. https://www.jstor.org/stable/2096842

[34] Mertz DB, Cawthon DA, Park T (1976) An experimental analysis of competitive indeterminacy in *Tribolium*. Proceedings of the National Academy of Sciences 73:1368–1372. https://doi.org/10.1073/pnas.73.4.1368

[35] Park T (1948) Experimental studies of interspecific competition. I. Competition between populations of the flour beetles *Tribolium confusum* Duval and *Tribolium castaneum* Herbst. Ecological Monographs 18:265–308. https://www.jstor.org/stable/1948641

[36] Park T (1954) Experimental studies of interspecific competition. II. Temperature, humidity, and competition in two species of *Tribolium*. Physiological Zoology 27:177–238. https://www.jstor.org/stable/30152164

[37] Park T (1957) Experimental studies of interspecific competition. III. Relation of initial species proportion to competitive outcome in populations of *Tribolium*. Physiological Zoology 30:22–40. https://www.jstor.org/stable/30166306

[38] Park T, Leslie PH, Mertz DB (1964) Genetic strains and competition in populations of *Tribolium*. Physiological Zoology 37:97–162. https://doi.org/10.1086/physzool.37.2.30152328

[39] Park T, Mertz DB, Petrusewicz K. (1961) Genetic strains of *Tribolium*: Their primary characteristics. Physiological Zoology 34:62–80. https://doi.org/10.1086/physzool.34.1.30152682

[40] Park T, Mertz DB, Grodzinski W, Prus T (1965) Cannibalistic predation in populations of flour beetles. Physiological Zoology 38:289–321. https://doi.org/10.1086/physzool.38.3.30152840

[41] Park T, Nathanson M, Ziegler JR, Mertz DB (1970) Cannibalism of pupae by mixed-species populations of adult *Tribolium*. Physiological Zoology 43:166–184. https://doi.org/10.1086/physzool.43.3.30155527

[42] Poincaré H (2017) The Three-Body Problem and the Equations of Dynamics: Poincaré's Foundational Work on Dynamical Systems Theory. Vol. 443. Cham, Switzerland, Springer.

[43] Pointer MD, Gage MJG, Spurgin LG (2021) *Tribolium* beetles as a model system in evolution and ecology. Heredity 126:869–883. https://doi.org/10.1038/s41437-021-00420-1

[44] Sokoloff A (1974) The Biology of *Tribolium*. Vol. 2. London, Oxford University Press.

[45] Stanley J (1932) A mathematical theory of the growth of populations of the flour beetle, *Tribolium confusum*, Duv. Canadian Journal of Research 6:632–671. https://doi.org/10.1139/cjr32-052

[46] Stanley J (1932) A mathematical theory of the growth of populations of the flour beetle, *Tribolium confusum*, Duv. II. The distribution by ages in the early stages of population growth. Canadian Journal of Research 7:426–433. https://doi.org/10.1139/cjr32-092

[47] Stanley J (1934) A mathematical theory of the growth of populations of the flour beetle, *Tribolium confusum*, Duv. III. The effect upon the early stages of population growth of changes in the nutritive value, palatability and density of packing of the flour medium. Canadian Journal of Research 11:728–732. https://doi.org/10.1139/cjr34-129

[48] Stanley J (1941) A mathematical theory of the growth of populations of the flour beetle, *Tribolium confusum*, Duv. IV. A modified theory descriptive of the relation between the limiting value of egg-populations in the absence of hatching, and the volume (or weight) of flour used in the culture. Ecology 22:23–37. https://www.jstor.org/stable/1930006

[49] Stanley J (1942) A mathematical theory of the growth of populations of the flour beetle, *Tribolium confusum*, Duv. V. The relation between the limiting value of egg-populations in the absence of hatching and the sex-ratio of the group of adult beetles used in a culture. Ecology 23:24–31. https://www.jstor.org/stable/1930869

[50] Stanley J (1943) A mathematical theory of the growth of populations of the flour beetle, *Tribolium confusum*, Duv. VI. Egg populations in which the initial number of eggs is above the limiting value. Ecology 24:323–328. https://www.jstor.org/stable/1930534

[51] Stanley J (1949) A mathematical theory of the growth of populations of the flour beetle, *Tribolium confusum*, Duv. VII. A study of the "re-tunnelling" problem. Ecology 30:209–222. https://doi.org/10.2307/1931187

[52] Stanley J (1964) A mathematical theory of the growth of populations of the flour beetle, *Tribolium confusum*, Duv. VIII. A further study of the "re-tunnelling" problem. Canadian Journal of Zoology 42:201–227. https://doi.org/10.1139/z64-019

[53] Stanley J (1966) A mathematical theory of the growth of populations of the flour beetle, *Tribolium confusum*, Duv. IX. The persistence of tunnels in the flour mass. Journal of Theoretical Biology 13:379–411. https://doi.org/10.1016/0022-5193(66)90028-2

[54] Verhulst, PF (1838) Notice Sur La Loi Que La Population Suit Dans Son Accroissement. Correspondence Mathematique et Physique 10:113–121.

[55] Volterra V (1927) Variazioni e Fluttuazioni Del Numero d'individui in Specie Animali Conviventi. Vol. 2. Societá Anonima Tipografica. Venezia, C. Ferrari.

CHAPTER 5

The LPA Model

In this chapter, our goal is to derive a model specifically designed to describe the dynamics of a population whose life cycle is represented by three stages. It is constructed so as to apply specifically to species of flour beetles (*Tribolium* spp.) and their characteristics when cultured in a laboratory setting, as described in Chapter 4. We derive the basic model in Section 5.1 and consider stochastic versions in Section 5.2. Using some historical data we parameterize and validate the model in Section 5.3.

5.1 THE DETERMINISTIC LPA MODEL

The first step in constructing a dynamic model is to specify the state variables, that is, the quantities to be measured and whose changes over time are to be followed. Larval, pupal, and adult stages of flour beetles can be accurately counted in laboratory cultures (Chapter 4). We chose the number of individuals in each of these stages as our state variables and denoted these numbers, at time t, by L_t, P_t, and A_t respectively. We will census these numbers at equally-spaced time units. The goal of our deterministic model is to predict the values of these variables at time $t+1$ after they have been counted at time t. That is to say, we want formulas to calculate L_{t+1}, P_{t+1}, and A_{t+1} in terms of L_t, P_t, and A_t which can then be used to construct (iteratively) the numbers in the three life cycle stages for all future times $t = 1, 2, \ldots$ once their "initial" values L_0, P_0, and A_0 are known. Mathematically, this procedure defines a (semi-) dynamical system whose solutions L_t, P_t, and A_t deterministically yield the unique future history of the population that follows from any initial condition. Models such as these that track population size only at discrete times are called *discrete-time (semi-) dynamical systems*, for which there exist a considerable amount of mathematical machinery for the study of solutions and their properties.

Mathematical formulas for L_{t+1}, P_{t+1}, and A_{t+1} can be derived straightforwardly with the following basic methodology for discrete-time dynamical systems. If x_t is a state variable known at time t, then each individual present in the census at time $t+1$ falls into one of two categories: those who were not present at time t and those who were present at time t. Thus,

$$x_{t+1} = \text{recruits} + \text{survivors}, \tag{5.1}$$

in which a recruit might be a newborn or an individual transferring from another state variable. A survivor is an individual who did not die and did not transfer to another state. The modeler's task is to prescribe mathematical expressions for these two components based on the biological mechanisms and processes of interest.

The simplest (lowest dimensional) model of type (5.1) involves only one state variable, for example, the number of adults. In such a model, any interactions among any life-cycle stages, for example intraspecific competition for resources and mates, cannibalism (intraspecific

DOI: 10.1201/9781003467120-5

predation), etc., would not be described by the model, at least not explicitly. The Ricker model in Chapter 3 is an example of such a one-dimensional model. However, in order to account for the significant life-cycle stages for our experimental populations of *Tribolium* and the interactions among them (of which cannibalism is of high importance), we need to construct a higher-dimensional model. Our model distinguishes three classes of individual beetles. Let L denote the number of feeding larvae; P the number of last instar (non-feeding) larvae, pupae, and callow adults; and A the number of sexually mature adults in a habitat of size V. For convenience, we refer to the stages as "larvae," "pupae," and "adults." Importantly, these state variables can be counted accurately under standard laboratory conditions.

Discrete-time models with equally-spaced census times $t = 0, 1, 2, 3, \cdots$ require specification of an appropriate unit of time. For our *Tribolium* model, we choose the unit of time according to the following considerations. Results of Park ([18], Table 10) suggest that for *T. castaneum* the feeding larval stage lasts about 14 days. Moffa and Costantino estimated the time from egg to adulthood as approximately 27 days ([16], Table 1). This includes egg and pupal stages of two to four days duration and an additional two to four days spent as a callow adult. Thus, the durations of the L-stage and P-stage are roughly equal, namely, approximately 14 days under standard laboratory conditions. We therefore took the unit of time in the model (and between our census data points) to be two weeks.

Note that we omitted an egg stage from the model. Although eggs can be and sometimes are counted in flour beetle studies, an inordinate amount of time is required to count eggs. Counting only larvae, pupae, and adults allows many more cultures to be maintained in any given experiment. The egg stage is fairly short in duration (two to four days [20]) and consequently most eggs laid within a two week period become larvae by the end of the period. Although inclusion of an egg stage could produce an improved model, a three-staged L, P, and A model turns out, as we will see, to be an accurate predictor of *Tribolium* dynamics.

With this choice of model time step, the application of the basic accounting procedure (5.1) to the larval stage $x_t = L_t$ has no survivors and the recruits are all newborn individuals. Thus, if b is the per adult production of eggs per unit of time, we have

$$L_{t+1} = bA_t.$$

In the case of pupae $x_t = P_t$, there are again no survivors in (5.1), and newcomers are surviving larvae who, after a unit of time, necessarily pupate. If μ_l is the fraction of larvae that die per unit time, then

$$P_{t+1} = (1 - \mu_l)\, L_t.$$

Finally, (5.1) applied to the adult population $x_t = A_t$ yields

$$A_{t+1} = \left(1 - \mu_p\right) P_t + (1 - \mu_a)\, A_t,$$

where μ_p and μ_a are the fraction of pupae and adults that die per unit time, respectively. Taken together, these three equations

$$\begin{aligned} L_{t+1} &= bA_t \\ P_{t+1} &= (1 - \mu_l)\, L_t \\ A_{t+1} &= \left(1 - \mu_p\right) P_t + (1 - \mu_a)\, A_t \end{aligned} \tag{5.2}$$

constitute the *linear LPA model.* (The model is linear because the right-hand sides of the equations are linear in the state variables.)

The linear LPA model (5.2) is an example of a classic Leslie-Lewis matrix model [11, 12]. The Fundamental Theorem of Demography (sometimes called the Strong Ergodic Theorem) implies, among other things, that the population goes extinct (all state variables tend to 0 as t increases without bound) if the inherent net reproduction number

$$R_0 \triangleq b\frac{(1-\mu_p)(1-\mu_l)}{\mu_a}$$

is less than 1. Here the symbol $\triangleq$ means "defined to be." Furthermore, all state variables increase (exponentially) without bound if $R_0 > 1$ [1, 3]. Hence, the linear Leslie model (5.2) cannot predict the long-term (asymptotic) dynamics of flour beetle populations, which typically persist under laboratory conditions but are of course bounded. In order to do that, the model must include the controlling mechanism of cannibalism (see Chapter 4).

Treating cannibalism as a random contact event between a cannibal and victim and following the type of reasoning used in Chapter 3 in the derivation of the Ricker model, we derived a model for cannibalism in *Tribolium* as follows. Consider the cannibalism of eggs by adults. Laboratory experiments (Chapter 4) show that the probability of contact between an individual adult and an egg, during any fixed time interval, is approximately inversely proportional to the habitat volume V [2].

If we assume that, during a small unit of time Δt, the probability an egg is contacted and eaten by an adult is approximately equal to $c_{ea}\Delta t/V$ for a constant $c_{ea} > 0$, then the probability the egg escapes cannibalism by the adult during Δt is approximately equal to $1 - c_{ea}\Delta t/V$. The probability the egg will survive two Δt units of time is approximately

$$\left(1-\frac{c_{ea}}{V}\Delta t\right)\left(1-\frac{c_{ea}}{V}\Delta t\right)=\left(1-\frac{c_{ea}}{V}\Delta t\right)^{2},$$

and so on. The probability it will survive a full unit of time, i.e., $1/\Delta t$ time units of length Δt, is approximately

$$\left(1-\frac{c_{ea}}{V}\Delta t\right)^{1/\Delta t}.$$

Under the assumption that encounters between the egg and adults are independent events, the survival probability of one egg over one unit of time in the presence of two adults is approximately

$$\left(1-\frac{c_{ea}}{V}\Delta t\right)^{1/\Delta t}\left(1-\frac{c_{ea}}{V}\Delta t\right)^{1/\Delta t}=\left(1-\frac{c_{ea}}{V}\Delta t\right)^{2/\Delta t}.$$

In the presence of A adults, the probability that an egg survives adult cannibalism is approximately

$$\left(1-\frac{c_{ea}}{V}\Delta t\right)^{A/\Delta t},$$

which, as $\Delta t \to 0$, approaches the exponential $\exp(-c_{ea}A/V)$.

Similar exponential terms describe the probabilities of surviving cannibalism that occur among the other life-cycle stages. The dominant cannibalistic interactions in populations of *T. castaneum* are egg cannibalism by larvae and adults and pupae cannibalism by adults [15]. Introducing these survival probabilities into the linear Leslie model (5.2), we obtain the model equations

$$\begin{aligned}
L_{t+1} &= bA_t e^{-\frac{c_{el}}{V}L_t} e^{-\frac{c_{ea}}{V}A_t}\\
P_{t+1} &= (1-\mu_l)\,L_t\\
A_{t+1} &= (1-\mu_p)\,P_t e^{-\frac{c_{pa}}{V}A_t} + (1-\mu_a)\,A_t\,.
\end{aligned}$$

In some of the later chapters of this book, habitat volume V appears explicitly in the model equations. Otherwise, however, we set $V = 1$ to be the volume occupied by 20 g of flour medium.

Under laboratory protocols described in Chapter 4, virtually all pupae of *T. castaneum* survive to emerge as adults in the absence of cannibalism. Therefore, for this species we take $\mu_p = 0$.[1] We have arrived at the (nonlinear) *LPA model*

$$\begin{aligned} L_{t+1} &= bA_t e^{-c_{el}L_t} e^{-c_{ea}A_t} \\ P_{t+1} &= (1-\mu_l)\, L_t \\ A_{t+1} &= P_t e^{-c_{pa}A_t} + (1-\mu_a)\, A_t. \end{aligned} \tag{5.3}$$

In this model, $b > 0$ is the per adult number of eggs laid per unit time (two weeks), and μ_l and μ_a are larval and adult survival probabilities per unit time ($0 \leq \mu_l, \mu_a < 1$), respectively. We refer to the positive coefficients c_{el}, c_{ea}, and c_{pa} as the "cannibalism coefficients."

5.2 STOCHASTIC LPA MODELS

Our next goal is to connect the LPA model (5.3) to data. To do this, and also to carry out statistical inferences, we need to modify the model to account for inevitable deviations of data from model predictions. This requires specification of the stochastic mechanisms affecting the population. Ecologists have identified two broad types of stochasticity important in biological populations [1, 14]: *environmental stochasticity*, which involves random variations in population numbers due to extrinsic forces that affect all individuals in the population (e.g., weather, resource availability, habitat conditions, etc.), and *demographic stochasticity*, which is due to independent random fluctuations that occur among vital rates or characteristics of individual organisms (e.g., fertility, survival probabilities, physical attributes, etc.).

A large number of statistical techniques are available for stochastic processes of the form

$$\mathbf{n}_{t+1} = \mathbf{h}(\mathbf{n}_t) + \mathbf{E}_t, \tag{5.4}$$

where $\mathbf{n}_t$ is a vector of (possibly transformed) state variables, $\mathbf{E}_t$ is a normal random vector with mean 0 and (symmetric) variance-covariance matrix denoted $\mathbf{\Sigma}$. Covariances among the components of $\mathbf{E}_t$ at any given time t are given by off diagonal elements in $\mathbf{\Sigma}$. However, we expected covariances *between* times to be small by comparison and therefore we assumed $\mathbf{E}_1, \mathbf{E}_2, \mathbf{E}_3, \cdots$ are uncorrelated. Such processes are called *nonlinear autoregressive (NLAR) models.* The nonlinear function $\mathbf{h}$ defines the deterministic "skeleton" $\mathbf{n}_{t+1} = \mathbf{h}(\mathbf{n}_t)$ that describes the model dynamics of $\mathbf{n}_t$ in the absence of noise [22].

It turns out that birth and survival processes derived for environmental and demographic stochasticity do not have the additive NLAR form (5.4) [5]. However, in each case it is possible to transform the state variables in such a way that the stochastic process for the resulting random variable $\mathbf{n}_t$ does have (at least approximately) the additive structure (5.4). Such transformations for environmental stochasticity and demographic stochasticity are logarithmic and square root transformations, respectively [5, 6, 7]. This leads to the *environmental stochastic LPA model*

$$\begin{aligned} L_{t+1} &= bA_t e^{-c_{el}L_t - c_{ea}A_t} e^{E_{1t}} \\ P_{t+1} &= (1-\mu_l)\, L_t e^{E_{2t}} \\ A_{t+1} &= \left(P_t e^{-c_{pa}A_t} + (1-\mu_a)\, A_t\right) e^{E_{3t}} \end{aligned} \tag{5.5}$$

[1] This assumption is corroborated by model parameter estimates using data. Therefore, for simplicity we assume *a priori* that $\mu_p = 0$.

and the *demographic stochastic LPA model*

$$\begin{aligned}
L_{t+1} &= \left(\sqrt{bA_t e^{-c_{el}L_t - c_{ea}A_t}} + E_{1t}\right)^2 \\
P_{t+1} &= \left(\sqrt{(1-\mu_l)\,L_t} + E_{2t}\right)^2 \\
A_{t+1} &= \left(\sqrt{P_t e^{-c_{pa}A_t} + (1-\mu_a)\,A_t} + E_{3t}\right)^2,
\end{aligned} \tag{5.6}$$

in which each E_{it} is a normally distributed random variable with mean 0 and variance-covariance matrix $\mathbf{\Sigma}$, and for which $E_{i1}, E_{i2}, E_{i3}, \ldots$ are uncorrelated.[2]

The stochastic model equations (5.5) and (5.6) have a number of useful mathematical and statistical advantages. First, on the logarithmic or square root scales, respectively, they have the form (5.4) with $\mathbf{n}_t = \mathbf{W}_t$ where

$$\mathbf{W}_t = \begin{pmatrix} \ln L_t \\ \ln P_t \\ \ln A_t \end{pmatrix}, \quad \mathbf{h}(\mathbf{W}_t) = \begin{pmatrix} \ln\left(bA_t e^{-c_{el}L_t - c_{ea}A_t}\right) \\ \ln\left((1-\mu_l)\,L_t\right) \\ \ln\left(P_t e^{-c_{pa}A_t} + (1-\mu_a)\,A_t\right) \end{pmatrix}$$

or

$$\mathbf{W}_t = \begin{pmatrix} \sqrt{L_t} \\ \sqrt{P_t} \\ \sqrt{A_t} \end{pmatrix}, \quad \mathbf{h}(\mathbf{W}_t) = \begin{pmatrix} \sqrt{bA_t e^{-c_{el}L_t - c_{ea}A_t}} \\ \sqrt{(1-\mu_l)\,L_t} \\ \sqrt{P_t e^{-c_{pa}A_t} + (1-\mu_a)\,A_t} \end{pmatrix},$$

respectively, and the vector

$$\mathbf{E}_t = \begin{pmatrix} E_{1t} \\ E_{2t} \\ E_{3t} \end{pmatrix}$$

has a multivariate normal distribution with mean $\mathbf{0}$. Many statistical estimation, testing, and evaluation methods and tools are available for such nonlinear autoregressive models [22].

Second, the state variables of the nonlinear deterministic model on the logarithmic scale are the expected values of the transformed state variables at time $t+1$ conditioned on their values are time t. That is to say, for the environmental stochastic model the expected values of L_{t+1}, P_{t+1}, and A_{t+1}, given L_t, P_t, and A_t, are equal to the right hand sides of the equations in the LPA model (5.3). (The same is true of the demographic stochastic model.) Thus, the stochastic versions retain the essential dynamic properties of their deterministic skeletons.

Third, as we will see in the following Section 5.3, the stochastic models (5.5) and (5.6) allow parameter estimation and easy computer simulation and exploration.

5.3 PARAMETER ESTIMATION AND MODEL VALIDATION

We "calibrate" the LPA model by obtaining maximum likelihood estimates of the model's parameters from data. A likelihood function gives the chance that an outcome of a proposed stochastic mechanism would result in the observed data, relative to all other possible outcomes. It is a fundamental tool in statistical inference and represents a crucial connection between model and data [21]. The stochastic LPA models (5.5) and (5.6) provide explicit likelihood functions. Observational data obtained from our experiments with *Tribolium* have

[2]In the demographic LPA model, there is a chance that an expression appearing inside the squaring function will be negative, in which case it should be set equal to zero.

the form of a time sequence of triples l_t, p_t, and a_t which is a realization of the joint stochastic state variables L_t, P_t, and A_t in the model. For the environmental stochastic model, we let

$$\mathbf{w}_t = \begin{pmatrix} \ln l_t \\ \ln p_t \\ \ln a_t \end{pmatrix} \tag{5.7}$$

denote the vector of observations on the log scale. (For the demographic environmental we replace the logarithm ln by the square root $\sqrt{\cdot}$.) Let $\boldsymbol{\theta}$ denote the vector of unknown parameters in the function $\mathbf{h}$ (i.e., the parameters in the deterministic model) together with the additional unknown parameters in the variance-covariance matrix $\boldsymbol{\Sigma}$. For a time series of observations $\mathbf{w}_0$, $\mathbf{w}_1$, $\cdots$, $\mathbf{w}_q$, The likelihood function $L(\boldsymbol{\theta}, \boldsymbol{\Sigma})$ is

$$L(\boldsymbol{\theta}, \boldsymbol{\Sigma}) = \prod_{t=1}^{q} p\left(\mathbf{w}_t | \mathbf{w}_{t-1}\right),$$

where $p\left(\mathbf{w}_t | \mathbf{w}_{t-1}\right)$ is the joint transition probability density function (PDF) for $\mathbf{W}_t$ conditional on $\mathbf{W}_{t-1} = \mathbf{w}_{t-1}$ and evaluated at $\mathbf{w}_t$. It has a multivariate normal PDF with a mean vector of $\mathbf{h}\left(\mathbf{w}_{t-1}\right)$ and a variance-covariance matrix $\boldsymbol{\Sigma}$:

$$\begin{aligned} p\left(\mathbf{w}_t | \mathbf{w}_{t-1}\right) &= \\ & |\boldsymbol{\Sigma}|^{-1/2} \left(2\pi\right)^{-3/2} \exp\left(-\frac{1}{2}\left(\mathbf{w}_t - \mathbf{h}\left(\mathbf{w}_{t-1}\right)\right)^T \boldsymbol{\Sigma}^{-1} \left(\mathbf{w}_t - \mathbf{h}\left(\mathbf{w}_{t-1}\right)\right)\right), \end{aligned}$$

where $\mathbf{v}^T$ denotes the transpose of a column vector $\mathbf{v}$. Most statistical calculations utilize the log-likelihood

$$\ln L(\boldsymbol{\theta}, \boldsymbol{\Sigma}) = \sum_{t=1}^{q} \ln p\left(\mathbf{w}_t | \mathbf{w}_{t-1}\right),$$

that is,

$$\begin{aligned} \ln L(\boldsymbol{\theta}, \boldsymbol{\Sigma}) = &-\frac{3}{2} q \ln 2\pi - \frac{q}{2} \ln |\boldsymbol{\Sigma}| \\ &- \frac{1}{2} \sum_{t=1}^{q} \left(\mathbf{w}_t - \mathbf{h}(\mathbf{w}_{t-1})\right)^T \boldsymbol{\Sigma}^{-1} \left(\mathbf{w}_t - \mathbf{h}(\mathbf{w}_{t-1})\right) \end{aligned} \tag{5.8}$$

The maximum likelihood (ML) estimates of the parameters in $\boldsymbol{\theta}$ and $\boldsymbol{\Sigma}$ are those values that maximize the likelihood function $L(\boldsymbol{\theta}, \boldsymbol{\Sigma})$ or, as is done in most statistical calculations, the log-likelihood $\ln L(\boldsymbol{\theta}, \boldsymbol{\Sigma})$. The ML estimates of the parameters have several desirable statistical properties. As the sample size increases, they are what is called *asymptotically efficient* (i.e., the variances approach the theoretically lowest bound), *unbiased* (i.e., the bias approaches zero), and *normally distributed* (which allows for the construction of approximate confidence intervals) [21]. For nonlinear time series models, the sample size is the number of observations—to be precise, the number q of one-step residuals—which are, by model assumption, independent. Theorems about these and other ML properties generally require the stochastic model to have a stationary distribution [22]. Typically, NLAR models of the form (5.4) have a stationary distribution when every trajectory of the underlying deterministic model $\mathbf{n}_{t+1} = \mathbf{h}(\mathbf{n}_t)$ has bounded trajectories and hence a bounded attractor. This is the case for the deterministic LPA model (5.3) [3, 5].

In [6] we calculated, using the environmental stochastic LPA model, the ML parameter estimates from data obtained from the control cultures (which consisted of four replicates) in an experiment using *T. castaneum* reported in the masters thesis of R. A. Desharnais

TABLE 5.1 Maximum likelihood (ML) parameter estimates for the LPA model (5.3) based on control cultures of the Desharnais experiment, using the environmental stochastic model (5.5). The ML estimates of the entries in the 3×3 variance-covariance matrix $\mathbf{\Sigma} = (\sigma_{ij})$ for the environmental stochastic LPA model (5.5) are $\hat{\sigma}_{11} = 0.2771$, $\hat{\sigma}_{12} = \hat{\sigma}_{21} = 0.02792$, $\hat{\sigma}_{13} = \hat{\sigma}_{31} = 0.009796$, $\hat{\sigma}_{22} = 0.4284$, $\hat{\sigma}_{23} = \hat{\sigma}_{32} = -0.008150$, and $\hat{\sigma}_{33} = 0.01112$.

Parameter	ML estimate	95% confidence interval
b	11.68	(6.2000, 22.2000)
μ_a	0.1108	(0.0700, 0.1500)
μ_l	0.5129	(0.4300, 0.5800)
c_{ea}	0.01097	(0.0040, 0.0180)
c_{el}	0.009264	(0.0081, 0.0105)
c_{pa}	0.01779	(0.0154, 0.0207)

at the University of Rhode Island (these data are part of the Beetle Team Tribolium Data Archive which is available from Dryad at https://doi.org/10.5061/dryad.qjq2bvqmp). The results for the six LPA model parameters, as well as the entries in the variance-covariance matrix $\mathbf{\Sigma}$, appear in Table 5.1.

We also calculated ML parameter estimates using other replicate cultures in the experiment. Furthermore, we calculated parameter estimates using the method of conditional least squares [7, 10, 22]. The estimates are similar to those in Table 4.1. See [5, 6].

Model fit is evaluated by analyzing the log-scale residuals, defined to be the differences between the (three) logarithmic state variables from the data and the estimated one-step expected values obtained from the parameterized environmental LPA model (5.5). Specifically, using vector notation for the three state variables, the log-scale residuals are $\mathbf{e}_{t+1} = \mathbf{w}_{t+1} - \mathbf{h}(\mathbf{w}_t)$ where $\mathbf{w}_t$ is the log-transformed data vector (5.7) and

$$\mathbf{h}(\mathbf{w}_t) = \begin{pmatrix} \ln\left(b a_t e^{-c_{el} l_t - c_{ea} a_t}\right) \\ \ln\left((1-\mu_l) l_t\right) \\ \ln\left(p_t e^{-c_{pa} a_t} + (1-\mu_a) a_t\right) \end{pmatrix}$$

is the vector of forecasted state variables obtained using the parameters given by the ML estimates in Table 5.1. If the model "fits," then the residuals $\mathbf{e}_1, \mathbf{e}_2, \ldots, \mathbf{e}_q$ (calculated from the *same* data used to estimate the parameters) should behave (approximately) like uncorrelated observations from a trivariate normal distribution. Model *validation* involves analyzing residuals calculated from data that were *not* used in the parameter estimation (to see how well the fitted model predicts novel data).

5.3.1 Model Fit

The residuals $\mathbf{e}_t$ can be inspected visually for accuracy by plotting the data of each state variable with the prediction obtained from the LPA model (5.3) using the preceding data point. For example, Figure 5.1 shows times series plots of the data from one of the replicates of the control cultures used to parameterize the model (Table 5.1) together with the one-step forecasts of the parameterized model. This plot, together with similar plots of the data from the other three replicates show, visually at least, that the one-step model predictions are reasonably accurate.

For further analysis of the model fit, we note that the residuals $\mathbf{e}_t$ for each individual state variable should have small autocorrelations and approximate univariate normal

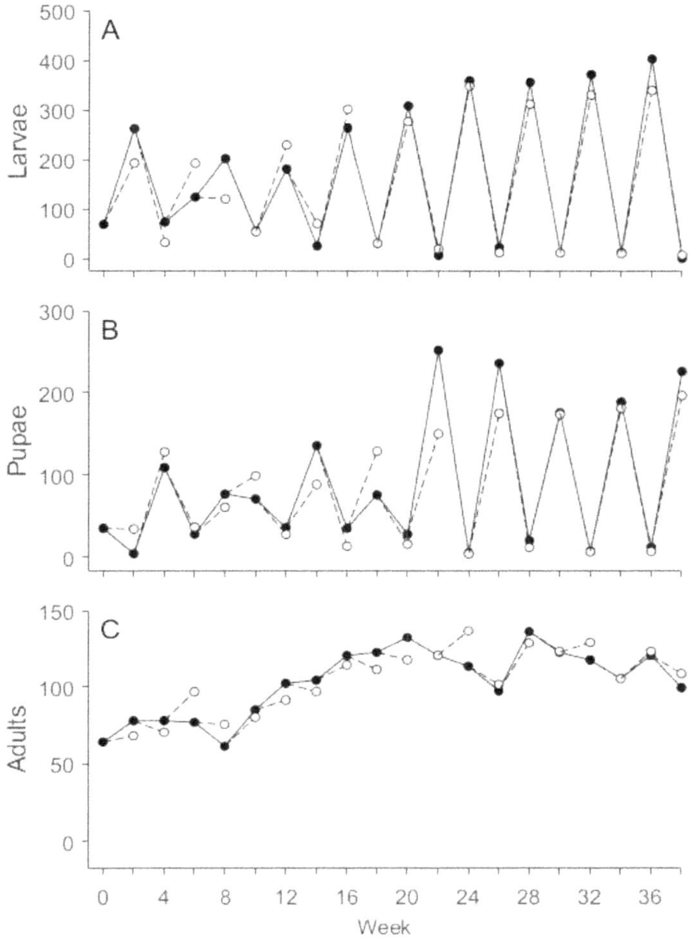

Figure 5.1 Time-series data (solid circles) and one-step forecasts (open circles) from replicate 1 of the Desharnais experimental data ([5], Appendix). Forecasts are calculated from the deterministic LPA model (5.3) using the parameter estimates in Table 5.1.

distributions. In addition to standard normal probability plots, statistical tests for (univariate) normality such as the Lin-Mudholkar test are useful [13, 22]. Standard univariate autocorrelation tests are informative as well. These statistical analyses, reported in [6], show that, overall, the residuals are not large in magnitude and do not vary systematically with the sizes of the state variables. We conclude that overall the residuals conform adequately to the multivariate normal model (5.5). In this sense, the parameters appearing in Table 5.1 produce a good "fit" of the data.

5.3.2 Model Validation

Certainly a parameterized model should fit the data used to parameterize it. But that is just a first step. Can the model successfully predict data that were not used for parameterization?

In our case, can the LPA model (5.3) successfully predict larval, pupal, and adult numbers for populations whose numbers were not used to estimate the parameters in Table 5.1?

Besides the control cultures that produced the data used for parameter estimation above, the Desharnais experiment included nine other cultures [8]. Using the data from these other treatments, one can perform a residual analysis like that carried out above (in Section 5.3.2), using *the same parameter estimates given in Table 5.1.* From the results of such an analysis one can evaluate the predictive capability of the LPA model. We will not describe the details of the analysis here, but all deviations from normality and autocorrelation were small and could be traced to a small number of outliers in each time series [6]. We concluded that, overall, the residual analysis using the prediction data supported the predictive accuracy of the stochastic LPA model (5.5).

5.4 PREDICTED DYNAMICS

The deterministic LPA model (5.3) and the ML parameter estimates in Table 5.1 provide a deterministic model description of flour beetle population dynamics in the Desharnais experiment. Furthermore, the estimates in Table 5.1 of the variance-covariance matrix $\mathbf{\Sigma}$ provide a (environmental) stochastic model (5.5) description of the population dynamics. In this section, we briefly consider what these models have to say.

A mathematical analysis of the dynamics implied by deterministic LPA model (5.3) begins with the consideration of solutions

$$\mathbf{x}_t = \begin{pmatrix} L_t \\ P_t \\ A_t \end{pmatrix}$$

that remain constant over time, that is, *equilibrium solutions*. Some fundamental facts about general discrete-time dynamic models, of which (5.3) is but one example, are the following [3, 4, 9]. The *extinction equilibrium* (also known as the *trivial equilibrium*)

$$\mathbf{x}_t = \begin{pmatrix} 0 \\ 0 \\ 0 \end{pmatrix}$$

is (locally asymptotically) stable and the population is predicted to go extinct as t increases without bound if the net reproduction number

$$R_0 = b\frac{1-\mu_l}{\mu_a} \tag{5.9}$$

satisfies $R_0 < 1$. (R_0 is the expected number of larval stage offspring that an individual beetle will produce over the course of its life span.) If, on the other hand, $R_0 > 1$, then there exists a *survival equilibrium* (*non-trivial equilibrium*)

$$\mathbf{x}_t = \begin{pmatrix} L \\ P \\ A \end{pmatrix}$$

with positive values for all three state variables L, P, and A. This phenomenon is called a *transcritical bifurcation* at $R_0 = 1$ where the survival equilibrium collapses to the extinction equilibrium.

For each $R_0 > 1$ close to one, the bifurcating survival equilibrium is stable and the model predicts an approach to this survival equilibrium state. For larger values of $R_0 > 1$,

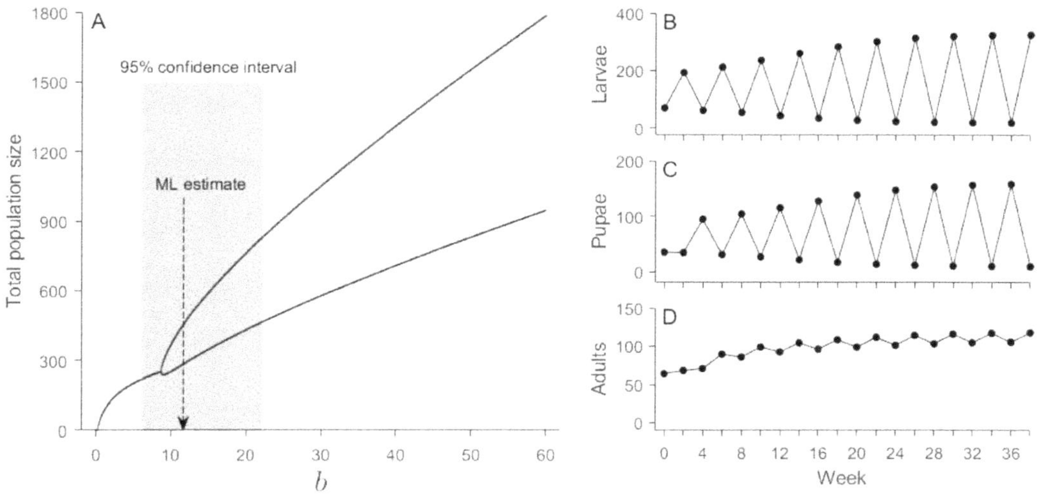

Figure 5.2 Deterministic LPA model (5.3) predictions. A. Bifurcation diagram showing total population size $L + P + A$ of the attractor as b is increased and all other parameters are held fixed at the ML estimates in Table 5.1. The extinction equilibrium loses stability at $b_{cr} \approx 0.2275$, creating a survival equilibrium for $b > b_{cr}$ which is stable until a period-doubling bifurcation occurs at approximately $b \approx 8.737$. The shaded region shows the confidence interval $6.2 < b < 22.2$ for the ML estimate $b = 11.68$ in Table 5.1. B–D. Model time series from all three state variables for the estimated value of $b = 11.68$ showing the approach to an oscillation of period two.

however, the survival equilibrium can lose stability and other bifurcations can occur that create new (nonequilibrium) attractors. One possibility for the new attractor is a two-cycle solution of the LPA model, as we saw in Chapter 3 with the Ricker model. This is called a *period-doubling bifurcation*. For higher-dimensional models, such as the LPA model, other types of attractors can be created by this bifurcation. Moreover, further loses of stability and bifurcations can occur with further increases in R_0. We saw that the Ricker equation displays a cascade of bifurcations that results in chaotic dynamics, and therefore we expect this can happen with the LPA model, as well. We will be concerned with such routes to chaos in later chapters. Here we restrict our attention to the dynamics of the LPA model with parameter values estimated in Table 5.1.

Note that we can manipulate the value of R_0 by manipulating any one of the inherent vital rates b, μ_a, or μ_l appearing in its formula (5.9). For example, the extinction equilibrium loses stability as b increases through the critical value $b_{cr} = \mu_a/(1 - \mu_l)$ where $R_0 = 1$, creating survival equilibria for $b > b_{cr}$. This bifurcation, as well as a subsequent period-doubling bifurcation as the survival equilibrium destabilizes, occurs in the LPA model when the other parameters in the model are given by the estimates in Table 5.1, as seen in the bifurcation diagram shown in Figure 5.2. According to this diagram, the deterministic LPA model (5.3) predicts, for the estimated value of $b = 11.68$ in Table 5.1, that the experimental beetle population will approach a two-cycle oscillation, whose two state variables can be

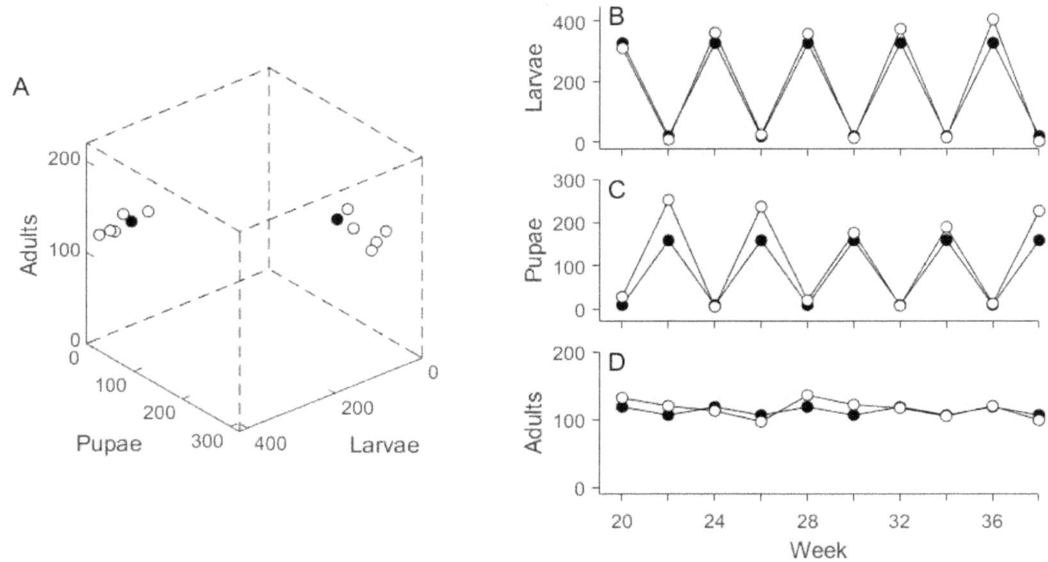

Figure 5.3 Model predictions and data from replicate 1 in the Desharnais experiment for $t \geq 10$ ($t \geq 20$ weeks). A. In state space, data (open circles) cluster about the model-predicted two-cycle (5.10) (shown as solid circles). B-D. Data time series (open circles) and two-cycle model prediction (solid circles).

calculated numerically from the model equations to be

$$\begin{pmatrix} L_1 \\ P_1 \\ A_1 \end{pmatrix} \approx \begin{pmatrix} 18.4 \\ 158.9 \\ 106.6 \end{pmatrix}, \quad \begin{pmatrix} L_2 \\ P_2 \\ A_2 \end{pmatrix} \approx \begin{pmatrix} 326.2 \\ 8.9 \\ 118.6 \end{pmatrix}. \tag{5.10}$$

A comparison between this prediction and the data from the experiment is shown in Figure 5.3. In those plots one sees, after transient data from the first 20 weeks has passed, that the data well matches the model-predicted two-cycle oscillation, both in time series plots and in the three-dimensional state space.

More patterns found in the Desharnais experimental data can be explained by using the stochastic LPA models (5.5) and (5.6). As we shall discover and emphasize in this book, stochasticity can allow data trajectories to visit and revisit all of the entities in state space, illuminating the deterministic structure of state space and providing a mix of dynamic patterns in time series.

For further exploration, see the examples for this section in the online LPA Simulator (https://lpasim.org/?c=5&s=4).

5.5 CONCLUSIONS

In this chapter we developed the discrete-time LPA model (5.3) for the deterministic dynamics of a population with three life-cycle stages (larvae, pupae, and adults). Stochastic versions (5.5) and (5.6) provide means by which to account for deviations of population data from the LPA model predictions (from one census time to the next) due to environmental and demographic stochasticity. We discussed statistical methods for parameterizing and for validating the ability of a model to fit and successfully predict data. As an example, we applied these methods to data obtained from a controlled, replicated laboratory experiment involving flour beetles.

While the analysis of the Desharnais experimental data using our LPA models leads to some interesting conclusions about flour beetle dynamics, such conclusions are not the main purpose of this chapter. Instead, the main purpose is to illustrate our modeling methodology and to demonstrate that the LPA model can provide accurate descriptions of flour beetle population dynamics under normal laboratory conditions. Having done this, we have gained some confidence in the LPA model for prediction purposes—a confidence that will allow us to proceed with other, more elaborate and complicated experiments based on this model's predictions.

Our next step is to put the predictive capability of the model to stronger tests. Can the LPA model predict the dynamics of the beetle populations under conditions different from those used in the Desharnais experiment (upon which the validation of the model was based)? Is the model accurate under other laboratory protocols designed to make specific changes in population parameters? Can model-predicted changes in dynamics be documented to occur in flour beetle populations by means of carefully controlled experiments?

To answer such questions, we need to design feasible, accurate, and controlled experiments to test model predictions and analyze the resulting data. A variety of such interdisciplinary projects populate this book. These projects accomplish several things. First, they further validate the prediction capability of the LPA model. Second, they demonstrate that many interesting and sometimes new and unexpected dynamic patterns (including chaos) predicted by this mathematical model actually occur in populations of *Tribolium*. And third, explanations for puzzling and sometimes subtle patterns observed in data can be provided by the model.

Bibliography

[1] Caswell H (2006) Matrix Population Models: Construction, Analysis and Interpretation. Second Edition. Sunderland, Massachusetts, Sinauer Associates, Inc.

[2] Costantino RF, Cushing JM, Dennis B, Desharnais RA, Henson SM (1998) Resonant population cycles in temporally fluctuating habitats. Bulletin of Mathematical Biology 60:247–275. https://doi.org/10.1006/bulm.1997.0017

[3] Cushing JM (1998) An Introduction to Structured Population Dynamics. CBMS-NSF Regional Conference Series in Applied Mathematics. Volume 71. Philadelphia, Society for Industrial and Applied Mathematics.

[4] Cushing JM (2024) Matrix Models for Population, Disease, and Evolutionary Dynamics. Student Mathematical Library. Volume 106. Providence, Rhode Island, American Mathematical Society.

[5] Cushing JM, Costantino RF, Dennis B, Desharnais RA, Henson SM (2003) Chaos in Ecology: Experimental Nonlinear Dynamics. Theoretical Ecology Series. Volume 1. New York, Academic Press.

[6] Dennis B, Desharnais RA, Cushing JM, Costantino RF (1995) Nonlinear demographic dynamics: Mathematical, models, statistical methods, and biological experiments. Ecological Monographs 65:261–281. https://doi.org/10.2307/2937060

[7] Dennis B, Desharnais RA, Cushing JM, Henson SM, Costantino RF (2001) Estimating chaos and complex dynamics in an insect population. Ecological Monographs 71:277–303. https://doi.org/10.1890/0012-9615(2001)071[0277:ECACDI]2.0.CO;2

[8] Desharnais RA, Costantino RF (1980) Genetic analysis of a population of *Tribolium*: VII. Stability: Response to genetic and demographic perturbations. Canadian Journal of Genetics and Cytology 22:577–589. https://doi.org/10.1139/g80-063

[9] Elaydi SN, Cushing JM (2024) Discrete Mathematical Models in Population Biology: Ecological, Epidemic, and Evolutionary Dynamics. Springer Undergraduate Texts in Mathematics and Technology. Chicago, Springer.

[10] Klimko LA, Nelson PI (1978) On conditional least squares estimation for stochastic processes. Annals of Statistics 6:629–642. https://doi.org/10.1214/aos/1176344207.

[11] Leslie PH (1945) On the use of matrices in certain population mathematics. Biometrika 33:183–212. https://doi.org/10.2307/2332297

[12] Lewis EG (1942) On the generation and growth of a population. Sankhya 6:93–96.

[13] Lin CC, Mudholkar GS (1980) A simple test for normality against asymmetric alternatives. Biometrika 67:455–461. https://doi.org/10.1093/biomet/67.2.455

[14] May RM (2001) Stability and Complexity in Model Ecosystems. Princeton Landmarks in Biology. Princeton, Princeton University Press.

[15] Mertz DB (1969) Age-distribution and abundance in populations of flour beetles. I. Experimental studies. Ecological Monographs 39:1–31. https://doi.org/10.2307/1948563

[16] Moffa AM, Costantino RF (1997) Genetic analysis of a population of *Tribolium*. VI. Polymorphism and demographic equilibrium. Genetics 87:785–805. https://doi.org/10.1093/genetics/87.4.785

[17] Olsson DM, Nelson LS (1975) The Nelder-Mead simplex procedure for function minimization. Technometrics 17:45–51. https://doi.org/10.1080/00401706.1975.10489269

[18] Park T (1932) Studies in population physiology: The relation of numbers to initial population growth in the flour beetle *Tribolium confusum* Duval. Ecology 13:172–181. https://doi.org/10.2307/1931067

[19] Press WH, Flannery BP, Teukolsky SA, Vetterling WT (1986) Numerical Recipes: The Art of Scientific Computing. Cambridge, Cambridge University Press.

[20] Sokoloff A (1974) The Biology of *Tribolium*. Volume 2. Oxford, Oxford University Press.

[21] Stuart A, Ord JK (1991) Kendall's Advanced Theory of Statistics. Volume 2. Classical Inference and Relationships. Fifth Edition. London, Griffin.

[22] Tong H (1999) Nonlinear Time Series: A Dynamical System Approach. Oxford, Oxford University Press.

III

Models as Scientific Hypotheses

CHAPTER 6

Transitions between Attractors

In Chapter 5 we developed a discrete-time, stage-structured model for the dynamics of a population with three life-cycle categories: a larval stage, a pupal stage, and an adult stage. We parameterized and validated the model by using time series data from laboratory experiments with the flour beetle species *Tribolium castaneum*. In this chapter, we turn our attention to a study based on certain predictions of that parameterized model. Specifically, the parameterized model predicts significant changes in dynamic behavior if the adult stage death rate μ_a is changed from its value estimated in Table 5.1 in Section 5.3 of Chapter 5.

We focus on a sequence of bifurcations that occur when an attractor destabilizes and creates a new type of attractor, such as we saw theoretically in Chapter 3 with the simpler Ricker model. Such a bifurcation sequence provides an excellent experimental opportunity to document the occurrence of model-predicted dynamics in a real biological population. Moreover, such an experimental project, if successful, would have two important consequences for our long-term goals. First, it would represent an additional validation of the LPA model in the sense that it would further substantiate the model's ability to *predict* dynamics that can be documented by controlled laboratory experiments. Second, this preliminary bifurcation experiment would serve as the starting point for other projects that empirically document the model's prediction of more complicated dynamical attractors and patterns such as the route to chaos addressed in Chapter 7.

6.1 BIFURCATIONS IN THE LPA MODEL

Recall that the LPA model for the dynamics of the larval, pupal, and adult stages of an insect population is given by the three difference equations

$$\begin{aligned} L_{t+1} &= bA_t e^{-c_{el}L_t - c_{ea}A_t} \\ P_{t+1} &= (1-\mu_l)\,L_t \\ A_{t+1} &= P_t e^{-c_{pa}A_t} + (1-\mu_a)\,A_t. \end{aligned} \tag{6.1}$$

Table 5.1 in Chapter 5 contains estimates for the six parameters appearing in these equations, derived using data obtained from the Desharnais experiment [2] with cultures of *T. castaneum*. For those point estimates, the LPA model predicts a stable (attracting) two-cycle, that is, the repetitive oscillation of the three state variables with period two (four weeks).

However, if the adult death rate μ_a is changed from the estimated value $\mu_a = 0.1108$ in Table 5.1 of Chapter 5, then the model predicted attractor can change and can do so quite

DOI: 10.1201/9781003467120-6

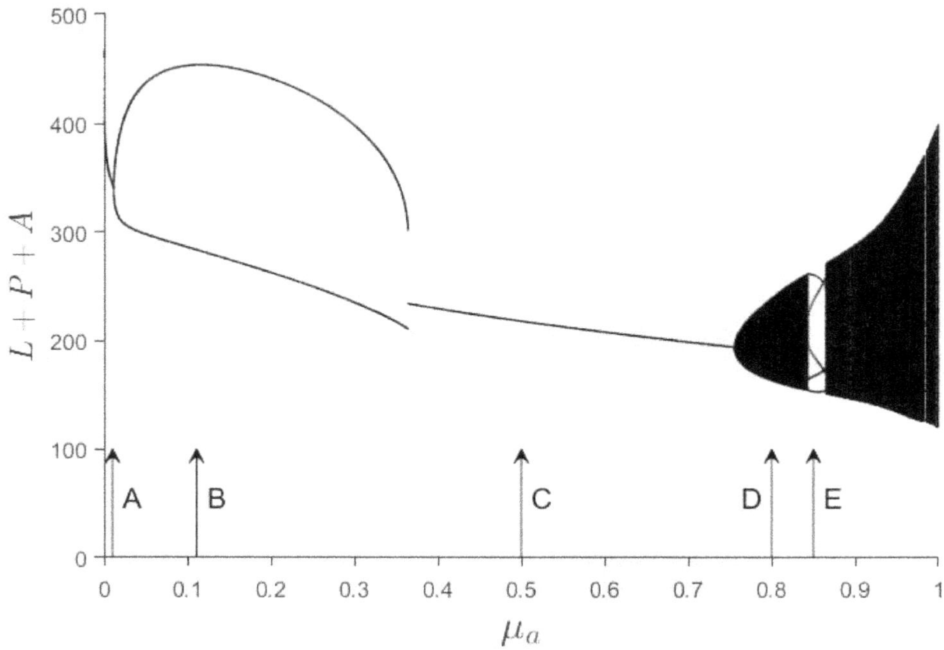

Figure 6.1 Bifurcation diagram for the LPA model (6.1) with parameter values from Table 5.1 (Chapter 5) when the adult death rate μ_a is varied from 0 to 1. Arrows mark values of μ_a for which attractors and sample time series are shown in Figure 6.2. A. Equilibrium for $\mu_a = 0.01$. B. Two-cycle for $\mu_a = 0.1108$. C. Equilibrium for $\mu_a = 0.5$. D. Invariant loop for $\mu_a = 0.8$. E. Five-cycle for $\mu_a = 0.85$.

significantly. This is seen in the bifurcation diagram shown in Figure 6.1, in which the value of μ_a ranges over its allowable interval from 0 to 1. This diagram was constructed as follows. For a selected value of the adult death rate μ_a and all other parameters set at the estimates in Table 5.1, a solution[1] $[L_t, P_t, A_t]$ of the LPA model (6.1) was simulated from initial conditions $[L_0, P_0, A_0] = [10, 10, 10]$ until it approached the attractor sufficiently closely so as to be indistinguishable from it. If the attractor is an equilibrium $[L^*, P^*, A^*]$, then it is represented in the bifurcation diagram by the single total population size $L^* + P^* + A^*$ point plotted above the selected value of μ_a. If, however, the attractor is, for example, a periodic solution with period two, that is, a two-cycle, consisting of two different points $[L_1^*, P_1^*, A_1^*]$ and $[L_2^*, P_2^*, A_2^*]$, then it is represented in the bifurcation diagram by plotting the two total population sizes $L_1^* + P_1^* + A_1^*$ and $L_2^* + P_2^* + A_2^*$ above the selected value of μ_a. Both of these types of attractors are seen in Figure 6.1 for μ_a between 0 and approximately 0.75.

In Figure 6.1, one sees a change from an equilibrium to a two-cycle attractor as μ_a increases from 0. This is an example of a period-doubling bifurcation of the type we saw in the Ricker equation in Chapter 3. As μ_a increases further the reverse happens as the two-cycles disappear and equilibria return.[2] This sequence of attractor changes is demonstrated

[1]We use square-bracket notation $[L, P, A]$ to mean the column vector $(L, P, A)^T$, which is the transpose of the row vector (L, P, A).

[2]This re-equilibration is not a "reverse" period-doubling bifurcation; that is, it is not a period-doubling bifurcation that occurs as μ_a *decreases* through a critical value. Instead, a period-doubling bifurcation occurs for the *unstable* equilibrium as μ_a *increases* through a critical value at approximately 0.35, causing the unstable equilibrium to become stable and creating *unstable* two-cycles. However, as μ_a increases further, the stable two-cycles that emanate from the first period-doubling bifurcation and the newly bifurcating unstable two-cycles "collide" and eliminate each other in what is called a *saddle-node* (or *blue-sky*) *bifurcation*. The

by the first row of panels in Figure 6.2, which show representative time series approaching the attractors. (The time series plots show some initial transients, followed by a temporal gap in the graph, followed by the attractor.) In each panel, the attractor is also shown in three-dimensional $[L, P, A]$-state space. In state space, an equilibrium appears as a single point and a two-cycle as two points.

As the adult death rate μ_a increases further, we observe in the bifurcation diagram (Figure 6.1) the arrival of new attractors as μ_a increases to higher values. In particular, a *Neimark-Sacker* (or *discrete Hopf*) *bifurcation* creates a closed-loop attractor (*invariant loop*) that is best viewed in state space (see panel D in Figure 6.2). This loop is invariant in that an initial condition on the loop leads to a solution trajectory that remains on the loop for all future times. The loop is not itself a solution, that is, the loop does not consists of a single solution trajectory, in contrast to the predator-prey loops whose trajectories in state space are continuous curves arising from differential equation models. Rather, an invariant loop for a discrete-time model, such as the LPA model (6.1), contains infinitely many different solution trajectories. Some of these trajectories can be periodic cycles and others are oscillatory but aperiodic. In total, they fill out the loop in state space.

Sometimes, however, trajectories on an invariant loop approach a periodic cycle on the loop. If the loop is an attractor in state space, then solution trajectories not on the loop also approach this cycle, a scenario referred to as periodic-locking. Panels D and E in Figure 6.2 show an invariant loop and a period-locked five-cycle attractor, respectively. The interval of μ_a values greater than the invariant loop bifurcation point in the bifurcation diagram (Figure 6.1) clearly shows a period-locking window with five-cycle attractors. Not so evident in this figure, however, is that if one zooms into that region of the bifurcation diagram, many other period-locking windows are seen in a complicated fractal-like array of attractors present for these large values of the adult death rate.

Our take away message from the bifurcation diagram in Figure 6.1 is that the LPA model (6.1) with parameter values estimated in Table 5.1 (Chapter 5) predicts a complicated sequence of bifurcations and attractors as the adult death rate is increased over its allowable range from 0 to 1. This is an unexpected and non-intuitive consequence of a large adult death rate. Since it is not difficult to manipulate the adult death rate in laboratory cultures of flour beetles, this bifurcation scenario presents an excellent experimental opportunity to test the use of models in population theory. A successful test of these model predicted bifurcations would, in and of itself, be a noteworthy case study that demonstrates a successful, quantitative, and predictive relationship between a mathematical model and a real biological population. In fact, for us it opened the door to interdisciplinary, model-supported study of many other dynamic phenomena using experimental cultures of flour beetles and the LPA model, as shown in subsequent chapters. An example is chaos, discussed in Chapter 7.

In this chapter, we ask: Will laboratory flour beetle cultures display the deterministic LPA model-predicted bifurcations shown in Figure 6.1 if their adult death rates are manipulated?

For further exploration, see the examples for this section in the online LPA Simulator (https://lpasim.org/?c=6&s=1).

6.2 A BIFURCATION EXPERIMENT

To answer the question raised at the end of the previous section, we designed an experiment in which cultures of flour beetles were grown under standard laboratory conditions, except

result is that there are values of μ_a for which stable equilibria, stable two-cycles, and unstable two-cycles all simultaneously exist. This complicated, multiple attractor situation occurs on a very small interval of μ_a values and is barely visible in Figure 6.1. It therefore plays no role in our study.

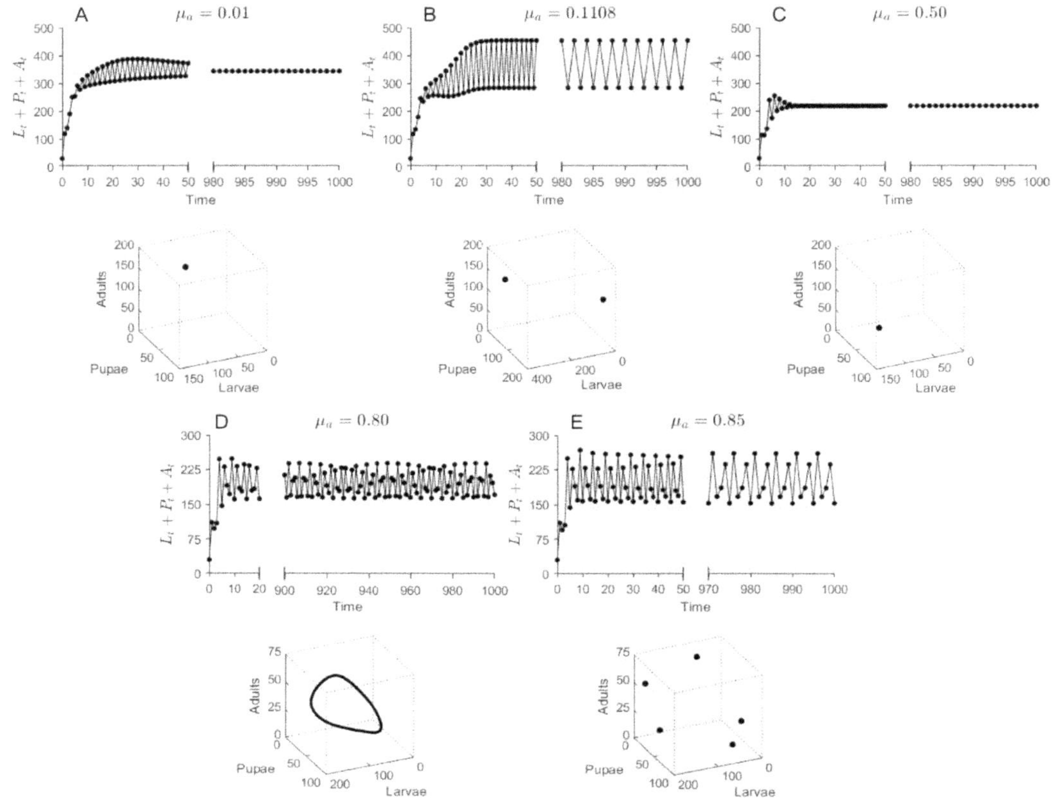

Figure 6.2 LPA model (6.1) time series for total population size with initial condition $[L_0, P_0, A_0] = [10, 10, 10]$. The time series plots show some initial transients, followed by a temporal gap in the graph, followed by the attractor. Each panel is associated with an arrow in the bifurcation diagram (Figure 6.1). Also shown, in each panel, is the model-predicted attractor in three-dimensional $[L, P, A]$ state space. A. Equilibrium for $\mu_a = 0.01$. B. Two-cycle for $\mu_a = 0.1108$. C. Equilibrium for $\mu_a = 0.5$. D. Invariant loop for $\mu_a = 0.8$. E. Five-cycle for $\mu_a = 0.85$.

that we manipulated the adult death rate to values selected to test the transitions between attractors predicted by the deterministic LPA model in Figure 6.1 [3]. Unfortunately, the corn oil sensitive (*cos*) strain of *T. castaneum* used in the Desharnais experiment, whose data were used for the model parameterization and validation in Chapter 5, was no longer available. Thus, we duplicated the experiment described above using two other strains—the so-called *SS* and *RR* strains of *T. castaneum.* Because of this change, and because we used different experimental protocols and laboratories, we performed a re-parameterization of the LPA model using the data from the new experiment. We also repeated model fit and validation analyses using the methods described in Chapter 5. The results of these analyses for the *SS* strain are discussed in the following sections. Parameter estimates and the sequence of dynamic bifurcations in Figure 6.1 did not change much, a testimony to the robustness of the LPA model's accuracy in describing the dynamics of the flour beetle populations. For a similar analysis utilizing the *RR* strain, see [1, 2, 3].

6.2.1 Experimental Design

We designed an experiment with 24 cultures of the *SS* strain of *T. castaneum*, each contained in a half-pint (237 mL) bottle with 20 g of standard media and kept in a dark incubator at 32°C. The cultures were initiated with 100 young adults, 5 pupae, and 250 small larvae. Every two weeks, the larvae, pupae, and adult stages were counted and returned to fresh media. This procedure continued for 36 weeks. Dead adults also were counted and removed. After 12 weeks, four cultures were assigned randomly to each of the adult death rates of $\mu_a = 0.04, 0.27, 0.50, 0.73$, and 0.96, leaving four cultures as unmanipulated controls. For 24 subsequent weeks, these manipulated death rates were enforced by removing or adding adults at the time of census in order to make the total number of adults who died during the previous time interval consistent with the desired treatment value of μ_a. (If the target death rate resulted in non-integer number of adults to be added or removed, we rounded the number to the nearest whole integer.)

This experimental design was repeated using cultures of the *RR* strain *T. castaneum* as well, although we describe here only the results obtained using the *SS* strain. Both the *RR* and *SS* data are available from Dryad (https://doi.org/10.5061/dryad.qjq2bvqmp).

6.2.2 Parameterization for the *SS* Strain

Using the *SS* data obtained from the experiment described in the previous section, we parameterized and validated the LPA model (6.1), repeating the procedures described in Chapter 5 [3].

For each treatment (including the control), we randomly designated two replicates for parameter estimation and withheld the other two replicates for model validation. We combined the time-series data from the 12 "estimation" populations into one log-likelihood estimation procedure as follows. For each population, each transition from $\mathbf{w}_t$ to $\mathbf{w}_{t+1}$ contributed a term $\ln p\left(\mathbf{w}_{t+1}|\mathbf{w}_t\right)$ to the log-likelihood function (5.8). For the experimentally manipulated populations, the value of μ_a in the term $\ln p\left(\mathbf{w}_{t+1}|\mathbf{w}_t\right)$ was fixed at the corresponding experimental value (either 0.04, 0.27, 0.50, 0.73 or 0.96). The value of μ_a in the control populations was estimated directly from the census counts of adults at time t and dead adults at time $t+1$. The resulting estimate of μ_a was included in the log-likelihood as fixed a constant for those populations. The control populations and experimental populations were given different variance-covariance matrices ($\mathbf{\Sigma}_C$ and $\mathbf{\Sigma}_T$, respectively) in the log-likelihood procedure, given that we expected the experimental manipulations to alter the variability of the adult stage.

Manipulations of adult death rate μ_a began at week 12, which was therefore the starting time ($t = 0$) in the parameter estimation procedure. Thus, each data time series was 24 weeks in length, producing 12 one-step transitions for each population. The estimation data contained 12 populations, yielding 144 transitions in the log-likelihood function.

The log-likelihood function contained 17 unknown parameters to be estimated: b, c_{el}, c_{ea}, c_{pa}, μ_l and six parameters each in the variance-covariance matrices Σ_C and Σ_T. We used the Nelder-Mead simplex algorithm to maximize numerically the log-likelihood function, and obtained the maximum likelihood (ML) parameter estimates in Table 6.1. The calculations were numerically well-behaved, and repeated maximizations from many different starting points converged to the same parameter values. (Confidence intervals for each parameter estimate also appear in this table, as calculated from the deterministic skeleton (6.1) using the profile likelihood method.)

TABLE 6.1 Maximum likelihood (ML) parameter estimates for the *SS* genetic strain from the experiment described in Section 6.2.1. The 95% confidence intervals were calculated from profile likelihoods. The control estimate for μ_a and its 95% confidence interval were calculated directly from the number of dead adults observed in the control cultures. The ML estimates for the variance-covariances matrix $\mathbf{\Sigma}_C = (\sigma_{ij})$ for the controls were $\hat{\sigma}_{11} = 0.0804$, $\hat{\sigma}_{12} = \hat{\sigma}_{21} = 0.0282$, $\hat{\sigma}_{13} = \hat{\sigma}_{31} = 0.00432$, $\hat{\sigma}_{22} = 0.159$, $\hat{\sigma}_{23} = \hat{\sigma}_{32} = 0.00341$, $\hat{\sigma}_{33} = 0.000199$. The estimates for $\mathbf{\Sigma}_T$ for the treatments were $\hat{\sigma}_{11} = 0.867$, $\hat{\sigma}_{12} = \hat{\sigma}_{21} = 0.0137$, $\hat{\sigma}_{13} = \hat{\sigma}_{31} = -0.00404$, $\hat{\sigma}_{22} = 0.222$, $\hat{\sigma}_{23} = \hat{\sigma}_{32} = 0.0101$, $\hat{\sigma}_{33} = 0.0109$.

Parameter	ML estimate	95% confidence interval
b	7.483	(5.4, 10.3)
μ_l	0.2670	(0.21, 0.32)
c_{ea}	0.009170	(0.0081, 0.1)
c_{el}	0.01200	(0.011, 0.013)
c_{pa}	0.004139	(0.0037, 0.0046)
μ_a (control)	0.003620	(0.0027, 0.0045)

To evaluate model fit on the estimation data, we substituted the ML parameter estimates from Table 6.1 into the conditional expected values

$$\begin{aligned} E\left(\ln L_{t+1} \middle| L_t = l_t,\ P_t = p_t,\ A_t = a_t\right) &= \ln\left(b a_t e^{-c_{el} l_t - c_{ea} a_t}\right) \\ E\left(\ln P_{t+1} \middle| L_t = l_t,\ P_t = p_t,\ A_t = a_t\right) &= \ln\left(\left(1-\mu_l\right) l_t\right) \\ E\left(\ln A_{t+1} \middle| L_t = l_t,\ P_t = p_t,\ A_t = a_t\right) &= \ln\left(p_t e^{-c_{pa} a_t} + \left(1-\mu_a\right) a_t\right). \end{aligned} \tag{6.2}$$

The conditioned one-step residuals on the log scale were the differences between the log-transformed observations at time $t+1$ and their estimated expected values conditioned on the observations at time t. We performed various diagnostic tests for autocorrelation and normality on these log-residuals, as described in Chapter 5, Section 5.3 (also see [3]). None of the *SS* series of residuals showed significant first-order autocorrelation, and just two (6%) of the *SS* series showed significant second-order autocorrelation. Most of the residuals in each series were well described by a normal distribution. Departure from normality was seen in six (17%) of the *SS* series, which can be traced to a small number of outliers.

6.2.3 Model Validation

To test the predictive capability of the parameterized model, we repeated the diagnostic residual tests for autocorrelation and normality on the validation data sets without re-parameterizing. We found no significant first-order or second-order autocorrelation present in any of the times series in the validation data sets. Departures from a normal distribution were seen in eight (22%) of the *SS* series, which again were caused by a small number of identifiable outliers.

Thus, our statistical analyses confirmed the accuracy of the LPA model in describing the data obtained from the bifurcation experiment. Plots of the both the estimation and validation data time series, together with each one-step model prediction, visually supported this conclusion. Figure 6.3 shows the four replicate data time series (solid circles) from the

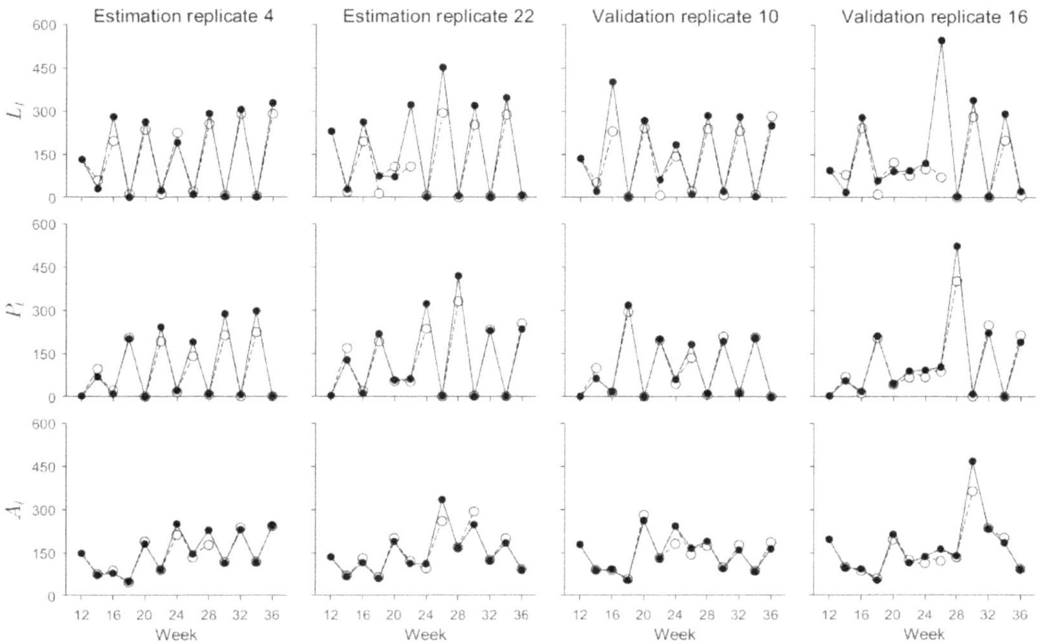

Figure 6.3 Census data (solid circles) and one-step predictions (open circles) for all four replicates of the $\mu_a = 0.5$ treatment for the *SS* strain in the bifurcation experiment. Replicates 4 and 22 were estimation data; replicates 10 and 16 were validation data.

$\mu_a = 0.5$ treatment along with their one-step predictions (open circles). Corresponding graphs for the rest of the treatments are shown in [3], for both the *SS* and *RR* strains.

It is noteworthy that the model evaluation results for both the *SS* and *RR* strains compared favorably with the same analyses performed in Chapter 5 on the *cos* strain of *T. castaneum*.

We now turn our attention to the model-predicted transitions (bifurcations) between attractors at the various values of μ_a. Was the predicted sequence of attractors observed in the laboratory?

6.2.4 Bifurcation and Attractor Results

The estimated parameters appearing in Table 6.1 for the *SS* genetic strain (obtained from the estimation subset of the data) differ somewhat from the estimates in Table 5.1 for the *cos* strain of *T. castaneum* obtained in Chapter 5. These new parameters yield a modified bifurcation diagram, displayed in Figure 6.4, which is qualitatively and quantitatively similar to the one in Figure 6.1 on which we based the experimental design. Both bifurcation diagrams have, at intermediate values of the adult death rate μ_a, a "two-cycle bubble" and, for high values of μ_a, a Neimark-Sacker bifurcation of invariant loops.

The predicted attractors can be calculated numerically by iterating the model equations, from selected initial conditions, sufficiently long and then ignoring initial transients. Table 6.2 lists the model-predicted attractors for the experimental values of μ_a. To compare the predicted attractors to the experimental data, we view them together (after initial data

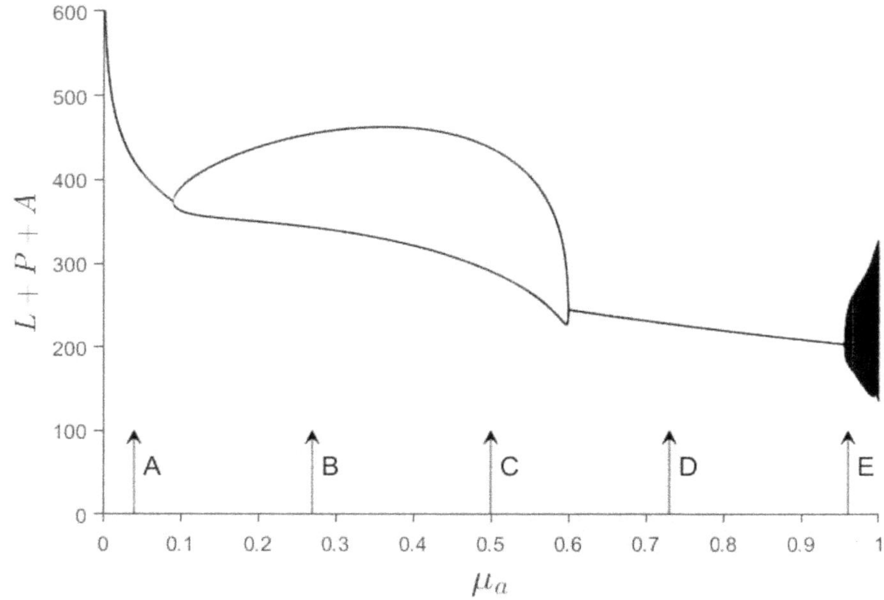

Figure 6.4 Bifurcation diagram for the LPA model (6.1) with parameter values from Table 6.1 when the adult death rate μ_a is varied from 0 to 1. Arrows indicate the values of μ_a used in the *SS* experimental treatments. The diagram predicts the following attractors: A. Equilibrium at $\mu_a = 0.04$. B. Two-cycle at $\mu_a = 0.27$. C. Two-cycle at $\mu_a = 0.50$. D. Equilibrium at $\mu_a = 0.73$. E. Invariant loop at $\mu_a = 0.96$. The corresponding attractors are shown in state space with experimental data in Figure 6.5.

TABLE 6.2 Deterministic LPA model-predicted attractors for the experimentally-manipulated values of μ_a, given the estimated parameters in Table 6.1.

μ_a	Stable attractor
0.04	Equilibrium: [62.55, 45.85, 313.3]
0.27	Two-cycle: [13.17, 175.7, 154.0]; [239.7, 9.657, 205.3]
0.50	Two-cycle: [12.43, 186.7, 91.25]; [254.7, 9.115, 173.6]
0.73	Equilibrium: [91.49, 67.07, 69.04]
0.96	Invariant loop

transients are removed) in three-dimensional $[L, P, A]$- state space, as shown in Figure 6.5. Equilibria appear as points in state space, two-cycles appear as two points, and invariant loops attractors appear as closed curves. The clusters of data for the predicted equilibrium cases ($\mu_a = 0.04$ and $\mu_a = 0.73$) are clearly different from the two separated clusters of data in the predicted two-cycle cases ($\mu_a = 0.27$ and $\mu_a = 0.50$). In all cases the data cluster near the model-predicted attractors. In particular, notice how the data for treatment $\mu_a = 0.96$ lie

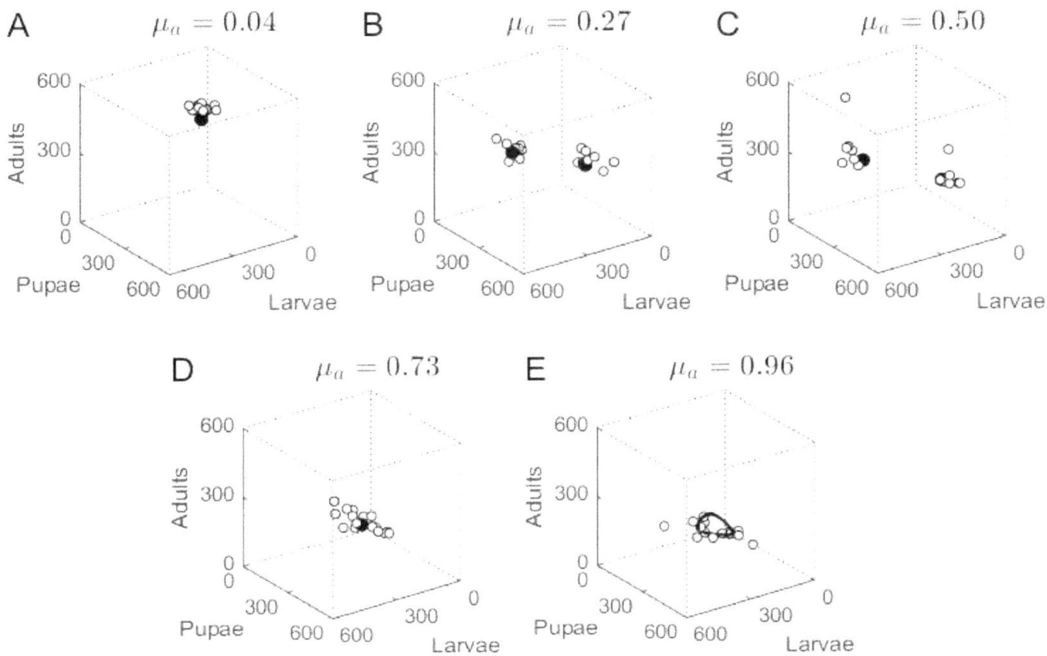

Figure 6.5 Attractors of the deterministic LPA model (6.1) with parameter estimates from Table 6.1 (solid circles and curves) and experimental data with transients removed (open circles) in state space. A. Equilibrium with $\mu_a = 0.04$. B. Two-cycle with $\mu_a = 0.27$. C. Two-cycle with $\mu_a = 0.50$. D. Equilibrium with $\mu_a = 0.73$. E. Invariant loop with $\mu_a = 0.96$.

close to the invariant loop. In these plots, the *SS* strain data appear remarkably consistent with the model-predicted attractors.

Similar results were obtained for the duplicated experiment with the *RR* strain [2, 3].

For further exploration, see the examples for this section in the online LPA Simulator (https://lpasim.org/?c=6&s=2).

6.3 CONCLUSIONS

In this chapter, we considered a bifurcation sequence predicted by the deterministic LPA model (6.1) as the adult death rate μ_a is varied. This sequence includes equilibria, period-doubling bifurcations that resulted in two-cycle attractors, and a Neimark-Sacker bifurcation involving aperiodic oscillations resulting from invariant loops. These transitions in dynamics motivated the implementation of experiments designed to document their occurrence in laboratory cultures of flour beetles.

The statistical analyses of the resulting data confirmed the accuracy of the model's description and prediction of the data. Residual analyses suggested that most of the systematic, predictable variability in the data was accounted for by the deterministic LPA model (6.1), and that the stochastic LPA model (5.5) in Chapter 5 adequately described the remaining variability. State space graphs of model-predicted attractors and data show quite convincingly that the beetle population dynamics did undergo the predicted transitions between attractors when the adult death rates were manipulated experimentally.

The success of the bifurcation experiment—documenting model-predicted dynamic changes by means of controlled and replicated laboratory experiments—gave us confidence to investigate other, more complicated model-predicted bifurcation sequences, including sequences involving chaotic attractors. We describe the "route-to-chaos" experiments in Chapter 7.

We emphasize the usefulness of viewing both model predictions and data in state space, as in Figure 6.5. For example, the structure of an attractor such as the invariant loop in Figure 6.5E would not be readily observable in time series plots of trajectories and data. The study of the geometry and stability properties of attractors in state space has been a standard tool in dynamical systems theory since the early work of many notable mathematicians and physicists, including Poincaré, Birkhoff, Gibbs, and others. The utility of this approach will become even more apparent in subsequent chapters as the dynamic patterns become more complicated, especially when coupled with stochasticity.

Bibliography

[1] Costantino RF, Cushing JM, Dennis B, Desharnais RA (1995) Experimentally induced transitions in the dynamic behaviour of insect populations. Nature 375:227–230. https://doi.org/10.1038/375227a0

[2] Cushing JM, Costantino RF, Dennis B, Desharnais RA, Henson SM (2003) Chaos in Ecology: Experimental Nonlinear Dynamics. Theoretical Ecology Series. Volume 1. New York, Academic Press.

[3] Dennis B, Desharnais RA, Cushing JM, Costantino RF (1997) Transitions in population dynamics: Equilibria to periodic cycles to aperiodic cycles. Journal of Animal Ecology 66:704–729. https://doi.org/10.2307/5923

CHAPTER 7

A Route to Chaos

In Chapters 5 and 6 we showed, using replicated and controlled laboratory experiments, that the LPA model

$$\begin{aligned} L_{t+1} &= bA_t e^{-c_{el}L_t - c_{ea}A_t} \\ P_{t+1} &= (1-\mu_l)\,L_t \\ A_{t+1} &= P_t e^{-c_{pa}A_t} + (1-\mu_a)\,A_t \end{aligned} \tag{7.1}$$

can accurately describe and predict the dynamics of the flour beetle *Tribolium castaneum*. The model successfully predicted characteristics of the long-term dynamics (attractors) of the larval, pupal, and adult stages, and, notably, predicted the *transitions* (bifurcations) between attractors as the adult mortality rate was varied. The predicted sequence of bifurcations used to design the experiment in Chapter 6 included equilibria, two-cycle oscillations, and aperiodic oscillations (resulting from an invariant loop attractor in state space), but did not involve chaos.

In this chapter we highlight experiments based on a different bifurcation sequence. Like the bifurcation sequence in Chapter 6, this "route-to-chaos" sequence contains equilibria, periodic cycles, and invariant loops—but it also contains exotic chaotic attractors. We experimentally tested this route to chaos in the laboratory. For one of the treatments predicted to result in chaotic dynamics, we maintained and censused the cultures for 484 weeks, or approximately 121 generations (over nine years).

7.1 A MODEL-PREDICTED ROUTE TO CHAOS

When drawing a bifurcation diagram to display a model's attractors as they depend on a model parameter, one must decide what component of the attractor to plot against the parameter. For a model with a single state variable, the choice is obvious. If a model has more than one state variable, then some scalar quantity must be derived from them to represent the attractor in the diagram. One can use just one of the state variables for this purpose, or one can use a linear combination of the state variables, for example a weighted average.

The LPA model has three state variables. Figure 7.1A shows a bifurcation diagram obtained from the LPA model that plots total population size $L+P+A$ on an attractor as the cannibalism coefficient c_{pa} ranges from 0 to 0.6. Here the adult death rate μ_a is fixed at 0.96 and the other parameters are the values shown in Table 7.1. From this diagram we obtain conjectures about the attractor for various values of the bifurcation parameter c_{pa}, given $\mu_a = 0.96$. The parameter estimates in Table 7.1 are the maximum likelihood (ML) estimates for the *RR* genetic strain of *T. castaneum*, obtained from the companion data set to the *SS* genetic strain used in Chapter 6. These estimates are reported in Table 1 of [8].

DOI: 10.1201/9781003467120-7

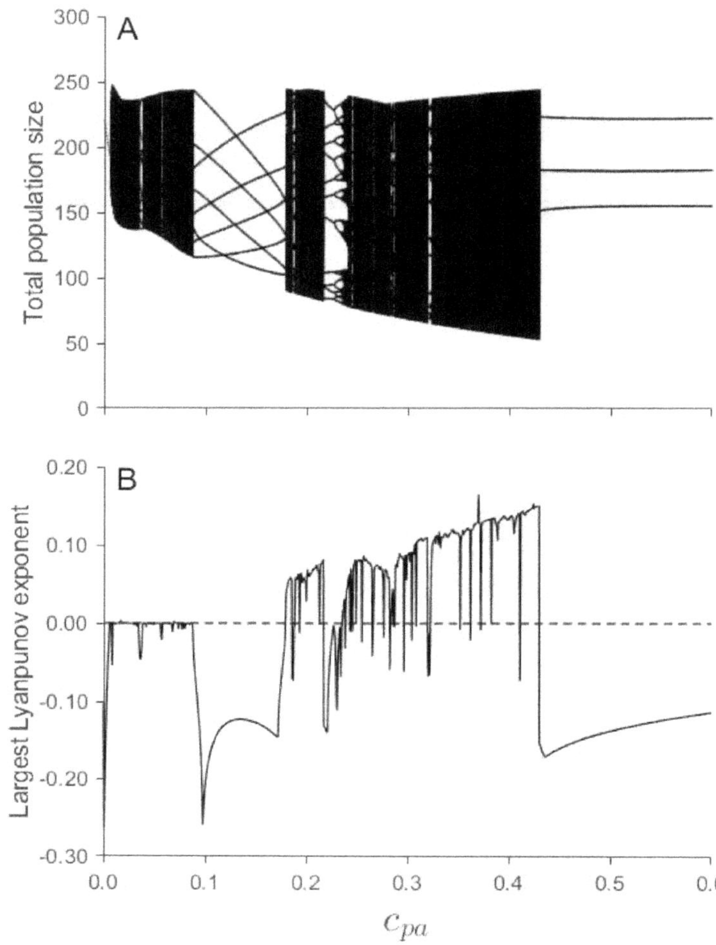

Figure 7.1 LPA model (7.1) predictions as c_{pa} ranges from 0 to 0.6, for $\mu_a = 0.96$. All other parameters are from Table 7.1. A. Bifurcation diagram. B. Corresponding plot of the dominant Lyapunov exponent (LE), a positive value of which indicates sensitivity to initial conditions.

If the model attractor at a certain value of c_{pa} is an equilibrium $[L_e, P_e, A_e]$,[1] then a single point $L_e + P_e + A_e$ would be plotted above it in the bifurcation diagram. One suspects the converse is true, that is, if a single point appears plotted above a value of c_{pa}, then the attractor for that value of c_{pa} would be an equilibrium. However, one must always be careful in interpreting bifurcation diagrams. For example, it is theoretically possible, although unlikely, that a two-cycle attractor could consist of the two distinct points $[L_1, P_1, A_1] \neq [L_2, P_2, A_2]$ with $L_1 + P_1 + A_1 = L_2 + P_2 + A_2$. Keeping this caution in mind, a finite number of points above a parameter value in the diagram suggests that the attractor is a cycle of some period. When there are so many points above the parameter that the plot looks like a solid line (or several line segments), there are several possibilities for the attractor: a periodic cycle of a very large period, or an aperiodic attractor of some

[1] We use square bracket notation $[L, P, A]$ to indicate the column vector $(L, P, A)^T$, which is the transpose of the row vector (L, P, A).

TABLE 7.1 Maximum likelihood (ML) parameter estimates for the RR genetic strain from the data observed in the experiment described in Section 6.2.1 of Chapter 6. The ML estimates for the variance-covariances matrix $\Sigma = (\sigma_{ij})$ for the controls are $\hat{\sigma}_{11} = 0.567$, $\hat{\sigma}_{12} = \hat{\sigma}_{21} = -0.145$, $\hat{\sigma}_{13} = \hat{\sigma}_{31} = 0.00247$, $\hat{\sigma}_{22} = 0.157$, $\hat{\sigma}_{23} = \hat{\sigma}_{32} = 0.00155$, $\hat{\sigma}_{33} = 0.000154$. For those treatments in which adult mortality was manipulated, the estimates are $\hat{\sigma}_{11} = 0.828$, $\hat{\sigma}_{12} = \hat{\sigma}_{21} = -0.0845$, $\hat{\sigma}_{13} = \hat{\sigma}_{31} = -0.00297$, $\hat{\sigma}_{22} = 0.175$, $\hat{\sigma}_{23} = \hat{\sigma}_{32} = -0.0202$, $\hat{\sigma}_{33} = 0.0206$. These estimates are reported in Table 1 of [8].

Parameter	ML estimate	95% confidence interval
b	7.876	(5.8, 10.7)
μ_l	0.1613	(0.10, 0.22)
c_{ea}	0.01114	(0.010, 0.012)
c_{el}	0.01385	(0.012, 0.015)
c_{pa}	0.004348	(0.0039, 0.0048)
μ_a (control)	0.004210	(0.0032, 0.0052)

kind, for example a chaotic attractor. One way to help determine what kind of attractor actually occurs at a specific parameter value is to plot time series of a solution and observe its characteristics, after discarding a suitable number of initial steps to rid the time series of the transient dynamics as it approaches the attractor. For this purpose, any or all of the components L_t, P_t, and A_t could be plotted against t or the points $[L_t, P_t, A_t]$ could be plotted in three-dimensional state space.

With these caveats in mind, we return to the bifurcation diagram in Figure 7.1A and observe that it implies a complex sequence of attractors as c_{pa} ranges from 0 to 0.6. The suggested attractors are as follows. For very small values of c_{pa} (less than approximately 0.01), the attractor, barely visible in Figure 7.1A, is an equilibrium. For values of c_{pa} larger than approximately 0.41, the attractor is a three-cycle. For c_{pa} between these two values, there is a complex sequence of attractors. On the interval from approximately 0.09 to 0.18 the attractor is an eight-cycle. Other small intervals similarly display a finite number of points and hence imply the attractor is a periodic cycle. These intervals, of which there are many, are referred to as *period-locking windows*. (Many more such windows are visible upon repeated magnification of the bifurcation diagram, in a fractal type phenomenon.) Notice how these period-locking windows appear to abruptly open and close along the diagram. They indicate a significant change in the attractor at these parameter values, from a periodic cycle to a more complicated aperiodic attractor, or vice versa.

Note the abrupt change from a complex aperiodic attractor (possibly chaotic attractor) to a three-cycle that occurs as c_{pa} increases through approximately 0.41. This significant change should be observable in controlled experiments. Points such as these are called *boundary crises.* [2]

In addition to the period-locking windows of cycle attractors, the bifurcation diagram in Figure 7.1A suggests significant c_{pa}-intervals in which there are aperiodic and possibly chaotic attractors.

For further exploration, see the examples for this section in the online LPA Simulator (https://lpasim.org/?c=7&s=1).

[2] A *boundary crisis* occurs when an attractor collides with the boundary of its basin of attraction as a parameter is varied.

7.2 CHAOS, SENSITIVITY TO INITIAL CONDITIONS, AND LYAPUNOV EXPONENTS

Mathematicians have identified and studied many kinds of complex patterns that can arise in simple dynamical systems. A special type called *chaos* has become in many ways an icon for complexity. But what exactly is chaos?

One can find many definitions of chaos, both formal and informal, throughout the scientific and mathematical literature [4]. Tien-Yien Li and James Yorke originally coined the term in their now famous study [18]. Edward Lorenz, who had encountered chaos a decade earlier in a different context [16], offered several informal definitions of chaotic dynamics in his book [17], including processes that "appear to proceed according to chance even though their behavior is in fact determined by precise laws" or "behavior that *is* deterministic, or is nearly so if it occurs in a tangible system that possesses a slight amount of randomness, but does not *look* deterministic." In his popular book on the subject [15], James Gleick provided several definitions of chaos formulated by various investigators, one of which (attributed to Philip Holmes) is: "complicated, aperiodic, attracting orbits of certain (usually low-dimensional) dynamical systems." Ian Stewart described chaos as "apparent randomness with a purely deterministic cause." As "[u]nruly behavior governed entirely by rules," it "inhabits the twilight zone between regularity and randomness" [21]. In a book by Kathleen Alligood, Tim Sauer, and James Yorke, a chaotic solution is described as a bounded solution "that forever continues to experience the unstable behavior that an orbit exhibits near a source, but that is not itself fixed or periodic" and "at any point of such an orbit, there are points arbitrarily near that will move away from the point . . . " [1].

There are numerous rigorous mathematical definitions of chaos. These definitions are not necessarily equivalent to one another, and none have become universally accepted [4, 19]. It is generally agreed, however, that a chaotic system should be generated by a deterministic and nonlinear rule whose solutions are bounded and exhibit *sensitivity to initial conditions*. This latter property means that a disturbance, no matter how small, in the state of the system becomes highly amplified over time. Or, put another way, trajectories initiating from two nearby states diverge and become significantly dissimilar as time passes. While these properties are central to the notion of chaos, they are not sufficient to define it [4]. Mathematically proving a system is chaotic is typically a formidable task.

For our purposes, it is not necessary to delve into these mathematical technicalities, and we take the following as a working definition: *A solution generated by a deterministic rule is chaotic if it is bounded, is neither periodic nor approaches a periodic state, and is sensitive to initial conditions.* The basic deterministic and nonlinear requirements and the boundedness of solutions are usually straightforward to establish mathematically. For example, it is shown in [6] that the solutions of the LPA model (7.1) are bounded. But we will, as do most researchers, rely on evidence obtained from computer simulations to establish the key property of sensitivity to initial conditions. This is usually done by estimating a diagnostic quantity called the (dominant) *Lyapunov exponent* (LE), a number whose positivity implies sensitivity to initial conditions.

Intuitively, one can think of the LE in the following way. In three-dimensional state space, consider an initial point $[L_0, P_0, A_0]$ and its next iterate $[L_1, P_1, A_1]$ as obtained from the LPA model (7.1). Imagine a tiny sphere, whose radius we define to be one, centered at $[L_0, P_0, A_0]$. Imagine iterating the entire sphere by plugging each point of the sphere into the right-hand sides of the LPA model equations. In this way the entire sphere is mapped, approximately, to an ellipsoid[3] surrounding the point $[L_1, P_1, A_1]$. This action might contract the sphere in some directions and expand it in others. In the directions

[3]This is because the action of the LPA model near the point $[L_0, P_0, A_0]$ is approximated by the linearization of the LPA model, and linear models map spheres into ellipsoids. See Chapter 3.

of contraction, nearby trajectories of the LPA model converge. In directions of expansion, nearby trajectories diverge. In a similar way, we can imagine iterating the initial point and its initial sphere t times into the future. The length r_t of the longest axis of the resulting (approximate) ellipsoid will grow exponentially, with exponential growth rate $\ln r_t$. The LE is the per unit time average of this exponential growth rate (namely, $t^{-1} \ln r_t$), in the limit as time increases:

$$\text{LE} = \lim_{t \to \infty} \frac{1}{t} \ln r_t. \tag{7.2}$$

If the LE of a trajectory is positive, then on average along the trajectory there is divergence of (at least some) nearby trajectories and hence sensitivity to initial conditions. This is the case if the trajectory is chaotic.[4] For a rigorous definition of LE and a discussion of this number, see [1]. For details on how we computed LE, see Chapter 11 and [9].

Figure 7.1B shows the LE of the LPA model, with parameter values from Table 7.1, corresponding to the bifurcation diagram shown in Figure 7.1A. We see there is a substantial interval, ranging from approximately 0.18 to 0.41, that contains c_{pa} values at which the LE is positive, indicating sensitivity to initial conditions and, as the hallmark signature of chaos, suggesting the existence of chaotic attractors in the LPA model. To date, however, there is no rigorous mathematical proof of chaotic attractors for the LPA model by any formal definition of chaos.

There are a number of other numerically-calculated evidences of chaotic attractors in the LPA model. For example, one axiom in a popular definition of chaotic attractor, due to Robert Devaney [2, 12], is that there exist infinitely many unstable periodic cycles on the attractor. In Chapter 9 we explore the existence and properties of several periodic cycles that appear in the LPA model's chaotic attractor and their influence in our experimental data. Another topological feature of a chaotic attractor in state space is its *fractal dimension*, which measures the "size" of the attractor. (A line has dimension one, a surface has dimension two, and so on.) A chaotic attractor is typically a *strange attractor*, meaning that it has a non-integer dimension. There are several definitions of dimension in topology, one of which is the *box dimension*. The attractor in Figure 7.1A located at $c_{pa} = 0.35$ has box dimension 1.26.[5]

Figure 7.1A predicts a distinctive sequence of bifurcations that includes chaotic attractors within a large interval of c_{pa} values, bookended on both sides by relatively uncomplicated attractors at both ends of the bifurcation sequence. Can this route to chaos be documented by means of manipulative experiments with cultures of flour beetles? Will the flour beetle cultures in the chaotic region show the properties of a chaotic attractor, bookended by cultures at both ends of the bifurcation sequence that show simple dynamics?

A key word in our working definition of chaos is "deterministic." Ordinary stochastic versions of non-chaotic population models such as the logistic can easily yield positive LEs if the LEs are calculated with the full stochastic trajectories rather than trajectories from the deterministic portion. The stochastic noise causes trajectories to revisit areas of state space where nearby trajectories diverge, such as the near-exponential growth behavior displayed by the logistic at low population abundances [10]. The original fascination with chaos in the 1970s and 80s arose from the possibility that random-seeming population fluctuations could be caused by low-dimensional deterministic (or near-deterministic) laws. We view the chaos definition as distinguishing between an invariant measure of a low-dimensional deterministic attractor and a stationary probability distribution of a stochastic process. The

---

[4]Chaotic attractors typically have in them a chaotic trajectory that recurrently comes arbitrarily close to every point of the attractor. Because of this, we can meaningfully associate the Lyapunov exponent with the attractor itself (although not every trajectory on a chaotic attractor is necessarily chaotic!).

[5]We thank Professors Jeffery Edmunds and Marek Rychlik for the calculation of this box dimension.

invariant measure seems like a probability distribution, but its predictions are not "random" in the sense of stochasticity arising from high-dimensional, unpredictable dynamics.

7.3 EXPERIMENTAL DESIGN

In November of 1995, we implemented an experiment designed to test the predictions contained in the bifurcation diagram in Figure 7.1A. In this *route-to-chaos experiment*, we used a protocol similar to that used in the bifurcation experiment described in Chapter 6.

Twenty-four cultures of the *RR* strain of *T. castaneum* were initiated with 250 L-stage individuals, 5 P-stage individuals, and 100 A-stage sexually mature young adults and were maintained in half-pint (237 ml) milk bottles with 20 g of standard media ($V = 1$) placed in a dark incubator at 32°C. Every two weeks the number of individuals in the L-stage, P-stage, and A-stage were counted and returned to fresh media. Dead adults also were counted and removed. We designated three cultures as unmanipulated controls. From the remaining 21 cultures, we randomly assigned three populations to each of seven treatments. In each treatment, we manipulated A-stage mortality μ_a by removing or adding adults at the time of census in such a way as to make the fraction of individuals that died between census counts equal to 0.96. However, in this experiment—unlike the bifurcation experiment in Chapter 6—we also manipulated the cultures to obtain a desired value of c_{pa}. This was done at census time $t+1$ by manipulating adult recruitment to equal the number $P_t \exp(-c_{pa} A_t)$, with values of c_{pa} set at one of the seven values 0.00, 0.05, 0.10, 0.25, 0.35, 0.50, and 1.00, designed to correspond to different attractors along the bifurcation diagram in Figure 7.1A. (The value $c_{pa} = 1.00$ is not shown in Figure 7.1A, but the predicted attractor is a three-cycle.)

To counter the possibility of genetic changes in life-history characteristics, at every other census the A-stage individuals returned to the population after the census came from separate stock cultures maintained under standard laboratory conditions. The results from this experiment are described in the next section.

7.4 PARAMETER ESTIMATION AND MODEL VALIDATION

In Chapters 5 and 6, we used the environmental stochastic LPA model (5.5). Environmental noise is the stochastic variation arising from extrinsic sources that affects many, if not all, individuals in the population. Demographic noise, on the other hand, arises from independent random contributions of births, deaths, and migrations of individual population members. Generally, demographic stochasticity becomes important at low population densities [20].

Although total population size does not become low across the bifurcation diagram in Figure 7.1A, the individual life-cycle stages do repeatedly attain low values. We can see this if we redraw the bifurcation diagram by plotting, say, the adult stage component of the attractor, as shown in Figure 7.2.

Therefore, for the route-to-chaos experiment, it is natural to consider whether the demographic stochastic version of the LPA model, discussed in Chapter 5,

$$\begin{aligned} L_{t+1} &= \left(\sqrt{bA_t e^{-c_{el}L_t - c_{ea}A_t}} + E_{1t}\right)^2 \\ P_{t+1} &= \left(\sqrt{(1-\mu_l)\,L_t} + E_{2t}\right)^2 \\ A_{t+1} &= \left(\sqrt{P_t e^{-c_{pa}A_t} + (1-\mu_a)\,A_t} + E_{3t}\right)^2, \end{aligned} \tag{7.3}$$

might be more appropriate, where the value inside the squaring function is set to zero if it is negative. Here $E_t = [E_{1t}, E_{2t}, E_{3t}]$ is a random noise vector assumed to have

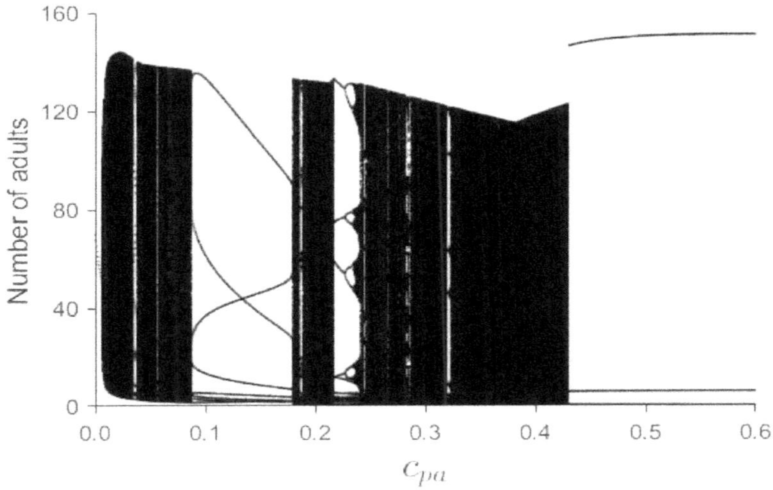

Figure 7.2 Bifurcation diagram from Figure 7.1A, redrawn with the adult stage component of the attractors plotted instead of total population size.

a joint normal probability distribution with a mean vector of zeros and a diagonal variance-covariance matrix denoted by $\mathbf{\Sigma}$. For demographic stochasticity, the noise variables would be uncorrelated with each other within a time unit (that is, the off-diagonal, covariances, of the matrix $\mathbf{\Sigma}$ would be small or zero) as well as uncorrelated through time. Note that the deterministic skeleton of the model—that is to say, the LPA model (7.1)—is obtained by setting $\mathbf{\Sigma} = \mathbf{0}$, or equivalently, by letting E_{1t}, E_{2t}, and E_{3t} equal zero.

To determine whether the demographic stochastic model (7.3) would be more appropriate for the route-to-chaos experiment, we compared the accuracy of the demographic (square root) and environmental (logarithmic) transformations using the data from the bifurcation experiment in Chapter 6. That analysis suggested that the model for demographic stochasticity was statistically superior [9]. This was not surprising, given that low (single-digit) values for larvae and pupae occurred in that experiment.

Therefore, despite the fact that we had used the environmental stochastic model (5.5) to design the route-to-chaos experiment, we re-fitted the LPA model using the demographic stochastic version (7.3) after obtaining 80 weeks of experimental data in the route-to-chaos experiment. These data are available in the data archive [3], and also can be found in the Appendix C in [7].[6]

The statistical properties of ML estimates do not hold if the model is a poor description of the underlying stochastic mechanisms producing the data. In particular, if the noise vector E_t in model (7.3) does not have a multivariate normal distribution, or is correlated through time, then the ML estimates could be biased. Thus, we also calculated the more robust conditional least squares (CLS) parameter estimates as a check on the ML estimates. CLS estimates are based on the sum of squared differences between the value of a variable observed at time $t+1$ and its expected (or one-step forecast) value, given the observed state of the system at time t.

[6]We will see in Section 7.5 that the predictions of the fitted environmental stochastic and demographic stochastic models are very similar.

For fitting the LPA model to a single time series, there are three such conditional sums of squares, specifically

$$Q_1(b, c_{el}, c_{ea}) = \sum_{t=0}^{q} \left(\sqrt{l_{t+1}} - \sqrt{ba_t \exp(-c_{el}l_t - c_{ea}a_t)} \right)^2$$

$$Q_2(\mu_l) = \sum_{t=0}^{q} \left(\sqrt{p_{t+1}} - \sqrt{(1-\mu_l)\, l_t} \right)^2$$

$$Q_3(c_{pa}) = \sum_{t=0}^{q} \left(\sqrt{a_{t+1}} - \sqrt{p_t \exp(-c_{pa}a_t) + (1-\mu_a)\, a_t} \right)^2.$$

Here l_t, p_t, and a_t, for $t = 0, 1, ..., q$, are the observed census counts for the three life stages. The conditional sums of squares are constructed on the square-root scale because that is the scale on which we assume noise is additive. The parameter μ_a was manipulated in the treatments and was estimated directly in the control cultures as a binomial probability from the counts of dead adults and the number of adults at risk of mortality at each census interval. The parameter c_{pa} was estimated for the control cultures by minimizing Q_3 with the value of μ_a fixed at its binomial estimate. For the other treatments, c_{pa} was manipulated, so Q_3 was not used.

Our 80-week data consisted of 24 population time series, with three replicates in each seven treatments and three replicates in the control. Each population began at time $t = 0$ and ended at time $t = 40$ (80 weeks). If Q_{ijk} is the conditional sum of squares for state variable i in the jth treatment in the kth replicate population, then the total conditional sum of squares for the ith state variable was calculated as

$$Q_i = \sum_{j=1}^{8} \sum_{k=1}^{3} Q_{ijk}.$$

CLS estimates minimize each of the total conditional sums of squares $Q_1(b, c_{el}, c_{ea})$, $Q_2(\mu_l)$, $Q_3(c_{pa})$. Because each parameter appears in only one Q_i, the three Q_i can be minimized independently or else the sum of the three can be minimized. For these minimizations, we used the Nelder-Mead simplex algorithm. The results appear in Table 7.2. For more details on the parameter estimation procedure and the bootstrapping procedure for constructing confidence intervals, see [7, 9].

A visual sense of the model's accuracy can be seen in the representative examples of stage-specific residuals shown in Figure 7.3. Details of the model validation analysis involving a generalized R^2, first- and second-order autocorrelation residual analyses, normality tests, and Q-Q plots appear in [7, 9]. The inferences we made about parameters are based on a CLS estimation that makes no particular assumptions concerning the distributional form of the noise. The diagnostic tests and plots mainly suggest there are no major systematic or symptomatic patterns in the departures of model and data [9]. This provides a high level of confidence in inferences obtained from the model. With that in mind, we turn our attention in the next section to the long-term dynamics and the attractors predicted by the model.

7.5 ATTRACTORS IN THE ROUTE-TO-CHAOS EXPERIMENT

The new CLS estimates (Table 7.2) for the route-to-chaos experiment are similar to the ML parameter estimates (Table 7.1) from the bifurcation experiment in Chapter 6. Because of this, the new bifurcation diagram predicted by the LPA model (7.1) using the CLS point

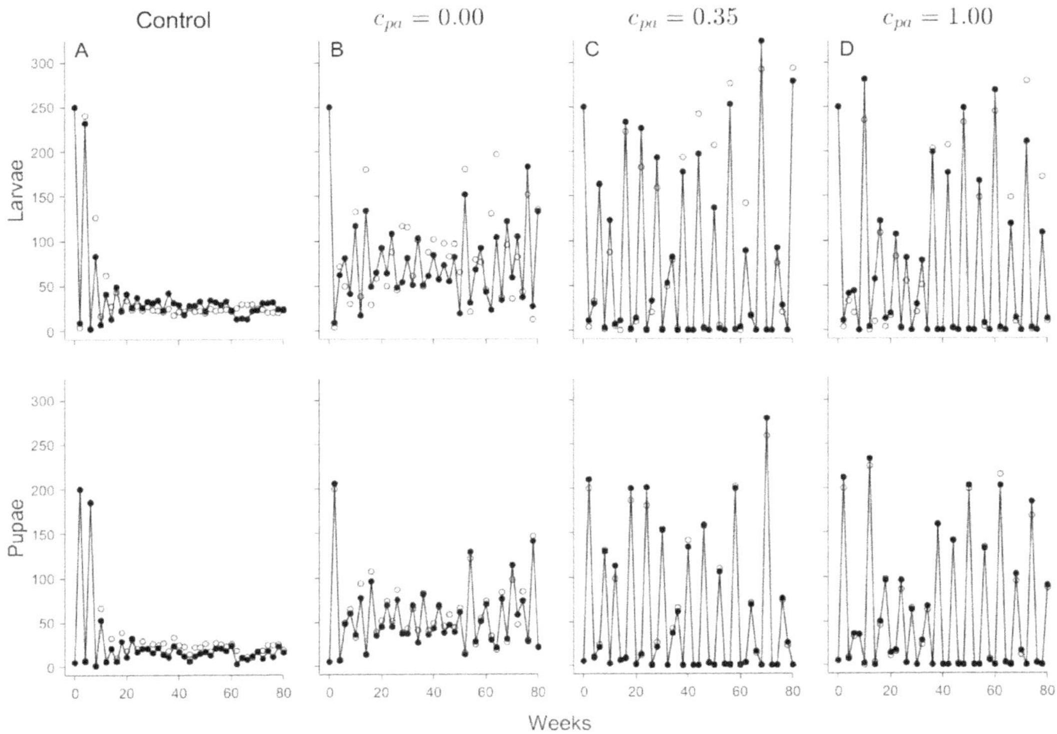

Figure 7.3 Sample L-stage and P-stage time-series data (solid circles) and one-step predictions (open circles) from the fitted LPA model (7.1). A. Control, replicate 11. B. Treatment with $c_{pa} = 0.00$, replicate 15. C. Treatment with $c_{pa} = 0.35$, replicate 13. D. Treatment with $c_{pa} = 1.00$, replicate 18. In all cases, $\mu_a = 0.96$ and the other parameters are given in Table 7.2.

estimates in Table 7.2 is similar, but not identical, to that predicted in Figure 7.1A using the ML estimates in Table 7.1. The new bifurcation diagram appears in Figure 7.4.

The prediction for the control remains a stable equilibrium. The prediction for the $c_{pa} = 0.0$ treatment is an invariant loop, although upon close examination it turns out to be two small, separated invariant loops, arising from a bifurcation caused by the destabilization of a two-cycle. As a result, the predicted dynamics for $c_{pa} = 0.0$ are quite similar to those of a two-cycle. Treatment $c_{pa} = 0.35$ and treatment $c_{pa} = 0.05$ predict chaotic attractors. Treatments $c_{pa} = 0.10$ and 0.25 are stable, long-period cycles that occur in period-locked windows nestled among more complex regions of the bifurcation diagram. The treatment $c_{pa} = 0.50$ is still a stable three-cycle, although the treatment $c_{pa} = 1.00$ is now a six-cycle, albeit one that is very close to a period-doubling bifurcation of the three-cycle. Table 7.3 summarizes these attractor types predicted by the deterministic LPA model (7.1) for the control and experimental treatments, with the CLS parameter point estimates in Table 7.2.

The (few) differences among the attractors from the preliminary bifurcation diagram in Figure 7.1A used to design the experiment and those in the re-fitted bifurcation diagram in Figure 7.4A are not surprising. The attractor types for $c_{pa} < 0.50$ are not robust on large parameter intervals in these bifurcation diagrams, a typical situation for regions in parameter space that contain complex dynamics. In such a region of parameter space, parameter

TABLE 7.2 Conditional least squares (CLS) estimates of the LPA model parameters and their 95% bootstrapped confidence intervals obtained from the 80-week, route-to-chaos experiment data. An asterisk indicates an estimate for the control cultures only. The CLS estimates for the variance-covariances matrix $\Sigma = (\sigma_{ij})$ for the controls are $\hat{\sigma}_{11} = 1.621$, $\hat{\sigma}_{12} = \hat{\sigma}_{21} = -0.1336$, $\hat{\sigma}_{13} = \hat{\sigma}_{31} = -0.01339$, $\hat{\sigma}_{22} = 0.7375$, $\hat{\sigma}_{23} = \hat{\sigma}_{32} = -0.0009612$, $\hat{\sigma}_{33} = 0.01212$. For those treatments where adult mortality and recruitment were manipulated the estimates are $\hat{\sigma}_{11} = 2.332$, $\hat{\sigma}_{12} = \hat{\sigma}_{21} = 0.007097$, $\hat{\sigma}_{22} = 0.2374$, $\hat{\sigma}_{13} = \hat{\sigma}_{31} = \hat{\sigma}_{23} = \hat{\sigma}_{32} = \hat{\sigma}_{33} = 0$.

Parameter	CLS estimate	95% confidence interval
b	10.45	(10.04, 10.77)
μ_l	0.2000	(0.1931, 0.2068)
μ_a^*	0.007629	(0.006769, 0.008489)
c_{el}	0.01731	(0.01611, 0.01759)
c_{ea}	0.01310	(0.01285, 0.01340)
c_{pa}^*	0.004619	(0.004446, 0.004792)

TABLE 7.3 Deterministic attractors predicted by the deterministic LPA model (7.1) and deterministic Lyapunov exponents (LE) with 95% bootstrapped confidence intervals (in parentheses). For the control treatment, the parameters are the point estimates in Table 7.2 except for $\mu_a = 0.96$. For the other treatments, the parameters are the point estimates in Table 7.2 except for $\mu_a = 0.96$ and the displayed value of c_{pa}.

Treatment	Attractor	Largest Lyapunov exponent
control	equilibrium	$-0.046\ (-0.047, -0.045)$
$c_{pa} = 0.00$	invariant loop	$0.000\ (-0.036, 0.000)$
$c_{pa} = 0.05$	chaotic	$0.029\ (-0.227, 0.029)$
$c_{pa} = 0.10$	26-cycle	$-0.109\ (-0.115, -0.0004)$
$c_{pa} = 0.25$	eight-cycle	$-0.076\ (-0.090, -0.071)$
$c_{pa} = 0.35$	chaotic	$0.096\ (-0.066, 0.100)$
$c_{pa} = 0.50$	three-cycle	$-0.089\ (-0.101, -0.080)$
$c_{pa} = 1.00$	six-cycle	$-0.003\ (-0.025, -0.0003)$

intervals can contain complicated mixtures of periodicity, aperiodicity, and chaos. For example, a plot of the LE in Figure 7.4B for the interval $0.3 < c_{pa} < 0.5$ shows a predominantly positive value, indicating sensitivity to initial conditions and chaotic attractors. However, negative LEs are densely interlaced with the positive ones, indicating no sensitivity to initial conditions and attracting cycles. In this region, small changes in point estimates of parameters inevitably lead to significant changes in attractor type, although complex dynamics will lurk in the background, influencing transients. Given that parameter estimates come with confidence intervals, it is important to categorize the types of attractors and their influence within those confidence intervals. A detailed study appears in [7, 9].

We can visualize the agreement of model and experimental data in state space. Figure 7.5 shows the estimated attractors of the deterministic LPA model (listed in Table 7.3)

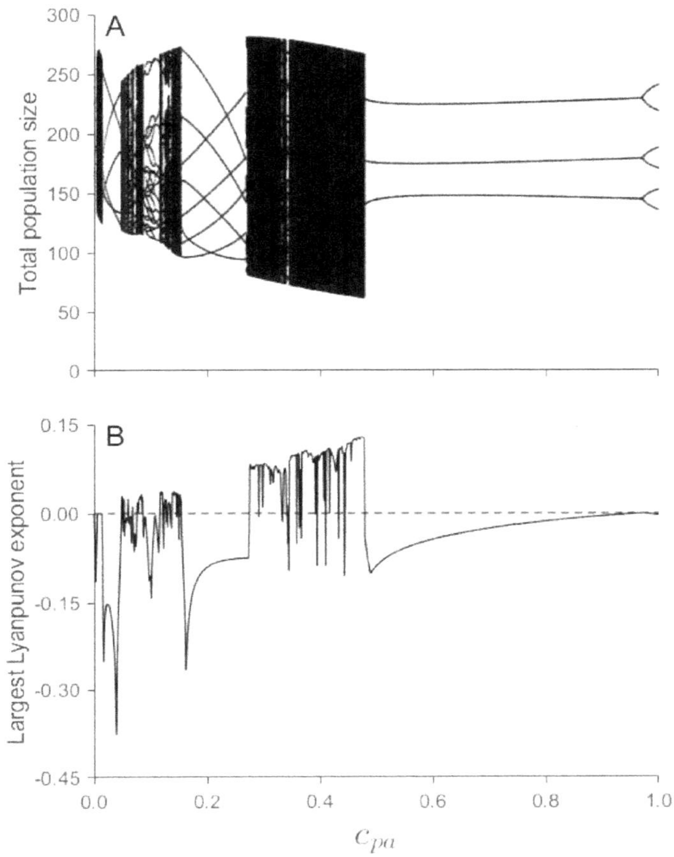

Figure 7.4 LPA model (7.1) predictions as c_{pa} ranges from 0 to 1, for $\mu_a = 0.96$, with all other parameters from Table 7.2. A. Bifurcation diagram. B. Corresponding plot of the dominant Lyapunov exponent (LE), a positive value of which indicates sensitivity to initial conditions.

plotted in three-dimensional state space together with the data triples (after initial transient observations have been removed). The transitions in dynamic behavior of the experimental data from treatment to treatment mirror the model-predicted attractors extremely well. Note the tight cloud of points around the stable point equilibrium for the control treatment. Note the two clouds of points around the small invariant loops for the $c_{pa} = 0.00$ treatment. Particularly striking are the attractor and data matches for the treatments $c_{pa} = 0.50$ and $c_{pa} = 1.00$.[7] Of special interest are those treatments in the region of complex dynamics. Here we see the irregular, triangular-shaped clouds of data points that accompany the three-pointed cyclic or chaotic attractor in the $c_{pa} = 0.05, 0.10, 0.25$, and 0.35 treatments. Clearly the data shown in Figure 7.5 are influenced strongly by the deterministic forces present in the model.

For further exploration, see the examples for this section in the online LPA Simulator (https://lpasim.org/?c=7&s=5).

[7]The six-cycle attractor is near a period-doubling bifurcation of a three-cycle, and thus consists of three separate groups of two nearby points. In this region of parameter space, we cannot expect to distinguish the six-cycle from a three-cycle in data.

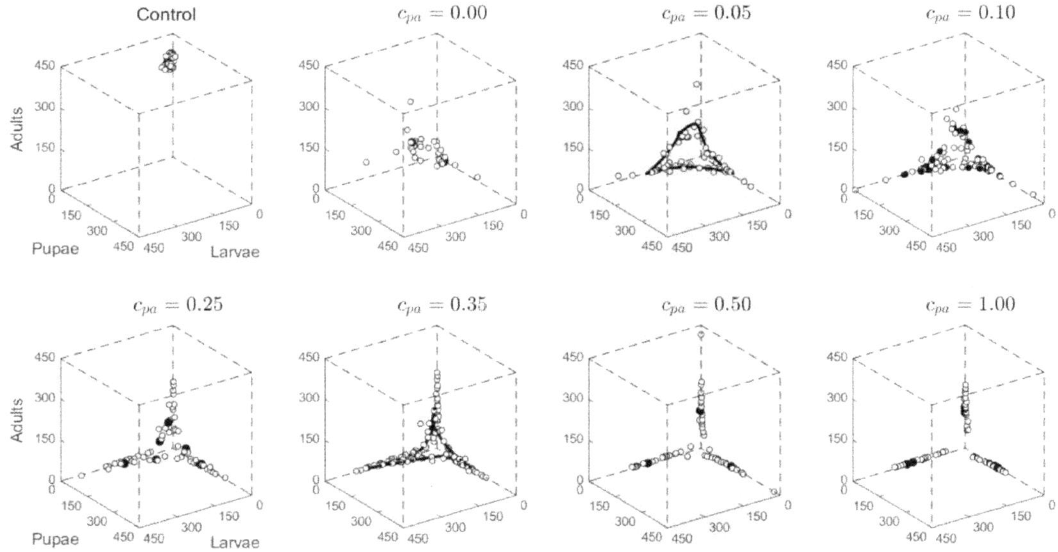

Figure 7.5 Experimental data $[L, P, A]$ triples (open circles) plotted in state space together with the LPA model (7.1) attractor (in black), using parameter values from Table 7.2 and $\mu_a = 0.96$. In each case, data from all three replicates are shown.

7.6 THE STOCHASTIC LPA MODEL AND THE INFLUENCE OF CHAOS

The data from the route-to-chaos experiment contain, of course, departures from the deterministic LPA model predictions (see Figures 7.3 and 7.5). These departures should be well described by the stochastic model. Does the variability predicted by the parameterized demographic stochastic LPA model (7.3) resemble the variability in the data?

Figure 7.6 contains plots of stochastic orbits from simulations using the estimated stochastic model (7.3) with parameter values given in Table 7.2 (and with initial transients removed). For the purpose of these simulations, we assumed the noise vector E_t has a multivariate normal distribution (recall that multivariate normality was not assumed for the estimation methods). The resemblance of these plots to those of the experimental data in Figure 7.5 is striking for all treatments. The stochastic model produced patterns of simulations around the estimated deterministic attractors that are very similar visually to the experimental data. In ecology, such concordance of model and data across experimental manipulations is rare.

The overall structure of the bifurcation diagram in Figure 7.4A remains robust throughout the parameter space bracketed by the confidence intervals of the estimated parameters [7, 8]. The c_{pa} interval ranging approximately from 0.5 to 0.95 robustly exhibits a single type of attractor (a three-cycle, Figure 7.4A) with negative LE (Figure 7.4B). The c_{pa} interval from approximately 0 to 0.5 contains a complicated mixture of periodicity, aperiodicity, and chaos. Zooming in on the bifurcation diagram in this latter region can reveal fractal-like nestings of period-locking windows, a situation typical in a chaotic regime. The dense embeddings of chaotic and periodic attractors, coupled with the real-world facts of parameter confidence intervals and ubiquitous process noise, begs the question of what it means to make an exact attractor identification for a real system in a complex regime.

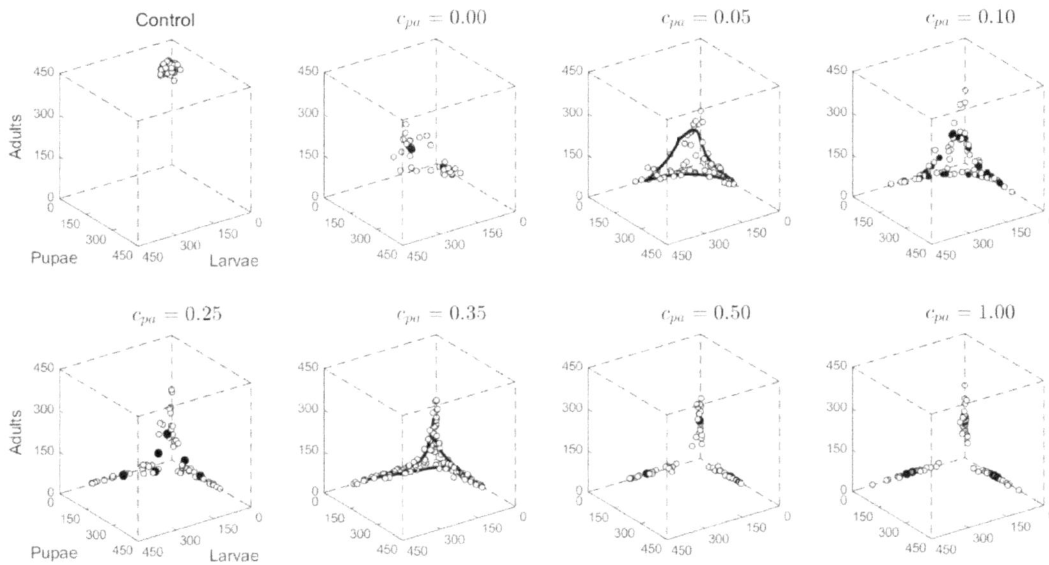

Figure 7.6 Simulations of the demographic stochastic LPA model (7.3) (open circles) plotted in state space together with the LPA model (7.1) attractors (in black), using parameter values from Table 7.2 with $\mu_a = 0.96$. In each case, three replicate simulations are shown, but with initial transients withheld. The simulations begin from the experimental initial conditions $[L, P, A] = [250, 5, 100]$. Compare to the experimental data in Figure 7.5.

We will return in detail to this question in Chapter 9. But the important point here is that, within these dense embeddings of period-locking windows, the complex invariant sets continue to lurk in the background, "momentarily" unstable (as one minutely tweaks c_{pa}) but still influencing the transients. Thus, the simulations of the stochastic demographic LPA model displayed in Figure 7.6 show the clear influence of the deterministic attractors, including the chaotic attractor; and likewise do the experimental population data displayed in Figure 7.5.

Relevant to this ambiguity of attractor identification in these treatments of the experiment is the concept of a *stochastic Lyapunov exponent* (SLE), which is designed to measure sensitivity to initial conditions for stochastic models [13, 14]. The SLE is a measure of the tendency of nearby stochastic orbits (suffering the same sequence of random perturbations) to diverge or converge. A positive value of SLE indicates a kind of average sensitivity to initial conditions associated with the stochastic model, although this sensitivity might arise from causes unrelated to deterministic chaos. This goes back to our working definition of chaos that is based on deterministic models. Indeed, a positive SLE does not imply that the deterministic skeleton has a chaotic attractor; see [10, 11] for examples. Nonetheless, SLEs are often estimated in studies of chaos, especially if sensitivity to initial conditions is a focal concern. The plot of the SLE for the parameterized demographic stochastic LPA model (7.3), shown in Figure 7.7, indicates a robust interval of c_{pa} values on which there is sensitivity to initial conditions. Compare this to the LE of the deterministic LPA model (7.1) displayed in Figure 7.4B. For more on the SLE in this experiment and its implications, see [7, 9].

For further exploration, see the examples for this section in the online LPA Simulator (https://lpasim.org/?c=7&s=6).

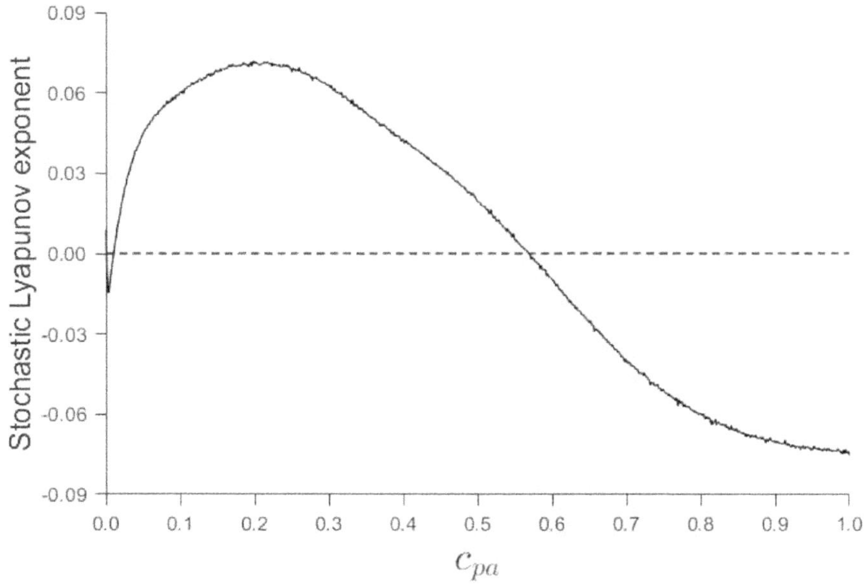

Figure 7.7 Stochastic Lyapunov exponent (SLE) associated with the bifurcation diagram in Figure 7.4.

7.7 THE CHAOTIC ATTRACTOR

Because the beetle populations in the route-to-chaos experiment described in this chapter did convincingly undergo the dynamic changes predicted by the parameterized LPA model (7.1), and because the deterministic model predicts chaotic dynamics for the treatments corresponding to $c_{pa} = 0.05$ and 0.35, we have some confidence in asserting that the beetle cultures in these treatments displayed chaotic dynamics—or, more accurately, that their dynamics were highly influenced by a chaotic attractor. To further investigate this conclusion, we continued the replicated cultures (and control) for the $c_{pa} = 0.35$ treatment beyond the 80 weeks of the experiment, for a total of 424 weeks (over 100 generations).

The best way to visualize attractors for the three-dimensional LPA model is to plot them in state space. An equilibrium attractor appears as a single point in state space; a two-cycle appears as two points; a three-cycle appears as three points; and so on for higher-period cycles. A chaotic attractor appears in state space as an infinite set of points whose geometric and topological properties can be extraordinarily complicated. The chaotic attractor associated with the $c_{pa} = 0.35$ treatment in the route-to-chaos experiment is shown in Figure 7.8. Although it has a roughly triangular appearance, its complex geometry is displayed in several magnified portions of the attractor.

A rigorous comparison of the experimental data with the deterministic chaotic attractor in state space is technically difficult. However, even without such statistical analyses, one can see from the state space plot in Figure 7.9 that the experimental data do cluster near the predicted deterministic chaotic attractor. In Chapter 9, we will focus on properties of the chaotic attractor and show they can be seen in the experimental data.

For further exploration, see the examples for this section in the online LPA Simulator (https://lpasim.org/?c=7&s=7).

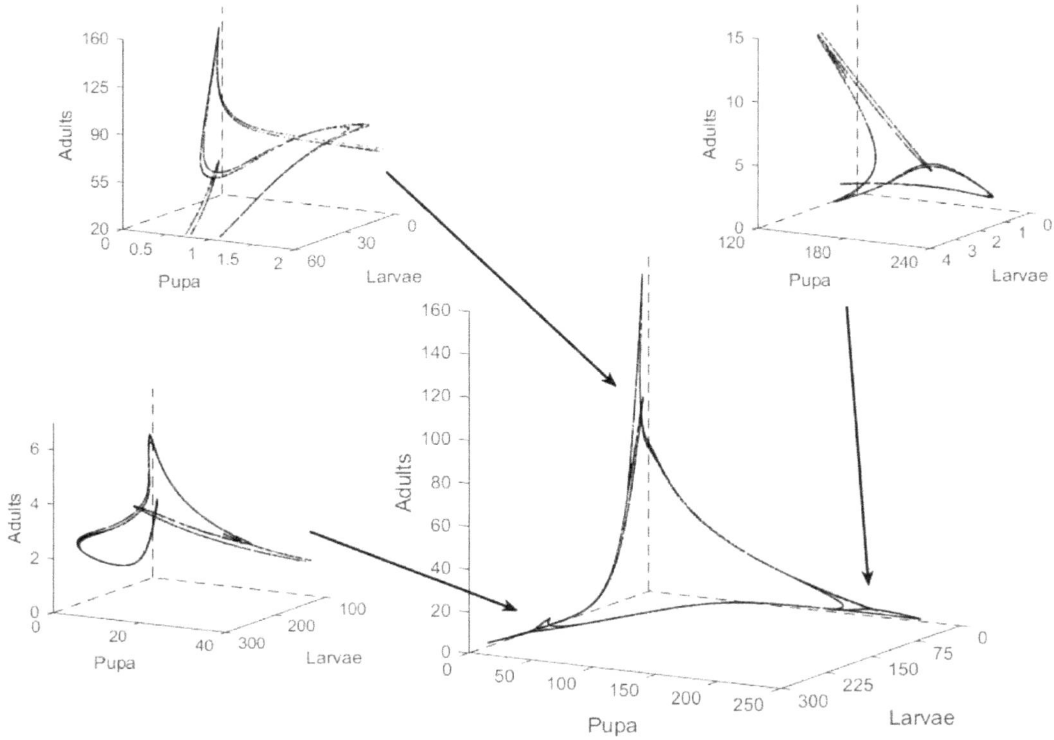

Figure 7.8 Chaotic attractor predicted by the deterministic LPA model (7.1) with treatment values $c_{pa} = 0.35$ and $\mu_a = 0.96$ and remaining parameter values given in Table 7.2. Magnifications of three portions of the attractor illustrate its geometric and topological complexity.

7.8 CONCLUSIONS

Our experimental approach in the route-to-chaos experiment described in this chapter was to investigate how the dynamics of the flour beetle populations responded to interventions. The fundamental goal in that approach was to determine whether or not the responses that resulted from changes in certain demographic parameters conformed to the bifurcations in an LPA model-predicted route to chaos. From this point of view, all the treatments at the various locations in the bifurcation diagram played a role in determining whether chaos was present in the experiment.

Although the state space evidence seen in Figure 7.9 offers solid evidence of the influence of the deterministic chaotic attractor on the beetle populations in the $c_{pa} = 0.35$ treatment, the temporal dynamic properties of a chaotic attractor are not easily discernible in state space and are not readily comparable to the time-series data. The time-series plots of the state variables from model trajectories on or near the chaotic attractor, besides possessing sensitivity to initial conditions, show irregular oscillations that resemble random or stochastic fluctuations. An example is shown in Figure 7.10 where the adult component of a solution that approaches the chaotic attractor displays oscillations that irregularly fluctuate between high and low values near zero. Similar irregular fluctuations are seen in the time series for the L and P components of the solution.

Despite the seemingly random fluctuations exhibited by such "chaotic" solutions, there are distinctive characteristics, properties, and dynamic patterns exhibited by the

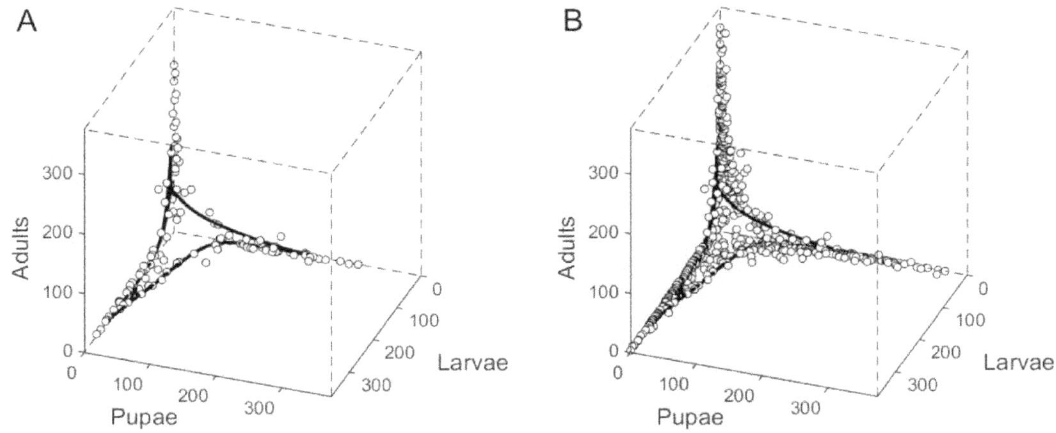

Figure 7.9 Data [L,P,A] triples in state space (open circles) cluster about the LPA model-predicted chaotic attractor (in black) for the treatment with $c_{pa} = 0.35$. All three experimental replicates are shown. Model parameters are from Table 7.2, except for the manipulated values of $\mu_a = 0.96$ and $c_{pa} = 0.35$. (A) Weeks 2–80. (B) Weeks 2–424.

deterministic chaotic attractor that one can expect to observe in the experimental data, if indeed the beetle populations are influenced by the chaotic attractor. Such properties include episodes of periodic cycles of specific periods (from the chaotic attractor), and sensitivity to initial conditions on certain "hot" areas of the attractor. These and other properties are

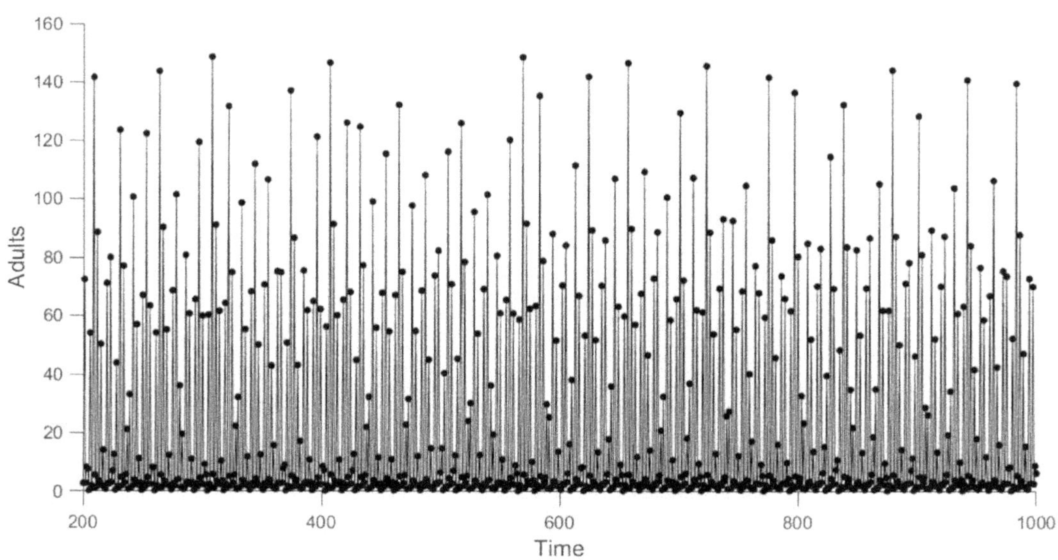

Figure 7.10 The adult component A_t from the solution with initial condition $[L_0, P_0, A_0] = [10, 10, 10]$ of the LPA model (7.1) with parameter values from Table 7.2 for the route-to-chaos experiment with $c_{pa} = 0.35$ and $\mu_a = 0.96$. The first 200 time steps are discarded so as to ignore transient behavior.

explored in subsequent chapters of this book in an effort to gain new insights into several issues involved in the detection of chaos in biological populations (and ecology in general). Parameterized stochastic LPA models play a key role in this program, as they allow for a wider exploration of state space with various implications for both transient and long-term dynamics.

Bibliography

[1] Alligood KT, Sauer TD, Yorke JA (1997) Chaos: An Introduction to Dynamical Systems. Berlin, Springer-Verlag. p. 106.

[2] Banks J, Brooks J, Cairns G, Davis G, Stacey P (1992) On Devaney's definition of chaos. American Mathematical Monthly 99:332–334. https://doi.org/10.2307/2324899

[3] Beetle Team *Tribolium* Data Archive. https://doi.org/10.5061/dryad.qjq2bvqmp

[4] Bishop R (2017) Chaos. In Zalta EN (ed.), The Stanford Encyclopedia of Philosophy, Spring 2017 Edition.

[5] Costantino RF, Desharnais RA, Cushing JM, Dennis B (1997) Chaotic dynamics in an insect population. Science 275:389–391. https://doi.org/10.1126/science.275.5298.389

[6] Cushing JM (2009) Matrix Models and Population Dynamics. In Lewis M, Chaplain AJ, Keener JP, Maini PK (eds.), Mathematical Biology, pp. 47-150. IAS/Park City Mathematics Series. Volume 14. Providence, Rhode Island, American Mathematical Society.

[7] Cushing JM, Costantino RF, Dennis B, Desharnais RA, Henson SM (2003) Chaos in Ecology: Experimental Nonlinear Dynamics. Theoretical Ecology Series. Volume 1. New York, Academic Press.

[8] Dennis B, Desharnais RA, Cushing JM, Costantino RF (1997) Transitions in population dynamics: Equilibria to periodic cycles to aperiodic cycles. Journal of Animal Ecology 66:704–729. https://doi.org/10.2307/5923

[9] Dennis B, Desharnais RA, Cushing JM, Henson SM, Costantino RF (2001) Estimating chaos and complex dynamics in an insect population. Ecological Monographs 71:277–303. https://doi.org/10.1890/0012-9615(2001)071[0277:ECACDI]2.0.CO;2

[10] Dennis B, Desharnais RA, Cushing JM, Henson SM, Costantino RF (2003) Can noise induce chaos? Oikos 102:329–340. https://doi.org/10.1034/j.1600-0706.2003.12387.x

[11] Desharnais RA, Costantino RF, Cushing JM, Dennis B (1997) Estimating chaos in an insect population. Response to Perry et al. Science 276:1881–1882. https://doi.org/10.1126/science.276.5320.1881

[12] Devaney RL (1989) Dynamics of simple maps. Proceedings of Symposia in Applied Mathematics 39:1–24. https://doi.org/10.1090/psapm/039/1010233

[13] Ellner S, Turchin P (1995) Chaos in a noisy world: New methods and evidence from time-series analyses. The American Naturalist 145:343–375. https://doi.org/10.1086/285744

[14] Ellner SP (2000) Defining chaos for real, noisy data: Local Lyapunov exponents and sensitive response to perturbations. In Perry JN, Smith RH, Woiwod IP, Morse DR (eds.), Chaos in Real Data: The Analysis of Nonlinear Dynamics from Short Ecological Time Series, pp. 1–32. Dordrecht, The Netherlands, Kluwer Academic Publishers.

[15] Gleick J (1988) Chaos: Making a New Science. New York, Viking Penquin Inc.

[16] Lorenz EN (1963) Deterministic nonperiodic flow. Journal of the Atmospheric Sciences 20:130–141. https://doi.org/10.1175/1520-0469(1963)020%3C0130:dnf%3E2.0.co;2

[17] Lorenz EN (1993) Essence of Chaos. Seattle, University of Washington Press.

[18] Li TY, Yorke JA (1975) Period three implies chaos. American Mathematical Monthly 82:985–992. https://doi.org/10.2307/2318254

[19] Robinson C (1995) Dynamical Systems: Stability, Symbolic Dynamics and Chaos. Boca Raton, Florida, CRC Press, Inc.

[20] Simberloff D (1988) The contribution of population and community ecology to conservation science. Annual Review of Ecology and Systematics 19:473–511. https://doi.org/10.1146/annurev.es.19.110188.002353

[21] Stewart I (2001) What Shape is a Snowflake? New York, W. H. Freeman and Company.

[22] Tong H (1999) Nonlinear Time Series: A Dynamical System Approach. Oxford, Oxford University Press.

CHAPTER 8

Lattice Effects

8.1 ANIMALS COME IN WHOLE NUMBERS

Counts of animals are integer valued. As a result, densities (counts per unit volume or area), although not necessarily integer valued, take on discrete values, as well. For example, in an area of 100 hectares, the possible population densities per hectare are 0, 0.01, 0.02, ... Thus, the observed values of state variables tracking population size exist on a discrete lattice of points in state space. In particular, for the *Tribolium* system, the observed count triples $[l_t, p_t, a_t]$ of larvae, pupae, and adults at time t exist on a lattice in three-dimensional state space [4].

Realistic deterministic models have bounded solutions. If such a bounded solution is required to "land" on discrete lattice points in state space, it must eventually repeat a value, because there are only finitely many lattice points on which to land in a bounded region of space. As soon as the solution of an autonomous model[1] repeats a value, it is locked into a cycle with period equal to the minimum number of time steps between the repeated values.

Given that chaos and other aperiodic attractors exist on a continuum of values in state space, we must ask: How are the nonlinear forces that create chaotic dynamics expressed on a discrete lattice in state space? This is the subject of the current chapter.

8.2 ILLUSTRATION USING THE RICKER MODEL

We begin our discussion of dynamics on lattices with an illustration using the Ricker model

$$y_{t+1} = by_t e^{-cy_t}, \tag{8.1}$$

where y_t is the population density at time t relative to a habitat volume $V > 0$. We can express the equation in terms of numbers of animals x in volume V by writing $y = x/V$, which yields

$$x_{t+1} = bx_t e^{-cx_t/V}. \tag{8.2}$$

Note that y_t is a solution of equation (8.1) if and only if $x_t = Vy_t$ is a solution of equation (8.2). This means that the dynamics of equation (8.2) are the same as the dynamics of equation (8.1), scaled up by factor V.

If $b = 17$ and $c = 1$, the Ricker model (8.2) has a chaotic attractor, which can be visualized by plotting x_{t+1}/V against x_t/V (Figure 8.1A). Suppose we confine the dynamics of the Ricker model to a lattice in state space by

$$x_{t+1} = \text{int}\left[bx_t e^{-cx_t/V}\right], \tag{8.3}$$

[1] An autonomous model is one that does not depend explicitly on time.

DOI: 10.1201/9781003467120-8

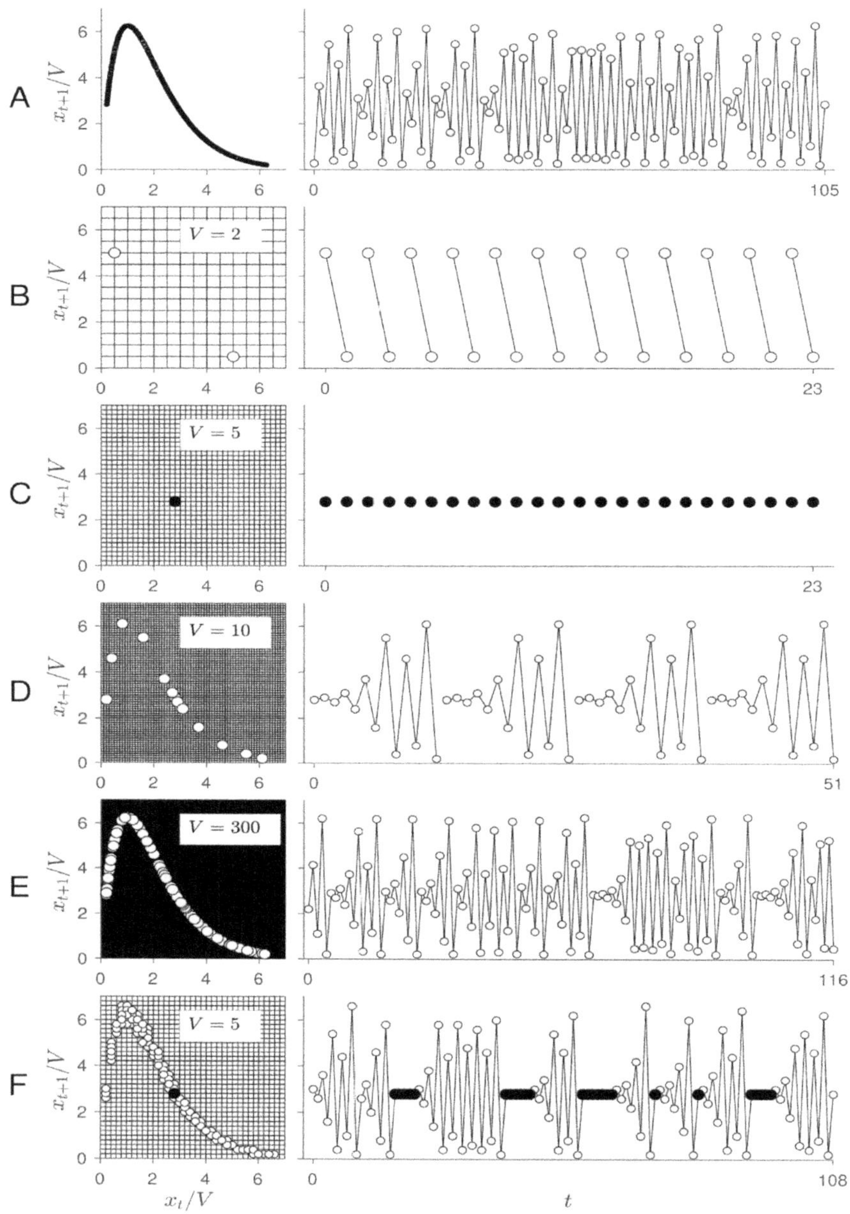

Figure 8.1 Ricker models with $b = 17$ and $c = 1$, showing state space plots with corresponding time series. A. Chaotic attractor of the Ricker model (8.2) with $V = 1$. B. Two-cycle attractor of the lattice Ricker model (8.3) with $V = 2$. C. One-cycle attractor of the lattice Ricker model with $V = 5$. D. 13-cycle attractor of the lattice Ricker model with $V = 10$. E. 117-cycle attractor of the lattice Ricker model with $V = 300$. F. Stochastic lattice model (8.4) with $V = 5$ and $\sigma = 0.03$. Realizations show episodes of the chaotic signal (open circles) interspersed with episodes of the lattice one-cycle (solid circles).

where "int[z]" rounds z to the nearest integer, with 0.5 rounding up. Note that we can decrease the mesh of the state space lattice by increasing V.

Figures 8.1B-E show attractors in state space for the lattice Ricker model (8.3), and a corresponding time series, for each of $V = 2, 5, 10, 300$. As V increases, the lattice mesh decreases, until in Figure 8.1E, the mesh is so fine that the background appears solid black. For $V = 2$, the chaotic dynamics collapse to a lattice two-cycle for equation (8.3) (Figure 8.1B). Interestingly, when V is increased to $V = 5$, the lattice two-cycle further reduces to an equilibrium (Figure 8.1C). For $V = 10$, the dynamics are a 13-cycle (Figure 8.1D), and for $V = 300$, a 117-cycle begins to trace out the underlying chaotic attractor (Figure 8.1E).

Figure 8.1F shows a realization of a stochastic version of the lattice Ricker model,

$$x_{t+1} = \text{int}\left[bx_t \exp\left(-\frac{cx_t}{V} + \sigma\varepsilon_t\right)\right], \tag{8.4}$$

with $V = 5$, where $\sigma > 0$ and the ε_t are standard normal random variables (mean zero and standard deviation one), uncorrelated in time. Even though the deterministic lattice cycle is an equilibrium (that is, a one-cycle) when $V = 5$ (Figure 8.1C), the introduction of noise on the lattice reveals the underlying chaotic signal (Figure 8.1F). Of particular interest is that the chaotic signal is punctuated by episodes of the lattice one-cycle (solid circles).

This example illustrates two ways in which an underlying chaotic mechanism can be expressed on a lattice:

- In a deterministic lattice system, if the lattice mesh is decreased toward zero (so that the lattice node density increases without bound), the lattice cycle periods eventually increase, and the cycles begin to "trace out" the limiting chaotic attractor. Lattice cycles of small period can reappear, however, even for very small lattice mesh.
- With the inclusion of noise, a lattice system's transients can trace out the underlying chaotic attractor even on a relatively coarse lattice. In this case, the chaotic signal can be punctuated by recurrent episodes of lattice cycles. We call these episodes "lattice effects."

Thus, the chaotic signal can be recovered on the lattice either deterministically, by decreasing the lattice mesh to zero, or stochastically on a coarse lattice, by including noise. In an experimental population system, the mesh cannot be taken to zero, but noise is ubiquitous. Thus, models predict that, *in experimental population systems, noise can illuminate underlying entities in the spatial continuum, allowing lattice trajectories to visit and trace out the chaotic attractor as stochasticity stirs the transients. The chaotic signal can be interspersed with episodes of lattice effects, which are signatures of the discrete nature of animal numbers.*

8.3 BEETLES ON A LATTICE

The LPA model

$$\begin{aligned} L_{t+1} &= bA_t \exp(-c_{el}L_t/V - c_{ea}A_t/V) \\ P_{t+1} &= (1-\mu_l)\,L_t \\ A_{t+1} &= P_t \exp\left(-c_{pa}A_t/V\right) + (1-\mu_a)\,A_t \end{aligned} \tag{8.5}$$

with $V = 1$ and experimental parameters

$$\begin{aligned} b = 10.67,\ \mu_l = 0.1955,\ c_{el} = 0.01647, \\ c_{ea} = 0.01313,\ c_{pa} = 0.35,\ \mu_a = 0.96 \end{aligned} \tag{8.6}$$

from the route-to-chaos experiment [2] predicts a chaotic attractor, shown in the continuum of state space and also as a time series in Figure 8.2A. The experimental beetle system in this chaotic regime produced time series in which larval, pupal, and adult numbers remained less than about 300 each. Thus, the experimental system in this regime exists on a lattice of about 2.7×10^7 nodes.

Consider the demographic stochastic lattice model

$$\begin{aligned} L_{t+1} &= \operatorname{int}\left[\left(\sqrt{bA_t \exp(-c_{el}L_t/V - c_{ea}A_t/V)} + E_{1t}\right)^2\right] \\ P_{t+1} &= \operatorname{int}\left[\left(\sqrt{(1-\mu_l)L_t} + E_{2t}\right)^2\right] \\ A_{t+1} &= \operatorname{int}\left[P_t \exp\left(-c_{pa}A_t/V\right)\right] + \operatorname{int}\left[(1-\mu_a)A_t\right], \end{aligned} \tag{8.7}$$

with experimental parameters (8.6). Here, the E_{it} are normal random variables with mean zero and variance-covariance matrix $\mathbf{\Sigma}$, and $V = 1$ corresponds to 20 g of flour. The quantities inside the squares are taken to be zero if negative. There is no random variable in the adult equation, because for the treatment under consideration from the route-to-chaos experiment, the adult equation was deliberately made exact (Chapter 7).

In this chaotic regime, the deterministic lattice model (8.7) with $\mathbf{\Sigma} = 0$ and $V = 1$ predicts a six-cycle lying close to the chaotic attractor (Figure 8.2B). For $V = 10$, that is in 200 g of flour, the period of the deterministic lattice cycle increases to 124, and the cycle traces out the chaotic attractor, as shown in Figure 8.2C. This latter scenario, however, in 200 g of flour, would involve counts of up to 3000 animals in a single life stage!

The inclusion of stochasticity in the lattice model (8.7) with $V = 1$, using variance-covariance entries in $\mathbf{\Sigma}$ of $\sigma_{11} = 2.332, \sigma_{22} = 0.2374, \sigma_{12} = \sigma_{21} = 0$ from [4], perturbs lattice trajectories away from the six-cycle and allows them to trace out the chaotic attractor (Figure 8.2D, open circles). Intermittently, however, the trajectories visit the lattice six-cycle (Figure 8.2D, solid circles). The episodes of six-cycle dynamics are shown in state space directly below the time-series episodes (Figure 8.2E). The reader should compare these six-cycle episodes to the lattice six-cycle in panel B.

Figure 8.3A shows a 304-week data time series from replicate 13 of the route-to-chaos experiment [2]. The episodes of transients resembling the chaotic attractor are shown with open circles, whereas the episodes of six-cycles are shown with solid circles. In Figure 8.3B, the time series is broken into four segments in state space, highlighting the lattice cycle episodes for weeks 58–92 and weeks 240–286. The reader should compare the six-cycle episodes in these data to the stochastic model predictions in Figure 8.2D,E.

For further exploration, see the examples for this section in the online LPA Simulator (https://lpasim.org/?c=8&s=3).

8.4 CONSTRUCTING LATTICE MODELS

Thus far in this chapter, we have restricted model dynamics to a lattice in an ad hoc way simply by rounding the state variables. In this section, we consider a more natural way of constructing lattice models by iterating the mode of discrete distribution models.

A population time series is the realization of a stochastic process in which the number of animals N_t is a discrete random variable. Suppose N_t has unimodal conditional distribution

$$N_{t+1} \sim P\left(n_t\right), \tag{8.8}$$

where the symbol $\sim$ denotes "distributed as" and $N_t = n_t$ is the observed number of animals at time t. Here $P(n_t)$ represents some probability distribution model for the random variable

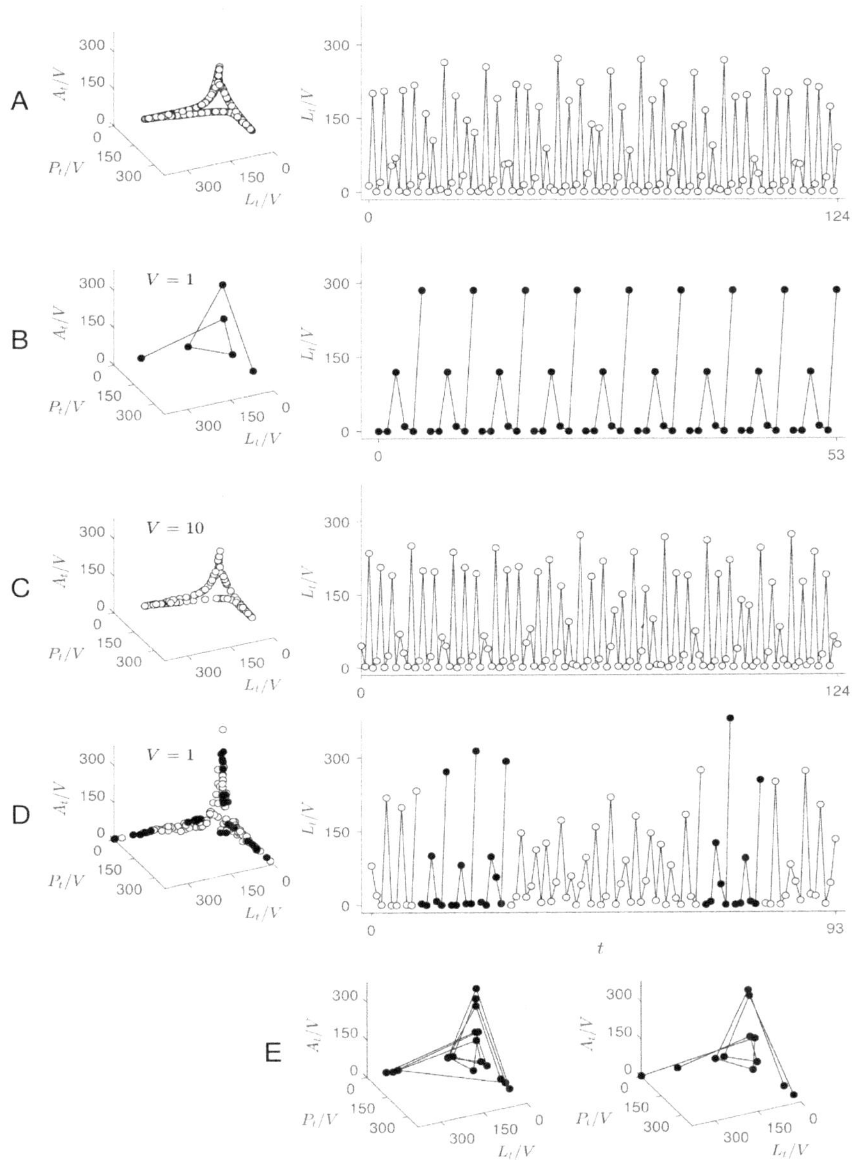

Figure 8.2 Dynamics of the LPA models with experimental parameters (8.6). A. Chaotic attractor of the LPA model (8.5) with $V = 1$. B. Six-cycle attractor of the lattice LPA model (8.7) with $\Sigma = 0$ and $V = 1$ (on the order of 10^7 lattice points). C. The 124-cycle attractor of lattice model (8.7) with $\Sigma = 0$ and $V = 10$ (on the order of 10^{10} lattice points), which resembles the chaotic attractor. D. Stochastic realization of model (8.7), with $V = 1$ and variance-covariance entries in Σ of $\sigma_{11} = 2.332$, $\sigma_{22} = 0.2374$, $\sigma_{12} = \sigma_{21} = 0$, exhibits episodes of the lattice six-cycle (solid circles) from panel B interspersed among episodes (open circles) that resemble the chaotic attractor from panel A. E. The six-patterns from panel D are shown in state space, where their resemblance to the six-cycle in panel B is apparent.

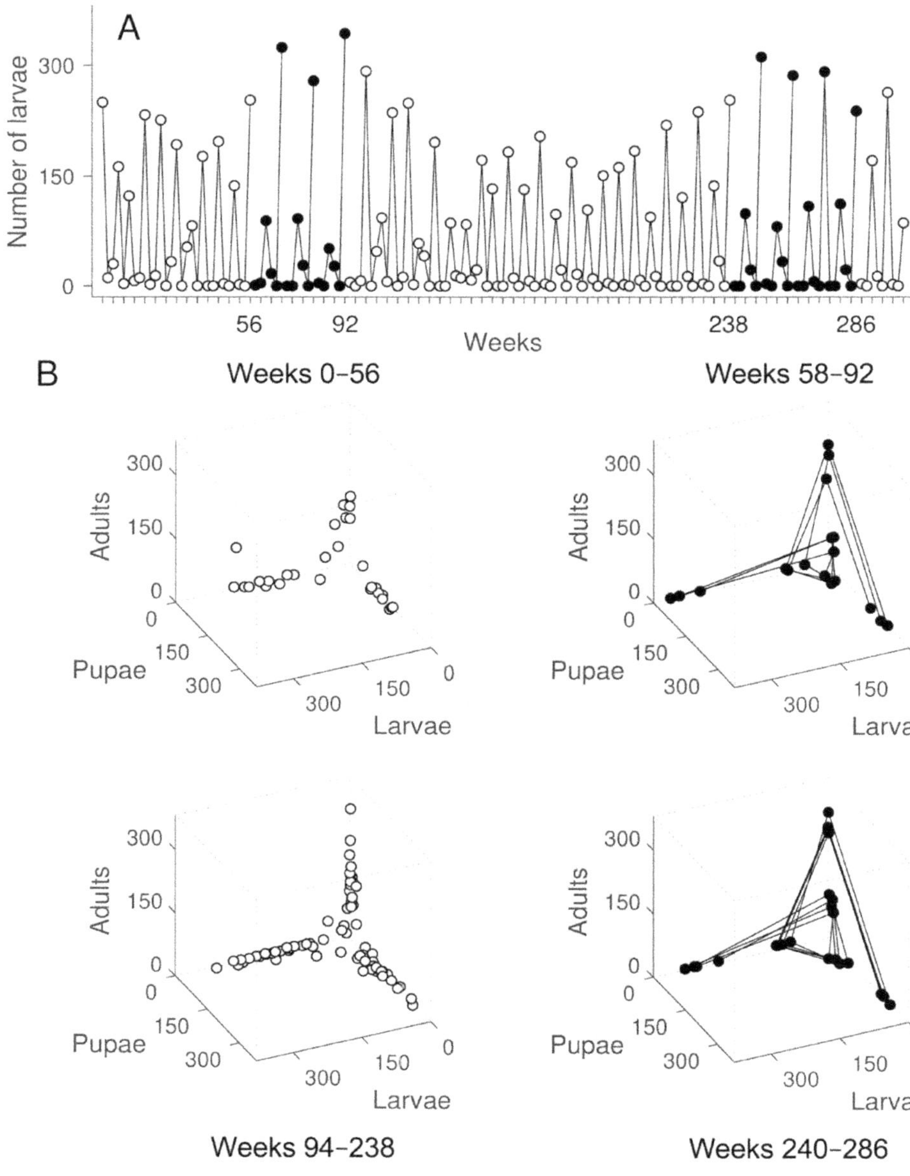

Figure 8.3 A 304-week data time series from replicate 13 of the route-to-chaos *Tribolium* experiment with experimental parameters (8.6) [2]. A. Selected temporal episodes of the lattice six-cycle (solid circles). The remaining data points (open circles) resemble the chaotic time series. B. Selected temporal episodes of the time series in panel A, shown in state space (on the order of 10^7 lattice points). Compare the six-cycle episodes in these data to the stochastic model predictions in Figure 8.2D,E.

N_{t+1} that has n_t appearing in it as a constant or vector of constants. If the time series is univariate, then the population size N_t has integer values. If the time series is multivariate, then N_t is a vector with integer entries. Denote the expected value of N_{t+1} given $N_t = n_t$ as

$$\mu(n_t) \mathrel{\hat{=}} E[N_{t+1} \mid N_t = n_t],$$

where the symbol $\mathrel{\hat{=}}$ means "defined to be."

One way to construct a deterministic skeleton for the stochastic process (8.8) is by iterating the conditional expectation as a deterministic model:

$$\begin{aligned} x_{t+1} &= \mu(x_t) \\ x_0 &= n_0. \end{aligned} \tag{8.9}$$

That is, the population size x_{t+1} at time $t+1$ is predicted to be the expected value of the distribution $P(x_t)$, where x_t was the prediction at time t. Model (8.9) takes on a continuum of values; it is not confined to a lattice in state space. This "mean model" cannot account for patterns that arise because of the discreteness of animal numbers.

We can, however, construct a deterministic skeleton with discrete states by using a different measure of central tendency. Given the observation of n_t animals at time t, the most probable next system state on the lattice is the mode

$$\lambda(n_t) \mathrel{\hat{=}} \text{mode}[N_{t+1} \mid N_t = n_t],$$

and a deterministic "mode model" iterates the conditional mode:

$$\begin{aligned} x_{t+1} &= \lambda(x_t) \\ x_0 &= n_0. \end{aligned} \tag{8.10}$$

If the mode is not unique, we can construct more than one mode model. The system states x_t of the mode model (8.10) are integer valued.

It turns out that the mean model (8.9) and the mode model (8.10) together can explain more of the low-dimensional patterns in the stochastic system (8.8) than either model could explain alone. Patterns in the stochastic realizations of (8.8) that are predicted by the mode model but not by the mean model are the result of lattice effects.

For example, consider the stochastic Ricker model

$$N_{t+1} \sim \text{Poisson}[bn_t \exp(-cn_t)], \tag{8.11}$$

where $b, c > 0$. The mean model is the familiar Ricker model

$$x_{t+1} = bx_t \exp(-cx_t), \tag{8.12}$$

whereas the mode model is[2]

$$x_{t+1} = \text{floor}[bx_t \exp(-cx_t)], \tag{8.13}$$

where "floor$[x]$" is the largest integer less than or equal to x.

When $b = 23$, $c = 1/43$, and $x_0 = 10$, the mean model (8.12) predicts a three-cycle (Figure 8.4A) whereas the mode model (8.13) predicts a five-cycle (Figure 8.4B). Simulations of the stochastic process (8.11) exhibit episodes of both three-cycles (open circles) and five-cycles (solid circles), as illustrated in Figure 8.4C. The episodes of five-cycles in the stochastic simulations are lattice effects.

---

[2]The mode of Poisson$[\mu]$ is floor$[\mu]$ if μ is not an integer. If μ is an integer, then the mode of Poisson$[\mu]$ is not unique, and is equal to both μ and $\mu - 1$. We can neglect this latter case, because for almost all values of b and c, there is no integer $x_t > 0$ for which $bx_t \exp(-cx_t)$ is an integer.

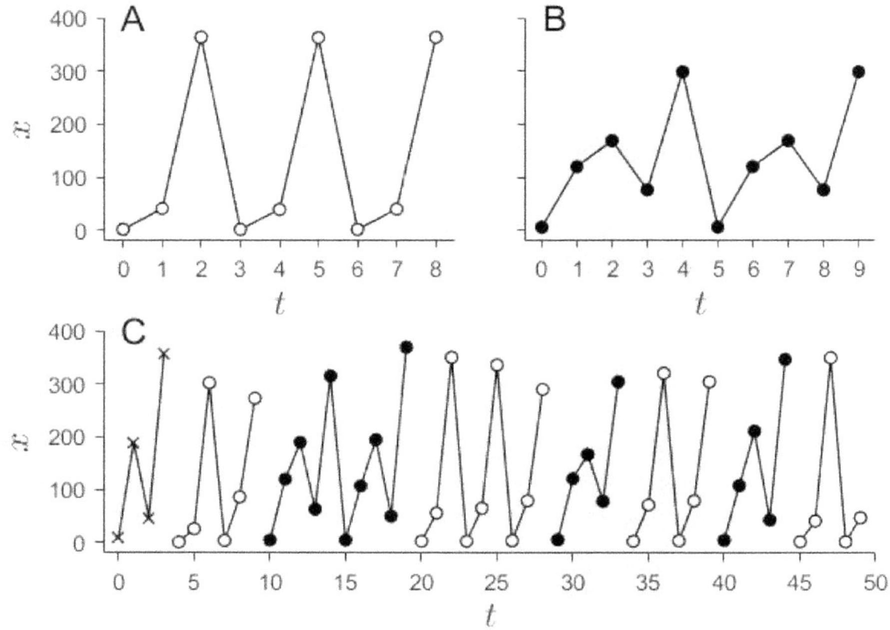

Figure 8.4 Ricker dynamics with $b = 23$, $c = 1/43$, and initial condition $x_0 = 10$. A. The mean model (8.12) predicts a three-cycle. B. The mode model (8.13) predicts a five-cycle. C. Realizations of the Poisson model (8.11) show recurring episodes of both the three-cycle (open circles) and the five-cycle (solid circles).

A stochastic model for *Tribolium* is [2]

$$\begin{aligned} L_{t+1} &\sim \text{Poisson}\left[ba_t \exp\left(-c_{el}l_t - c_{ea}a_t\right)\right] \\ P_{t+1} &\sim \text{binomial}\left[l_t, (1-\mu_l)\right] \\ R_{t+1} &\sim \text{binomial}\left[p_t, \exp\left(-c_{pa}a_t\right)\right] \\ S_{t+1} &\sim \text{binomial}\left[a_t, (1-\mu_a)\right] \\ a_t &\mathrel{\mathring{=}} r_t + s_t, \end{aligned} \tag{8.14}$$

where L_t is the number of feeding larvae, P_t is the number of non-feeding larvae, pupae and callow adults, R_t is the number of sexually mature adult recruits, and S_t is the number of surviving mature adults. The quantities l_t, p_t, r_t, and s_t are the respective abundances observed at time t. The total number of mature adults is $R_t + S_t$, and $a_t = r_t + s_t$ is the total number of mature adults observed at time t. The mean model for stochastic model (8.14) is the LPA model (8.5) with $V = 1$, and the mode model is [5]

$$\begin{aligned} L_{t+1} &= \text{floor}\left[bA_t \exp(-c_{el}L_t - c_{ea}A_t)\right] \\ P_{t+1} &= \text{floor}\left[(1-\mu_l)\left(L_t+1\right)\right] \\ A_{t+1} &= \text{floor}\left[(P_t+1)\exp\left(-c_{pa}A_t\right)\right] + \text{floor}\left[(1-\mu_a)\left(A_t+1\right)\right]. \end{aligned} \tag{8.15}$$

Figure 8.5 shows model simulations and 148 weeks of experimental data from replicate 13 of the route-to-chaos experiment [2], with experimental parameters (8.6), and initial condition $[250, 5, 100]$. The mean model (8.5) predicts chaos (Figure 8.5A), with a recurring

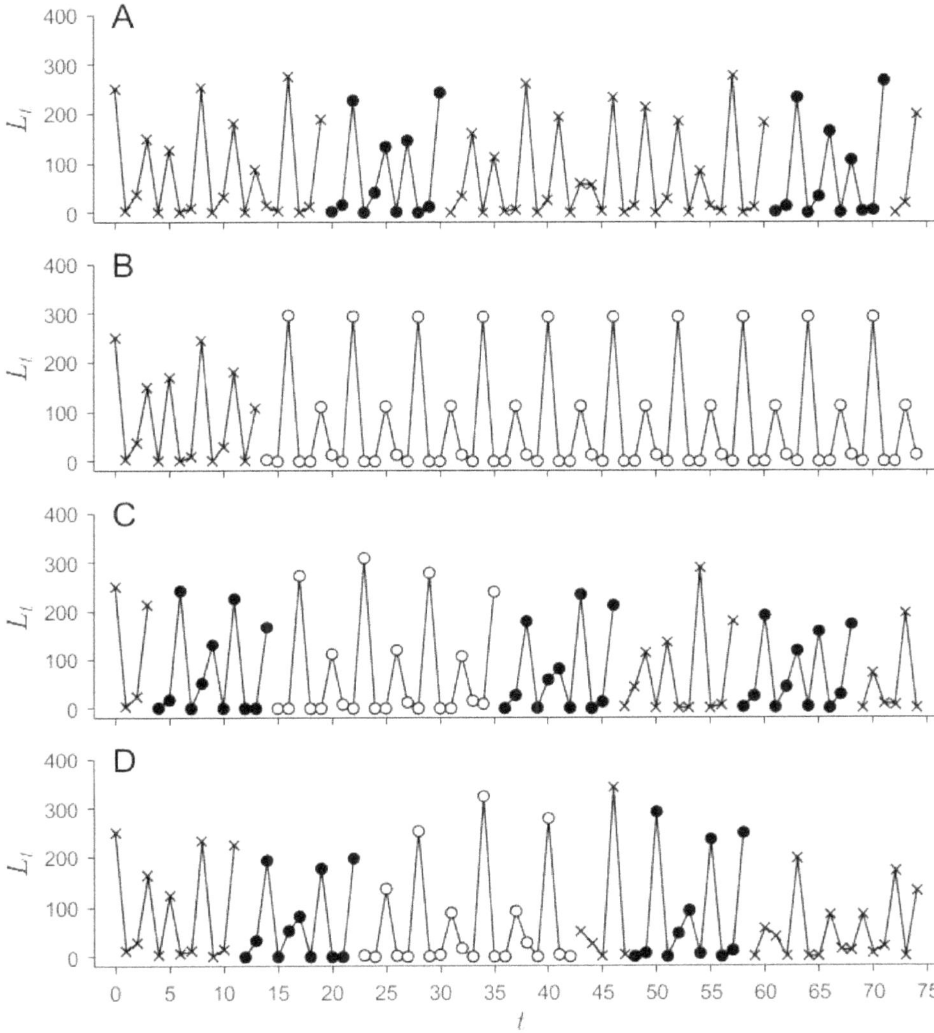

Figure 8.5 LPA model simulations with experimental parameters (8.6) and initial condition [250, 5, 100]; and 148 weeks of experimental *Tribolium* data from replicate 13 of the route-to-chaos experiment [2]. A. The mean model (8.5) predicts chaos, with recurrent episodes of an 11-cycle (solid circles) pattern on the chaotic attractor. B. The mode model (8.15) locks into a six-cycle (open circles). C. The Poisson-binomial model (8.14) exhibits episodes of both the 11-cycle (solid circles) and the lattice six-cycle (open circles). D. The experimental data also show episodes of both cycles.

11-cycle pattern (solid circles) [1, 7]. As we shall see in the next chapter, the 11-cycle pattern is related to an 11-cycle saddle on the chaotic attractor. The mode model (8.15) predicts a six-cycle (Figure 8.5B, open circles). Note that this is essentially the same six-cycle predicted by the LPA lattice model using "int," model (8.7), with $V = 1$ and $\mathbf{\Sigma} = 0$, shown in Figure 8.2B. Realizations of the stochastic process (8.14), as illustrated in Figure 8.5C, contain episodes of the 11-cycle from the chaotic attractor of the mean model (solid circles) as well as episodes of the dynamics of the mode model (open circles). The experimental data also show both types of episodes (Figure 8.5D).

For further exploration, see the examples for this section in the online LPA Simulator (https://lpasim.org/?c=8&s=4).

8.5 CONCLUSIONS

Animals come in discrete units, and that matters. It matters because biological populations are therefore restricted to a lattice in state space. Deterministic autonomous models restricted to a bounded lattice in state space can predict only cycles. If the nonlinear mechanisms in such a model produce complex dynamics in the continuum of state space, then the model's behavior when restricted to a lattice depends on the granularity of the lattice, which depends on population size. If population sizes are large, these lattice cycles might (or might not) approximate underlying complex dynamics. If population sizes are small, these lattice cycles typically are low-period, but are "weakly stable" in the sense that small perturbations on the lattice can cause complicated transients, as the system attempts to follow the underlying tendency of chaos on the continuum. Thus, in the presence of noise, these transients can dominate the time series as stochasticity repeatedly stirs the system. In fact, stochastic models predict transient episodes that mimic the underlying complex dynamics, punctuated by episodes that resemble the deterministic lattice cycles. *Furthermore, in this chapter, we have demonstrated these effects in experimental data. Chaotic forces in noisy populations are observed in data time series as transient visits to the underlying continuum attractors, punctuated by visits to low-period lattice cycles.*

What determines the period and form of lattice cycles? Consider again the Ricker stochastic process (8.11). The mean model (8.12) predicts a three-cycle (Figure 8.4A). It seems strange at first glance that the mode model (8.13) would lock onto a five-cycle (Figure 8.4B) rather than something resembling the mean model three-cycle. The model parameters, however, place the mean model in the big three-period-locking window of the Ricker bifurcation diagram, where an unstable chaotic set lurks in the background, influencing transients. Thus, it is not surprising that the mode model can lock onto dynamics that are quite different from the mean-model three-cycle—nor is it surprising that the stochastic model (8.11) shows richer dynamics than a simple noisy three-cycle (Figure 8.4C).

Similarly, for the LPA stochastic process (8.14), the mode model locks onto the lattice six-cycle (Figure 8.5B) likely because a similar six-pattern occurs within the recurrent 11-cycle pattern on the chaotic attractor, as predicted by the mean model (Figure 8.5A; first six solid circles of the 11-cycle pattern). The mean model also has a six-cycle "nearby" in parameter space (at $\mu_a = 1$). *Thus, the nature of a lattice cycle (its period and form) is determined by the arrangement of the stable and unstable invariant sets of the mean model in the underlying continuum—those existing nearby in state space and those existing nearby in parameter space.*

In Chapter 9, we explore in detail what it means for a stochastic discrete-state system to be influenced by chaos. We develop a tool called the "lag metric comparison" (LMC) to quantify the existence, length, and frequency of cycle episodes in stochastic data, and we use the LMC to suggest new ways to detect chaos in population data [7].

Bibliography

[1] Cushing JM, Henson SM, Desharnais RA, Dennis B, Costantino RF, King AA (2001) A chaotic attractor in ecology: Theory and experimental data. Chaos, Solitons, and Fractals 12:219–234. https://doi.org/10.1016/S0960-0779(00)00109-0

[2] Dennis B, Desharnais RA, Cushing JM, Henson SM, Costantino RF (2001) Estimating chaos and complex dynamics in an insect population. Ecological Monographs 71:277–303. https://doi.org/10.1890/0012-9615(2001)071[0277:ECACDI]2.0.CO;2

[3] Domokos G, Scheuring I (2002) Random perturbations and lattice effects in chaotic population dynamics. Science 297:2163a. https://doi.org/10.1126/science.297.5590.2163a

[4] Henson SM, Costantino RF, Cushing JM, Desharnais RA, Dennis B, King AA (2001) Lattice effects observed in chaotic dynamics of experimental populations. Science 294:602–605. https://doi.org/10.1126/science.1063358

[5] Henson SM, King AA, Costantino RF, Cushing JM, Dennis B, Desharnais RA (2003) Explaining and predicting patterns in stochastic population systems. Proceedings of the Royal Society of London B 270:1549–1553. https://doi.org/10.1098/rspb.2003.2420

[6] King AA, Desharnais RA, Henson SM, Costantino RF, Cushing JM, Dennis B (2002) Random perturbations and lattice effects in chaotic population dynamics. Reply to Domokos G, Scheuring L. Science 297:2163a. https://doi.org/10.1126/science.297.5590.2163a

[7] King AA, Costantino RF, Cushing JM, Henson SM, Desharnais RA, Dennis B (2004) Anatomy of a chaotic attractor: Subtle model-predicted patterns revealed in population data. Proceedings of the National Academy of Sciences 101:408–413. https://doi.org/10.1073/pnas.2237266100

CHAPTER 9

How Is Chaos Manifest in Noisy Populations?

In this chapter, we focus on some dominant temporal signatures associated with the chaotic attractor discussed in Chapters 7–8 and analyze the experimental data for corresponding patterns.

9.1 WHAT IS CHAOS IN A REAL POPULATION?

The experiments described in Chapters 6–7 tested the ability of the LPA model

$$\begin{aligned} L_{t+1} &= bA_t e^{-c_{el}L_t - c_{ea}A_t} \\ P_{t+1} &= (1-\mu_l)\,L_t \\ A_{t+1} &= P_t e^{-c_{pa}A_t} + (1-\mu_a)\,A_t \end{aligned} \tag{9.1}$$

to make accurate predictions. In particular, we designed long-term, replicated, and controlled experiments to test whether flour beetle population dynamics would correspond to a sequence of attractors in the LPA model-predicted route to chaos (shown in Figure 7.4A of Chapter 7) as selected vital rate parameters were manipulated. The fact that the predicted sequence of bifurcations was experimentally observed gave strong support to the conclusion that the LPA model (9.1) is capable of accurately describing and predicting flour beetle dynamics across these dynamic regimes.

Among the several different types of attractors predicted along the bifurcation diagram (Figure 7.4A) in the route-to-chaos experiment is the chaotic attractor at the parameter value $c_{pa} = 0.35$ (with $\mu_a = 0.96$). Can we say that the replicate beetle populations cultured in this treatment exhibited chaotic dynamics? Let's review three points from Chapters 7–8 that are relevant to this question.

First, in Chapter 7 we discussed the fact that estimated parameter values come with confidence intervals, and that within a chaotic regime, each parameter interval, no matter how small, typically is riddled densely with period-locking windows. In particular, this means that the deterministic LPA model (9.1) does not predict chaos for every value within the parameter confidence intervals for the treatment with $c_{pa} = 0.35$ and $\mu_a = 0.96$.

Second, in Chapter 7 we noted that chaos is a deterministic phenomenon [2], whereas populations are always subject to stochastic perturbations. The stochastic demographic LPA model (7.3), used in Chapter 7 for model parameterization and validation, does not have chaotic dynamics in a rigorous sense.

Third, in Chapter 8, we saw that population sizes are restricted to a finite lattice in state space, and that dynamics of a deterministic system on a bounded lattice must become

DOI: 10.1201/9781003467120-9

periodic. For example, the deterministic lattice version of the LPA model cannot predict chaotic dynamics.

What, then, does it mean for a real population to have chaotic dynamics? In light of the first two points, what does it mean for any real system to be chaotic?

The issue raised in the first point above is a purely deterministic matter. In Chapter 7, we discussed how, within these small period-locking windows that are packed in a chaotic regime, the chaotic set, although "momentarily" unstable (as the bifurcation parameter moves through the window), still lurks in the background, influencing transients. That is to say, although a stable cycle appears, the chaotic set remains nearby in parameter space, strongly affecting trajectories that do not start directly on the cycle.

When we include the reality of stochasticity in this picture, acknowledging that noise continually stirs up transients, we see that stochasticity is in fact what makes the chaotic signal robust across the parameter confidence interval within a chaotic regime. And, in Chapter 8 we saw that, even if we restrict the dynamics to a lattice in state space, the chaotic set in the underlying continuum still exerts its influence. Noisy transients trace out the chaotic set on the state space lattice, punctuated by episodes of lattice cycles.

The point is that stochasticity, rather than being a nuisance, reveals the underlying deterministic mechanisms of the model's skeleton [8].

With all these considerations in mind, for the $c_{pa} = 0.35$ (with $\mu_a = 0.96$) treatment in the route-to-chaos experiment in Chapter 7, we say the experimental dynamics are *influenced* by chaos rather than saying they *are* chaotic.

The influence of chaos in the $c_{pa} = 0.35$ (with $\mu_a = 0.96$) treatment was demonstrated in at least three ways in Chapters 7–8. First, when the data are plotted in three-dimensional state space, they do cluster closely around the predicted chaotic attractor, as seen in Figure 7.5 of Chapter 7. Second, cultures in the sequence of expermental treatments as c_{pa} was manipulated from 0 to 1 (with $\mu_a = 0.96$ fixed) showed complex fluctuations in the $c_{pa} = 0.35$ treatment, bookended by simple cyclic dynamics in the treatments at both ends of the c_{pa} range. Third, the model-predicted lattice six-cycle makes its episodic appearance in the experimental data (Chapter 8).

Are there additional ways to demonstrate the influence of chaos in these replicates? Are there other characteristics of the chaotic attractor that we can see in the experimental data?

9.2 SADDLE FLYBYS

Robert Devaney defines a chaotic invariant set to be a bounded invariant set of a deterministic dynamical system that has three properties [7]. The first is sensitivity to initial conditions, which we will address in detail in Chapter 11. The second is that the set contains a dense set of infinitely many periodic cycles. The third is a technical requirement called topological transitivity,[1] which we will not address here.

The periodic cycles embedded in a chaotic attractor are generally unstable saddles, which means they have both attracting directions and repelling directions in state space. A dynamic consequence is that some trajectories can move toward an unstable periodic cycle before moving away from it, a phenomenon called a *saddle flyby*.

In Chapter 3 we discussed saddle flybys of saddle equilibria (which can be viewed as cycles of period one). In a stochastic system, this can occur when a stochastic bump places a time series near the stable manifold of a saddle equilibrium, to which it approaches and then retreats. It turns out that equilibrium flybys were commonly observed in the route-to-

[1] It turns out that the latter two properties imply the former. Therefore, sensitivity to initial conditions can be removed from Devaney's definition of chaos and treated instead as a derivable property of a chaotic set.

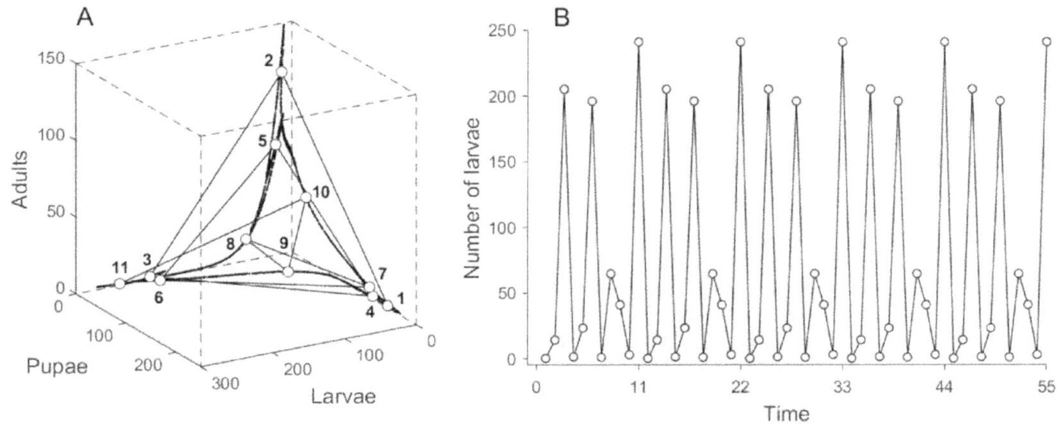

Figure 9.1 A dominant saddle 11-cycle on the chaotic attractor is predicted by the LPA model (9.1) in the route-to-chaos experiment with parameters from Table 7.2 (Chapter 7) except for the treatment values of $c_{pa} = 0.35$ and $\mu_a = 0.96$. A. The 11-cycle is shown on the chaotic attractor in state space. B. Five repetitions of the L-stage component of the 11-cycle, shown as a time series.

chaos experiments discussed in Chapter 7. There was, for example, an extraordinary saddle equilibrium flyby in one of the three replicates of the chaos treatment at $c_{pa} = 0.35$ (with $\mu_a = 0.96$) that occurred approximately during weeks 350–375 in this 424-week experiment. (It is important to note that the existence of the unstable equilibrium was predicted by the parameterized LPA model at the onset of the experiment approximately 6.5 years earlier!) In Chapter 12, we present a detailed discussion of this and other saddle equilibrium flybys observed in our experimental data. But what about flybys of saddle cycles embedded in a chaotic attractor?

Does the chaotic attractor predicted by the LPA model (9.1) for the route-to-chaos experimental treatment with $c_{pa} = 0.35$ and $\mu_a = 0.96$ contain dominant saddle cycles, and does the model predict flybys of those cycles? If so, do the experimental data show these saddle flybys? An answer in the affirmative will provide further evidence of the influence of the chaotic attractor on the dynamics of the flour beetle cultures.

9.3 A DOMINANT 11-CYCLE SADDLE

According to chaos theory, there are infinitely many periodic cycles on the chaotic attractor predicted for the $c_{pa} = 0.35$ (with $\mu_a = 0.96$) treatment of the route-to-chaos experiment described in Chapter 7. It is often the case, however, that one unstable (saddle) cycle has dominant influence on the dynamics. The dominant cycle on the chaotic attractor predicted in this treatment is an 11-cycle [5]. Figure 9.1A shows this saddle cycle in state space as it exists on the chaotic attractor, along with the corresponding L-stage component of the 11-cycle (Figure 9.1B). Flybys of the 11-cycle occur in chaotic time series generated by the deterministic LPA model (9.1), as shown in Figure 9.2.

The question is, can flybys of the saddle 11-cycle in Figure 9.1 be observed in the route-to-chaos experimental data?

To answer this question, we devised a metric to measure how close a time series, or a segment of a time series, is to a given periodic cycle. In the simple case of a period one

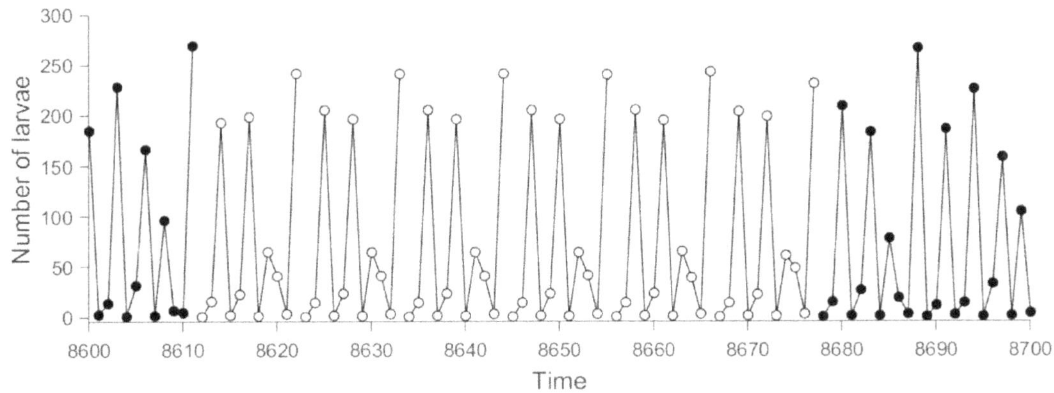

Figure 9.2 Portion of larval time series from a trajectory on the chaotic attractor, generated by the deterministic LPA model (9.1) with parameters from Table 7.2 except for the treatment values of $c_{pa} = 0.35$ and $\mu_a = 0.96$, and with initial condition $[1.8, 154.2, 2.4]$. At $t = 8612$ the trajectory lands near the 11-cycle saddle. This is followed by several repetitions of patterns similar to the 11-cycle (open circles), after which the trajectory moves away.

cycle, that is, an equilibrium $\mathbf{x}_e = [L_e, P_e, A_e]$,[2] we simply measure the distance from the time series triple $\mathbf{x}(t) = [L_t, P_t, A_t]$ to the equilibrium at each point in time t using the norm $\|\mathbf{x}\| = |L| + |P| + |A|$ (total population size). This distance $\|\mathbf{x}(t) - \mathbf{x}_e\|$ is a function of time t, and if the time series $\mathbf{x}(t)$ undergoes a flyby of a saddle equilibrium $\mathbf{x}_e$, then a plot of this distance will show a time interval during which $\|\mathbf{x}(t) - \mathbf{x}_e\|$ is close to 0 (before and after which it is not close to 0).

For a periodic cycle $\mathbf{x}_T(t)$ of period $T \geq 2$, we are interested in how close $\mathbf{x}(t)$ is to $\mathbf{x}_T(t)$ on average over an interval of length T. At a specified time $t \geq T$, we look at the distances $\|\mathbf{x}(t) - \mathbf{x}_T(t)\|$ to the cycle at the previous times over an interval equal to the period T, that is, for times $t - q$ for $q = 0, ..., T-1$. There are T phase shifts $\mathbf{x}_T(t+s)$ of the cycle $\mathbf{x}_T(t)$, $s = 0, ..., T-1$, to which the solution might be close. We define the *lag metric comparison* to be the average distance to phase shift s at time t, that is,

$$\text{LMC}(s,t) \mathrel{\mathring{=}} \frac{1}{T}\sum_{q=0}^{T-1} \|\mathbf{x}(t-q) - \mathbf{x}_T(t+s-q)\|,$$

where the symbol $\mathrel{\mathring{=}}$ means "defined to be." If we pick a phase s and plot LMC(s,t) for $T - 1 \leq t \leq n$ up to some final time n, then we can judge how close the time series $\mathbf{x}(t)$ gets to that phase of the cycle at times during this interval. The time n should be large (relative to the period T) so that there is plenty of opportunity to observe intervals over which LMC(s,t) is small, which indicate saddle cycle flybys. The lag metric LMC can be calculated for data time series or for model time series.

We used the LMC to look for saddle 11-cycle flybys in the experimental data. In particular, we computed the LMC for the 424-week data time series from replicate 13 in the route-to-chaos experiment (from the treatment with $c_{pa} = 0.35$ and $\mu_a = 0.96$) as compared to the LPA model-predicted saddle 11-cycle on the chaotic attractor. The larval component

[2]We use the notation $[L, P, A]$ to denote the column vector $(L, P, A)^T$, which is the transpose of the row vector (L, P, A).

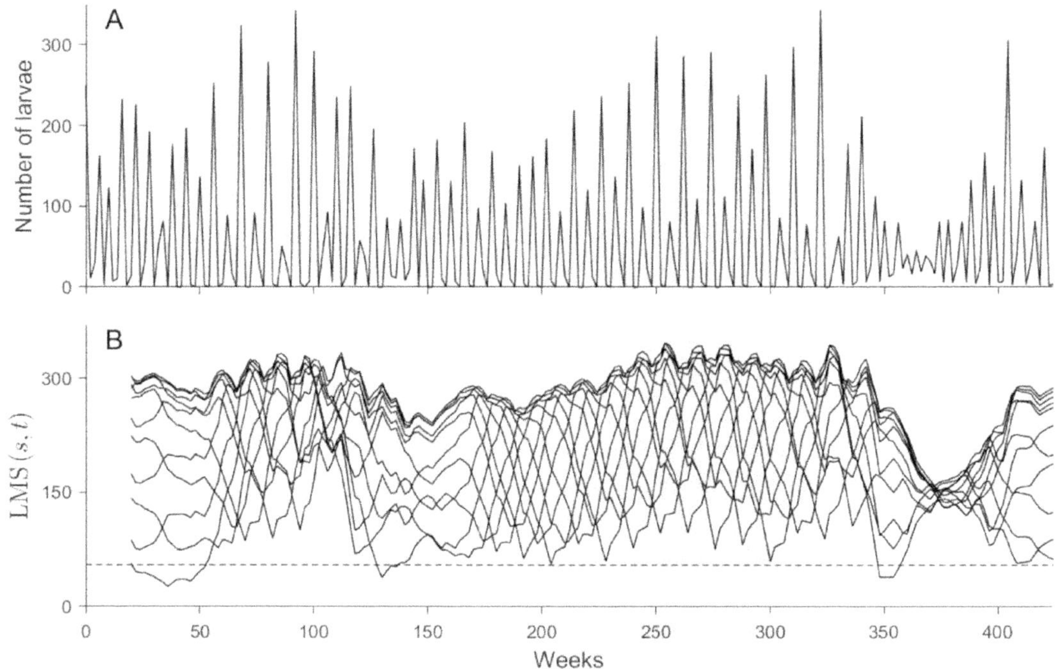

Figure 9.3 Experimental data (424 weeks) from replicate 13 of the route-to-chaos experiment described in Chapter 7, from the treatment with $c_{pa} = 0.35$ and $\mu_a = 0.96$. A. L-stage component of the time series data. B. Eleven lag metric comparisons (LMCs) showing the distance between the data and the model-predicted 11-cycle on the chaotic attractor. The threshold is set at a total population size of 55 (dashed line).

L_t of the experimental data time series is plotted in Figure 9.3A. The LMC for all 11 phases of the 11-cycle are shown in Figure 9.3B. The 11 different LMC metrics appear as 11 braided strands, which unravel when one of the strands dips below the arbitrary threshold of 55 (total population size; dashed line). These indicate times at which the data are close to the 11-cycle, that is, they indicate flybys of the 11-cycle saddle. Note that being "close" means being close over 11 consecutive census times, which is five and one half generations.

From the LMC plots in Figure 9.3B, we see that replicate 13 was close to the 11-cycle over three distinct time intervals during the nine-year experiment. The fact that the dominant saddle cycle on the chaotic attractor is thus observed in the time series data adds support for our assertion that the deterministic chaotic attractor has influenced the dynamics of this replicate. For more analysis of the influence of saddle cycles on the dynamics of the experimental beetle cultures see [4, 5, 6, 9].

9.4 LATTICE CYCLES

Numerical simulations of the deterministic lattice LPA model,[3] with parameter values from Table 7.2 but using the treatment values $c_{pa} = 0.35$ and $\mu_a = 0.96$, show three prominent

[3]Model (8.7) without stochasticity, that is, with $\mathbf{\Sigma} = 0$.

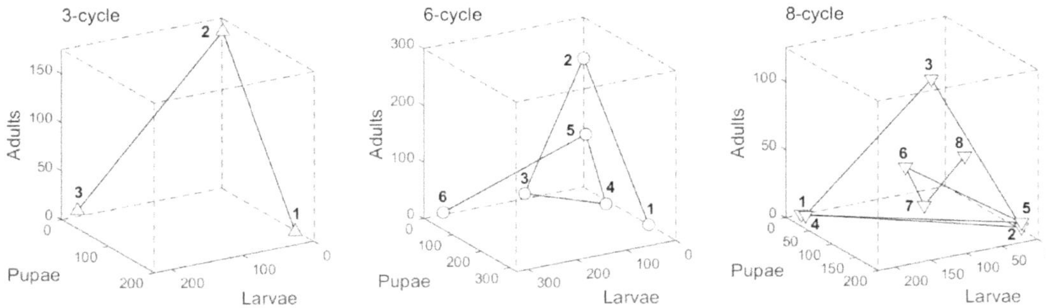

Figure 9.4 The three prominent lattice cycles predicted by the deterministic lattice LPA model (model (8.7) without stochasticity) for the parameter values in Table 7.2 except for the treatment values of $c_{pa} = 0.35$ and $\mu_a = 0.96$ in the route-to-chaos experiment.

cycles on the lattice with periods three, six, and eight (Figure 9.4). Notice that these lattice cycles bear a strong resemblance to segments of the 11-cycle which resides on the chaotic attractor. It is as if the lattice model trajectories attempt to follow the chaotic attractor with its dominant 11-cycle in state space, but being constrained to a lattice they become "stuck" on one of these lattice cycles. In this sense, we can view the dynamics on the lattice and these lattice cycles as being influenced by the 11-cycle and hence by the chaotic attractor.

Can the lattice cycles be observed in the chaotic ($c_{pa} = 0.35$ and $\mu_a = 0.96$) treatment of the route-to-chaos experiment? The LMCs for the same data time series used in Figure 9.3 are shown in Figure 9.5A-C for all three lattice cycles, with a threshold of 55. These plots show frequent and sometimes persistent episodes of each of these cycles.

9.5 ANATOMY OF CHAOS

Figure 9.5D shows the time duration spent by the data time series near the three lattice cycles as well as the continuous state space 11-cycle, according to the threshold level of 55 individuals. In this figure (Figure 9.5D), we see that initially the experimental data follow the saddle 11-cycle on the chaotic attractor. During this same interval, the lattice eight-cycle is also evident, but this is not surprising since these two cycles lie close together (see Figures 9.1 and 9.4). After about week 50, a six-cycle episode is observed. Over weeks 110 to 134, another 11-cycle/eight-cycle episode appears, which is followed by a three-cycle episode over weeks 150 to 172. At around week 200, a startlingly long 130-week six-cycle episode begins (over 65 generations)! Weeks 328 to 356 display another 11-cycle/eight-cycle episode. It should be noted that a one-cycle episode (the afore mentioned equilibrium flyby) is clearly evident in these data around week 364 (Figure 9.3A). This saddle equilibrium is not located on the chaotic attractor, but a stochastic event bumped the data time series near it (or its stable manifold) and a saddle equilibrium flyby resulted. This saddle equilibrium flyby is considered further in Chapter 12.

The breakdown in Figure 9.5D provides a kind of "anatomy" for the influence of the chaotic attractor on the experimental data.

For further exploration, see the examples for this section in the online LPA Simulator (https://lpasim.org/?c=9&s=5).

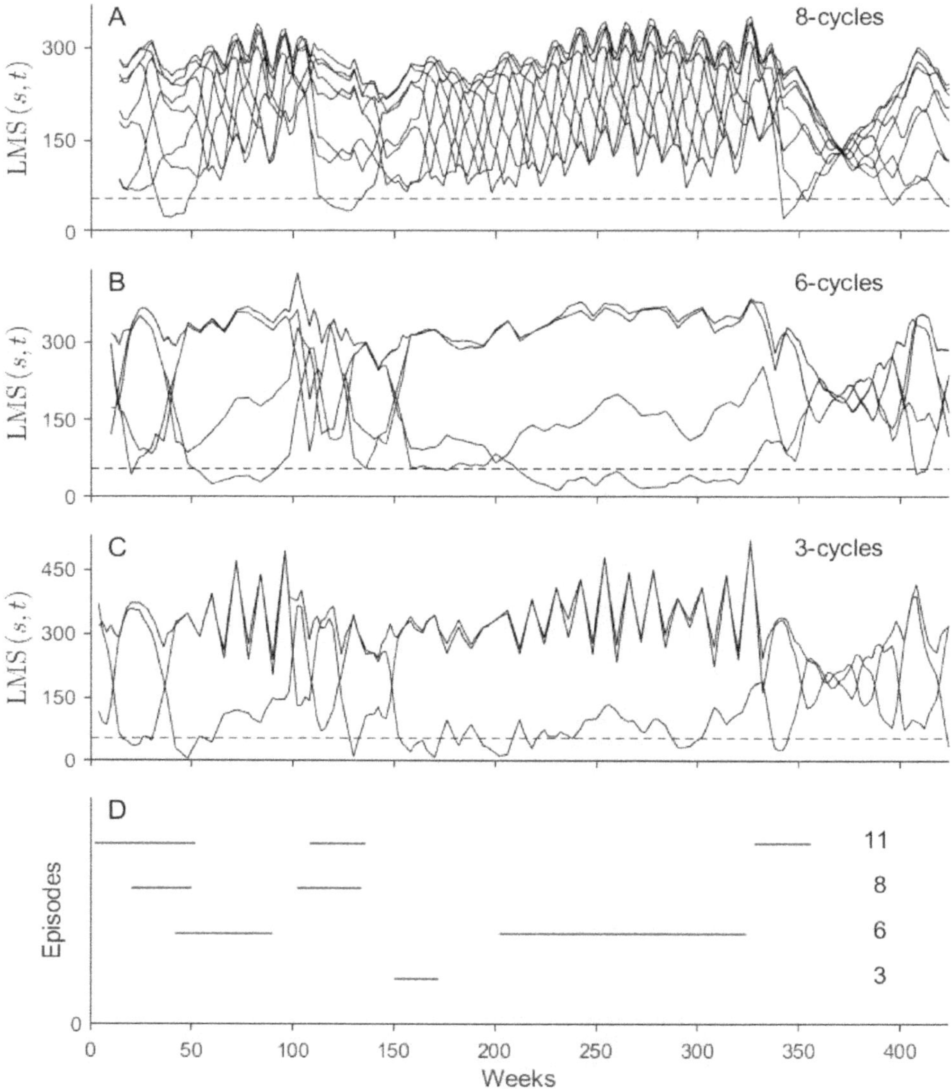

Figure 9.5 The lag metric comparisons (LMCs) for the experimental data in replicate 13 of the route-to-chaos experiment, as compared to the three lattice cycles shown in Figure 9.4. A. The LMCs for the lattice eight-cycle. B. The LMCs for the lattice six-cycle. C. The LMCs for the lattice three-cycle. D. Time duration spent by the data time series near each of these three lattice cycles as well as the continuous state space 11-cycle, according to the threshold level of 55 individuals.

9.6 CONCLUSIONS

A main focus of the route-to-chaos experiment described in Chapter 7 was the occurrence of LPA model-predicted chaotic attractors. In this chapter, we focused on the chaotic attractor predicted to occur in the experimental treatment with $c_{pa} = 0.35$ and $\mu_a = 0.96$. How should we expect this chaotic attractor to be manifest in the time series of a noisy, discrete-state system? What signals will mark the influence of underlying chaotic forces?

Mathematically, chaotic dynamics are not devoid of order but can display flybys of saddle cycles. Furthermore, if dynamics are confined to a lattice in state space, stochasticity can restore the signal of an underlying chaotic attractor as noise continually stirs up transients, and this signal will be episodically interrupted by lattice cycles whose forms and periods are characteristic of the interplay between the scale of the system (mesh of the lattice) and the nature of the chaotic attractor. Because of, rather than in spite of, the noisy nature of biological time series and the discrete nature of individual organisms, the influence of chaos is detectable in population time series. Chaotic mechanisms manifest themselves in noisy, discrete-state population systems as transient but recurrent and predictable cyclic patterns generated by chaotic attractors and their discrete-state analogues, woven together by stochasticity.

Bibliography

[1] Banks J, Brooks J, Cairns G, Davis G, Stacey P (1992) On Devaney's definition of chaos. American Mathematical Monthly 99:332–334. https://doi.org/10.2307/2324899

[2] Bishop R (2017) Chaos. In Zalta EN (ed.), The Stanford Encyclopedia of Philosophy, Spring 2017 Edition.

[3] Costantino RF, Desharnais RA, Cushing JM, Dennis B, Henson SM, King AA (2005) Nonlinear stochastic population dynamics: The flour beetle *Tribolium* as an effective tool of discovery. Advances in Ecological Research 37:101–141. https://doi.org/10.1016/S0065-2504(04)37004-2

[4] Cushing JM, Dennis B, Desharnais RA, Costantino RF (1998) Moving toward an unstable equilibrium: Saddle nodes in population systems. Journal of Animal Ecology 67:298–306. https://doi.org/10.1046/j.1365-2656.1998.00194.x

[5] Cushing JM, Henson SM, Desharnais RA, Dennis B, Costantino RF, King A (2001) A chaotic attractor in ecology: Theory and experimental data. Chaos, Solitons and Fractals 12:219–234. https://doi.org/10.1016/S0960-0779(00)00109-0

[6] Cushing JM, Costantino RF, Dennis B, Desharnais RA, Henson SM (2003) Chaos in Ecology: Experimental Nonlinear Dynamics. Theoretical Ecology Series. Volume 1. New York, Academic Press.

[7] Devaney, RL (2019) An Introduction to Chaotic Dynamical Systems. Second Edition. Boca Raton, Florida, CRC press.

[8] Henson SM, King AA, Costantino RF, Cushing JM, Dennis B, Desharnais RA (2003) Explaining and predicting patterns in stochastic population systems. Proceedings of the Royal Society of London B 270:1549–1553. https://doi.org/10.1098/rspb.2003.2420

[9] King AA, Costantino RF, Cushing JM, Henson SM, Desharnais RA, Dennis B (2003) Anatomy of a chaotic attractor: Subtle model predicted patterns revealed in population data. Proceedings of the National Academy of Sciences 101:408–413. https://doi.org/10.1073/pnas.2237266100

CHAPTER 10

Demographic Stochasticity

The random factors that cause fluctuations in population numbers are of two types: (1) changes due to extrinsic forces such as weather and climate, and (2) events that are intrinsic to the population such as variation in the numbers of offspring produced by individuals or the timing of deaths. These phenomena are called, respectively, *environmental stochasticity* and *demographic stochasticity* [11]. Fluctuations driven by environmental stochasticity affect the entire population as a whole, whereas fluctuations due to demographic stochasticity result from the sum of independent random events among individuals. Theoretical studies have shown that, for a population fluctuating around a stable equilibrium, the influence of demographic stochasticity diminishes as population density increases. In particular, for single-species population models with a single age/stage class, if the deterministic version of the stochastic model predicts a stable equilibrium, then demographic stochasticity scales as the inverse square root of population size [1, 2, 3, 6, 7, 10, 13, 14]. Robert May showed this is also true for models with exponential growth [11]. In ecology, this result is known as *inverse square root rule for demographic stochasticity*. It suggests that, for large populations, demographic stochasticity can be ignored. It is sometimes cited as a justification for using deterministic models of populations in a constant environment.

In this chapter, we examine numerically the robustness of the square root rule for two versions of a stochastic Ricker model. We then extend the investigation to structured population models and to complex dynamics, again showing numerically the inverse square root rule. Finally, we conduct an empirical test of the inverse square root rule using *Tribolium* populations cultured with the protocol from the chaotic treatment of the route-to-chaos experiment discussed in Chapter 7.

10.1 DEMOGRAPHIC STOCHASTICITY AND THE INVERSE SQUARE ROOT RULE

Demographic stochasticity involves the sum of the independent random contributions of the individuals that constitute the population. Consider a population at its stable equilibrium value N^*. If we let σ^2 represent the expected variance in the contribution of a single individual to the next time interval, then, the variance in population numbers for the next time interval will be the sum of the variances for each individual, $N^*\sigma^2$. Since the population is at its equilibrium, the expected mean number of individuals for the next time interval will be, approximately, N^*. The coefficient of variation, CV, is commonly used as a measure of dispersion for population density; it is equal to the standard deviation (square root of the variance) divided by the mean. For the population near its equilibrium,

$$CV = \frac{\sqrt{N^*\sigma^2}}{N^*} = \frac{\sigma}{\sqrt{N^*}}.$$

DOI: 10.1201/9781003467120-10

This suggests that, for a large population near its equilibrium N^*, demographic stochasticity can be ignored.

10.1.1 Example Using the Ricker Model

Consider the following deterministic discrete-time Ricker model:

$$N_{t+1} = N_t b \exp\left(-\frac{cN_t}{V}\right), \tag{10.1}$$

where N_t is the population size, b is the per capita number of offspring, c is a parameter describing density-dependent effects on reproduction with increases in population density, and V is the size of the habitat. Population density is given by the size of the population divided by the habitat size, N_t/V. By multiplying both sides of equation (10.1) by c/V, it is easy to see that parameters c and V have no effect on the population's dynamics. For values of $b < e^2 \approx 7.389$, the model has a stable equilibrium $N^*/V = (\ln b)/c$.

The deterministic model (10.1) can be converted to a mechanistic model with demographic stochasticity. There are many ways this could be done. To keep matters simple, we will assume that the reproductive output of each individual i is a Poisson-distributed random variable R_i with mean value b and that the survival of each offspring to time $t+1$ is an independent Bernoulli trial with a probability $p_t = \exp(-cN_t/V)$. Given N_t, the population number in the next generation is a random variable:

$$\begin{aligned} N_{t+1} &\sim \sum_{i=1}^{N_t} \mathrm{Bin}\,(R_i, p_t) \\ &\sim \sum_{i=1}^{N_t} \mathrm{Pois}\,(bp_t) \\ &\sim \mathrm{Pois}\,(N_t b p_t)\,, \end{aligned} \tag{10.2}$$

where $\mathrm{Bin}\,(R_i, p_t)$ represents a binomial distribution for R_i Bernoulli trials, each with probability p_t, and "$\sim$" means "distributed as." The final Poisson distribution comes from the fact that hierarchical binomial distribution $\mathrm{Bin}\,(R_i, p_t)$, where R_i is given by a Poisson distribution with mean b, yields a Poisson distribution with mean bp_t; and the sum of N_t independent identically distributed Poisson random variables, each with mean bp_t, follows a Poisson distribution with mean $N_t b p_t$ [9].

Mathematically, models such as (10.2) have no stationary probability distribution for population numbers. This is because, for any time interval, there is a non-zero probability that the population will go extinct. $N = 0$ is an *absorbing state*; a closed population cannot recover from extinction. From a practical point of view, in many cases, the expected time to extinction is enormously long and the population spends most of its time fluctuating around the deterministic attractor. Modelers often refer to the *quasi-stationary distribution* which is conditional upon non-extinction.

Figure 10.1 shows some simulations of the deterministic and stochastic Ricker models for $b = 1.2$ and $c = 0.001$, with population densities N_t/V plotted against time. The deterministic model shows population numbers smoothly approaching the stable equilibrium. For $V = 1$, the stochastic model shows significant variability around the equilibrium. For $V = 10$ and $V = 100$, the amplitude of the fluctuations decreases, and the behavior of the stochastic model begins to resemble the deterministic model.

We used the model (10.2) to demonstrate the inverse square root rule. Habitat size was varied from $V = 1$ to $V = 100$ using 201 evenly-spaced values on a logarithmic scale.

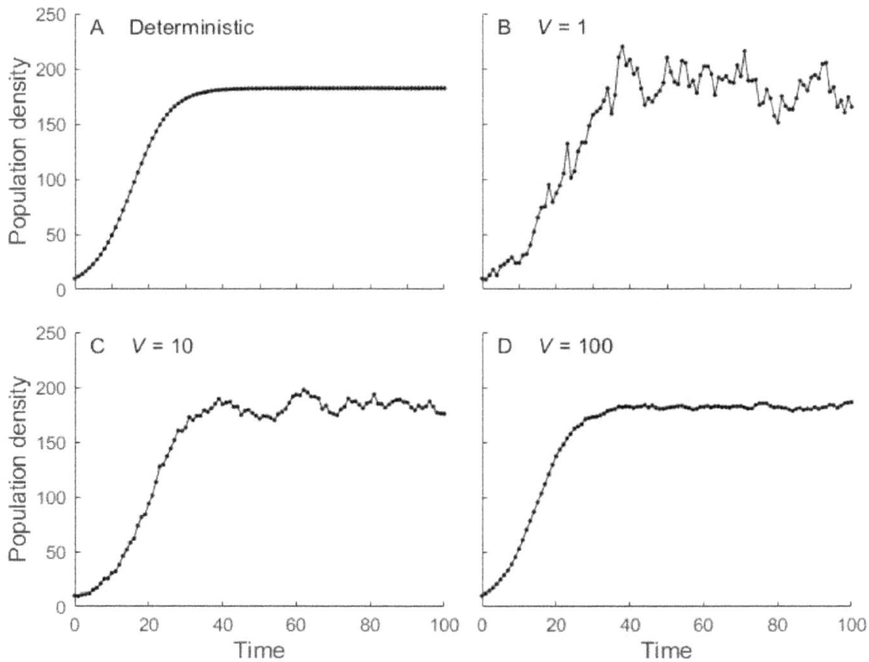

Figure 10.1 Time-series simulations of the Ricker model with $b = 1.2$ and $c = 0.001$. A. Trajectory of population density for the deterministic Ricker model (10.1). B–D. Realizations of the stochastic Ricker model (10.2) with $V = 1$, $V = 10$, and $V = 100$.

For each habitat size, the model was first iterated 100 times to approach the quasi-steady state distribution and then 1000 times to obtain an estimate of the mean and variance of the population numbers. Figure 10.2 shows the coefficient of variation CV in population numbers versus V on a log-log scale, along with the best-fit least-squares regression line. On the log-log scale, the CV values showed a strong linear trend with a slope close to -0.5. This implies that the coefficient of variation in population numbers decreases in proportion to $1/\sqrt{V}$ as predicted by the inverse square root rule.

10.1.2 Robustness of the Inverse Square Root Rule

One might argue that the choice of a Poisson distribution for demographic stochasticity is the reason model (10.2) supports the inverse square root rule, and that other models of demographic stochasticity could lead to different results. One might think the inverse square root rule is holding for the Poisson distribution only because the mean λ is equal to the variance, so that the coefficient of variation is $CV = \sqrt{\lambda}/\lambda = 1/\sqrt{\lambda}$. As the mean increases, the CV decreases according to the inverse square root rule. However, this argument is misleading. The inverse square root rule arises because, for a population of size N, demographic stochasticity is the result of summing the random variation in the individual contributions to the next time interval. Both the mean and the variance increase in proportion to N, and so the coefficient of variation decreases in proportion to $\sqrt{N}/N = 1/\sqrt{N}$. This is true for any probability distribution describing the contribution of an individual to the next generation. Indeed, as we mentioned in Chapter 2, demographic stochasticity is

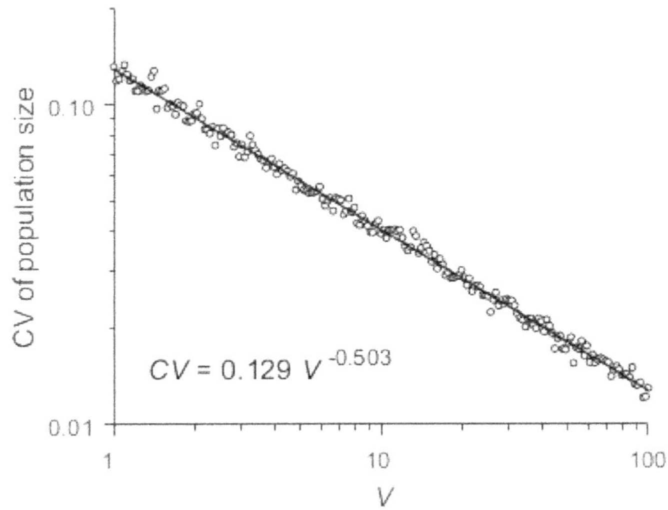

Figure 10.2 Demonstration of the inverse square root rule for the stochastic Ricker model (10.2) with $b = 1.2$ and $c = 0.001$ which yields a stable equilibrium for the deterministic Ricker model (10.1). For habitat sizes ranging from $V = 1$ to $V = 100$, the model was iterated 100 times to remove transients and then 1000 times to obtain the mean and variance of population numbers. The coefficient of variation (CV) of population numbers versus V is plotted on a log-log scale. The solid line is the linear least squares regression result for $\ln(CV)$ versus $\ln(V)$ which yielded the equation shown on the plot.

essentially an application of the law of large numbers from probability, in which the CV of a sample mean is inversely proportional to the square root of the sample size.

For example, consider a stochastic Ricker model with a negative binomial distribution, with mean b and variance σ^2, for the variation in the reproduction, R_i, for each individual i. For a negative binomial distribution, the variance always exceeds the mean, therefore it is often used in situations where one needs a distribution that is "overdispersed" relative to a Poisson distribution. It can be obtained as a compound distribution where it is assumed that the mean for the Poisson distribution has an underlying gamma probability distribution. In the case of the Ricker model, this might occur if there is random variation among individuals in b, the expected contribution to reproduction. The negative binomial has two parameters, one which can be used to specify the mean and another the variance. The stochastic model is given by

$$\begin{aligned} N_{t+1} &\sim \sum_{i=1}^{N_t} \text{Bin}\left(R_i, p_t\right) \\ &\sim \text{Bin}\left(\sum_{i=1}^{N_t} R_i, p_t\right) \\ &\sim \text{Bin}\left(X_t, p_t\right), \end{aligned} \tag{10.3}$$

where $X_t \sim \text{Nbin}\left(N_t b, N_t \sigma^2\right)$ represents a negative binomial distribution with mean $N_t b$ and variance $N_t \sigma^2$. This result comes from the fact that the sum of two or more binomial random variables, each with probability parameter p, is also a binomial random variable

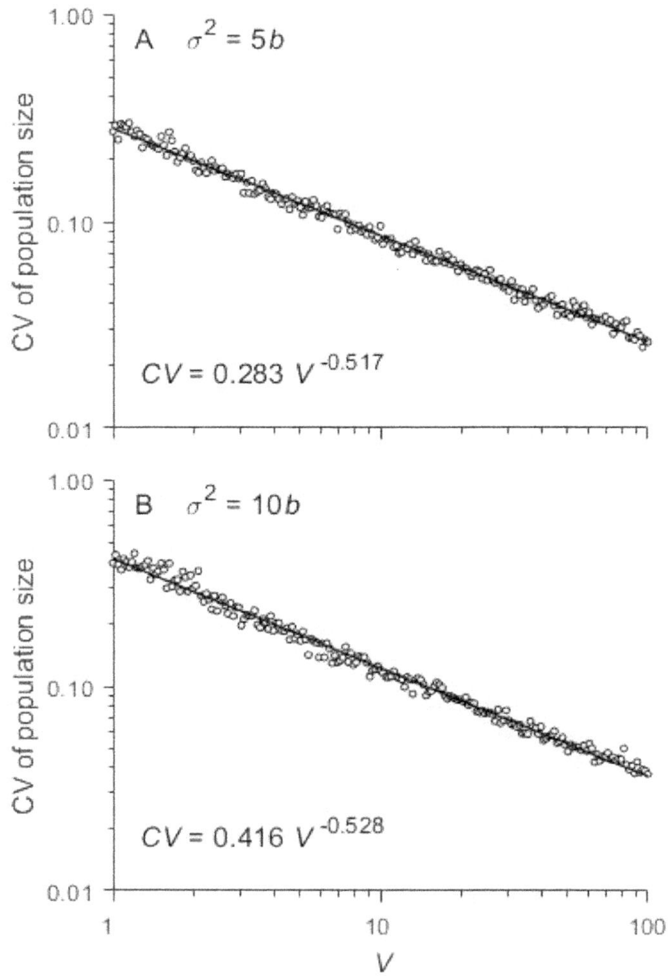

Figure 10.3 Demonstration of the inverse square root rule for the stochastic Ricker model (10.3) with $b = 1.2$ and $c = 0.001$ which yields a stable equilibrium for the deterministic Ricker model (10.1). For habitat sizes ranging from $V = 1$ to $V = 100$, the model was iterated 100 times to remove transients and then 1000 times to obtain the mean and variance of population numbers. The coefficient of variation (CV) of population numbers versus V is plotted on a log-log scale. The variance of the negative binomial distribution for reproduction was $5b$ in panel A and $10b$ in panel B. The solid lines are the linear least squares regression results for $\ln(CV)$ versus $\ln(V)$ which yielded the equations shown on the plots.

with probability parameter p; and the sum of N independent negative binomial random variables, each with mean b and variance σ^2, has a negative binomial distribution with mean Nb and variance $N\sigma^2$ [9].

Using the stochastic model (10.3), we repeated the simulation protocol for Figure 10.2 with negative binomial variances of $\sigma^2 = 5b$ and $\sigma^2 = 10b$. The results appear in Figure 10.3. In both these cases, the coefficient of variation from the model follows the inverse square root rule as habitat size, V, increases.

10.2 INVERSE SQUARE ROOT RULE FOR STRUCTURED POPULATIONS AND COMPLEX DYNAMICS

A limitation of the inverse square root rule is that it applies to populations with one age class fluctuating around a stable equilibrium. The mean for the quasi-stationary distribution of the population should approximate the equilibrium value, and so the coefficient of variation characterizes the overall magnitude of the population's long-term fluctuations near the equilibrium. However, if the population exhibits a cycle, or complex asymptotic dynamics such as an invariant loop or chaotic attractor, the variation in population numbers caused by deterministic behavior would be confounded with the variation due to demographic stochasticity. Additionally, if the population consisted of two or more age/stage classes, a simple coefficient of variation would not be applicable. How does one characterize the influence of demographic stochasticity in these situations?

One of the approaches we have proposed is to look at population density in state space and see how closely the population values cling to the predicted deterministic attractor [7]. Let $\mathbf{N}_t = [N_{1,t}, N_{2,t}, \ldots, N_{m,t}]$ represent an observed vector[1] of different age or stage classes at time t and let $A = \{\tilde{\mathbf{N}}_1, \tilde{\mathbf{N}}_2, \ldots, \tilde{\mathbf{N}}_\omega\}$ be the set of points comprising a deterministic ω-cycle attractor or a large finite set of points in the case of an aperiodic attractor. For an observed time series of length n, the mean minimum distance in state space between the data and attractor is defined as

$$d_S = n^{-1} \sum_{t=1}^{n} \min_{\tilde{\mathbf{N}}_i \in A} \left\| \mathbf{N}_t - \tilde{\mathbf{N}}_i \right\|,$$

where here we take

$$\|\mathbf{x} - \mathbf{y}\| := \sqrt{\sum_{j=1}^{m} (x_j - y_j)^2},$$

which is the Euclidean distance, in m-dimensional space, between two points $\mathbf{x} = [x_1, \ldots, x_m]$ and $\mathbf{y} = [y_1, \ldots, y_m]$. To compute d_S, for every observed data vector, one finds the nearest point on the attractor and computes the distance between the data point and the attractor. One then averages these distances over the time series. In practice, if the attractor is made up of an infinite number of points, one would need to sample a large enough number of points on the attractor so that the sample is representative of the attractor, that is, so that there are no significant gaps in state space. The mean minimum distance can be used to extend the inverse square root rule to structured populations and complex dynamics [7].

10.3 EXAMPLE USING THE LPA MODEL

Consider the following stochastic LPA lattice model that uses the Poisson and binomial distributions for demographic stochasticity:

$$\begin{aligned} L_{t+1} &\sim \text{Pois}\left(bA_t \exp\left(-\frac{c_{el}}{V}L_t - \frac{c_{ea}}{V}A_t\right)\right) \\ P_{t+1} &\sim \text{Bin}\left(L_t, (1-\mu_l)\right) \\ A_{t+1} &= \text{int}\left(P_t \exp\left(-\frac{c_{pa}}{V}A_t\right)\right) + \text{int}\left(A_t\left(1-\mu_a\right)\right), \end{aligned} \tag{10.4}$$

where int (x) refers to rounding x to the nearest integer. The random variables L_{t+1} and P_{t+1} are conditioned on the realized values L_t, P_t, and A_t from the previous time interval.

[1] We use the notation $[x, y]$ to denote the column vector $(x, y)^T$, which is the transpose of the row vector (x, y).

The derivation of the first equation for larvae follows the same reasoning as in the Ricker model (10.2), where b is the number of eggs produced per adult and eggs must survive cannibalism by both larvae and adults. The number of the pupae is described by a binomial distribution where each member of a cohort of L_t larvae is independently at risk of dying with probability μ_l. The data to be presented in the next section are from an experiment where we manipulated adult numbers using the expression from the deterministic LPA model with the numbers of new and surviving adults confined to integer values; this is reflected in the equation for A_{t+1}. For model (10.4), the mean minimum distance is computed using

$$d_S = n^{-1} \sum_{t=1}^{n} \min_{i\in[1,\omega]} \sqrt{\left(\frac{L_t}{V} - \frac{\tilde{L}_i}{V}\right)^2 + \left(\frac{P_t}{V} - \frac{\tilde{P}_i}{V}\right)^2 + \left(\frac{A_t}{V} - \frac{\tilde{A}_i}{V}\right)^2},$$

where $[\tilde{L}_1, \tilde{P}_1, \tilde{A}_1], [\tilde{L}_2, \tilde{P}_2, \tilde{A}_2], \ldots, [\tilde{L}_\omega, \tilde{P}_\omega, \tilde{A}_\omega]$ is the set of ω points that make up the attractor (a large finite sample of points in the case of an aperiodic attractor).

We conducted simulations to see if the mean minimum distance follows the inverse square root rule for complex population dynamics. We used model (10.4) with the maximum likelihood parameter values estimated in the route-to-chaos study [5] with $c_{pa} = 0.35$ and $\mu_a = 0.96$ which yields chaotic dynamics. Otherwise, our simulation protocol was the same as for stochastic Ricker model in Figure 10.2, with d_S replacing the coefficient of variation.

The results appear in Figure 10.4. For $V \geq 5$, the mean minimum distance from the chaotic attractor follows the inverse square root rule. However, for $V < 5$, there are systematic deviations from the inverse square root relationship. These are caused by lattice effects, which are prevalent when the habitat size is small (see Chapter 8). In this case, population numbers spend much of their time near the stable lattice cycles which are not on the chaotic attractor. As habitat size increases, the lattice cycles are closer to the continuous-state-space attractor and the inverse square rule applies. In Chapter 8, we examined a dominant lattice six-cycle for model (10.4) when $V = 1$. The asterisk on the vertical axis in Figure 10.4 marks the mean minimum distance between the lattice six-cycle and the chaotic attractor when $V = 1$.

We also tested the robustness of the inverse square root rule for the mean minimum distance from the simulation data to the attractor by substituting the negative binomial distribution for the Poisson distribution for the random production of eggs by adults. Following the derivation for the Ricker model (10.3), the equation for larval numbers in model (10.4) becomes

$$L_{t+1} \sim \text{Bin}\left(X_t, \exp\left(-\frac{c_{el}}{V} L_t - \frac{c_{ea}}{V} A_t\right)\right), \tag{10.5}$$

where $X_t \sim \text{Nbin}\left(A_t b, A_t \sigma^2\right)$. Using this model, we repeated the simulation protocol for Figure 10.4 with negative binomial variances of $\sigma^2 = 5b$ and $\sigma^2 = 10b$. The results appear in Figure 10.5. Once again, for both cases, the mean minimum distance from the chaotic attractor follows the inverse square root rule for $V \geq 5$ and the influence of lattice effects are obvious for $V < 5$. The inverse square root rule for d_S is robust to the details of how demographic stochasticity comes about.

For further exploration, see the examples for this section in the online LPA Simulator (https://lpasim.org/?c=10&s=3).

10.4 EXPERIMENTAL TEST OF THE INVERSE SQUARE ROOT RULE

We experimentally evaluated the scaling rule by comparing two sets of *Tribolium castaneum* populations with similar densities and the same deterministic dynamics, but different overall

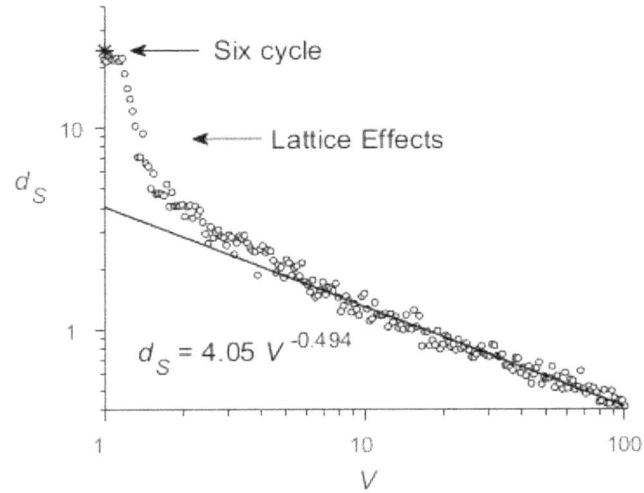

Figure 10.4 Demonstration of the inverse square root rule for the stochastic LPA model (10.4) with parameter values $b = 10.67$, $c_{el} = 0.01647$, $c_{ea} = 0.01313$, $c_{pa} = 0.35$, $\mu_l = 0.1955$, and $\mu_a = 0.96$. These parameter values yield chaotic population dynamics in the deterministic LPA model. To obtain sample points on the chaotic attractor, the deterministic LPA model (5.3) was iterated 1000 times to remove transients and then 10,000 times to obtain the attractor sample points. For habitat sizes ranging from $V = 1$ to $V = 100$, the model was iterated 100 times to remove transients and then 1000 times to compute d_S. The mean minimum distances, d_S, between the population numbers and the chaotic attractor versus V is plotted on a log-log scale. The solid line is the linear least squares regression result for $\ln(d_S)$ versus $\ln(V)$ in the range of $V = 5$ to $V = 100$ which yielded the equation shown on the plot. Lattice effects caused deviations from the inverse square root rule for $V < 5$. The asterisk on the y-axis is the mean minimum distance between the dominant lattice six-cycle for $V = 1$ and the chaotic attractor.

numbers. We maintained three replicate populations in 20 g of flour media ($V = 1$) and a second treatment of three populations in 60 g of media ($V = 3$). We subjected both sets of cultures to a laboratory protocol that fixed the adult mortality rate at $\mu_a = 0.96$ and the cannibalism rate of pupae at $c_{pa} = 0.35$ [4]. Previous work in the route-to-chaos experiment (Chapter 7) showed that these conditions produce chaotic population dynamics [4, 5]. The initial populations were composed of 250 larvae, five pupae, and 100 adults for the 20-g cultures and 114 larvae, five pupae, and 148 adults for the 60 g cultures. (The different initial stage numbers come from the fact that the 20 g cultures were also part of a separate study. However, all the populations quickly approached the chaotic attractor and initial numbers were excluded from the data analyses.) All cultures were maintained in a half-pint (237 mL) milk bottles and kept in a dark incubator at 32°C. Every two weeks the larval, pupal and adult stages were counted and returned to fresh media (95% bleached wheat flour and 5% dried brewer's yeast). To control for genetic change, at every other census, adults were replaced with an equal number from the laboratory stocks. This procedure was repeated for 108 weeks yielding a time series of 54 censuses for each replicate population.

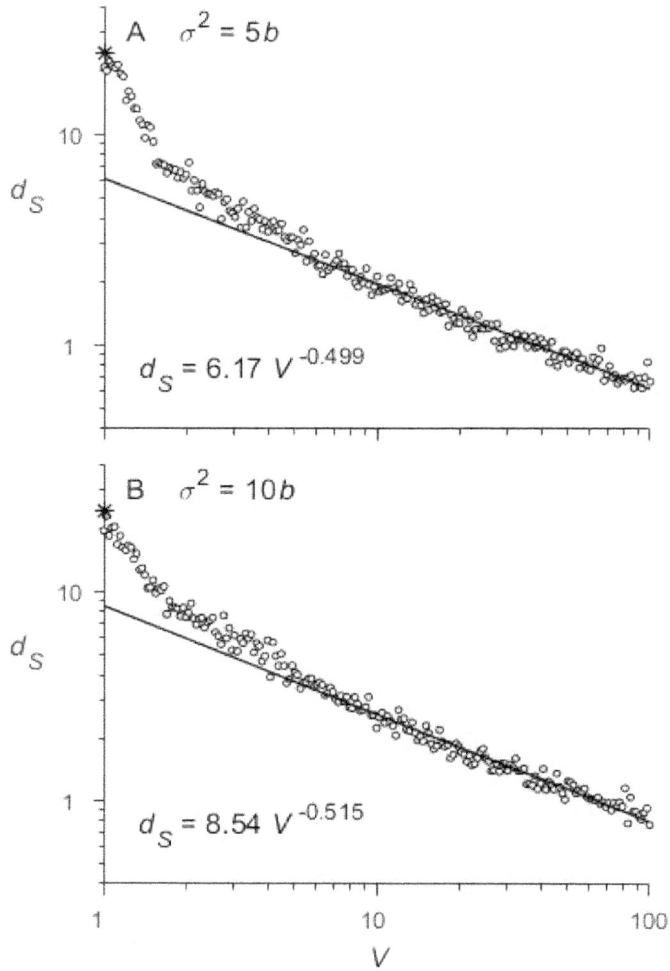

Figure 10.5 Demonstration of the inverse square root rule for the stochastic LPA model (10.5) with a negative binomial probability distribution for egg production. The parameter values and simulation protocol are the same as for Figure 10.4. The variance of the negative binomial distribution for reproduction was $5b$ in panel A and $10b$ in panel B.

The larval densities for the replicates from the three populations in each treatment appear in Figure 10.6. The solid circles connected by lines show the trajectories of each population. The open circles are the one-time-step values predicted by the deterministic LPA model (5.3) using the observed life stage values from the previous census. We used maximum likelihood parameters from the route-to-chaos experiment (Chapter 7) [5] using the stochastic model (10.4), with $\mu_a = 0.96$ and $c_{pa} = 0.35$ as determined by the experimental protocol.

Overall, the deviations between the observed and predicted values were smaller for the populations in the larger habitat ($V = 3$). Figure 10.7 shows the observed population values (open circles) together with the predicted chaotic attractor (solid points) in state

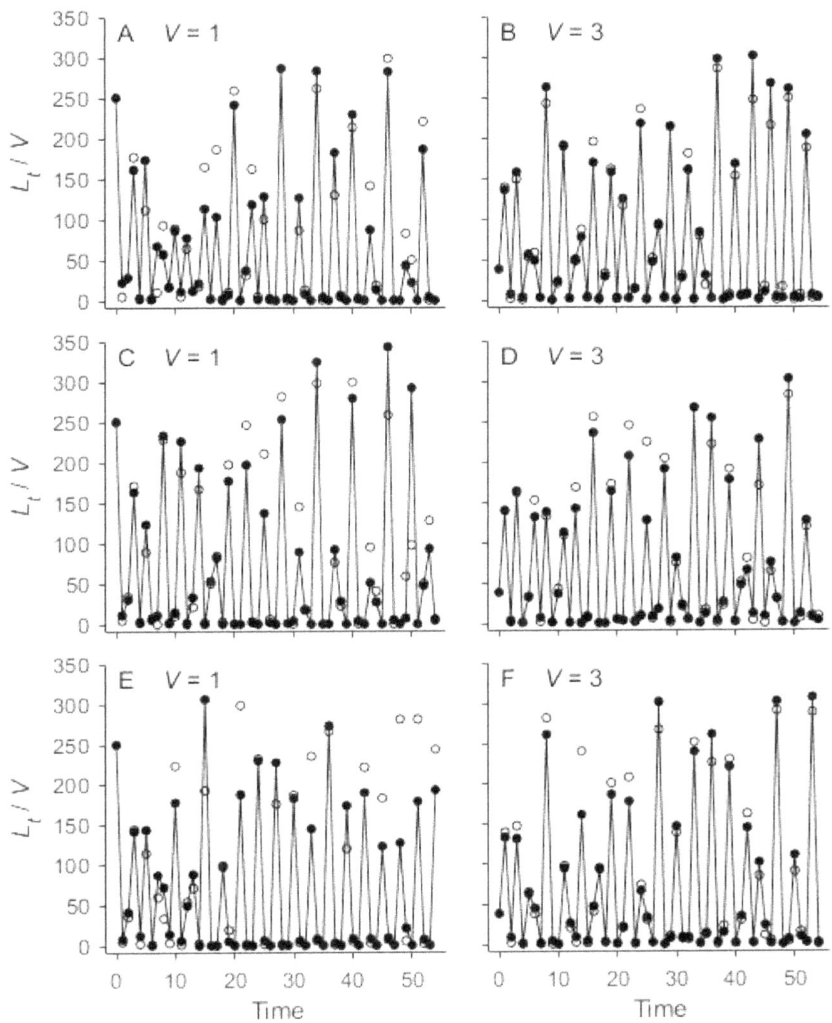

Figure 10.6 Time series of larval densities for six experimental populations. Closed circles are the observed densities and open circles are the one-step predictions for the deterministic LPA model (5.3) with parameter values as specified in Figure 10.4. On the left are the populations cultured in 20 g of medium ($V = 1$); on the right are the populations cultured in 60 g of medium ($V = 3$).

space. For $V = 3$, the data values cling more closely to the predicted attractor. For $V = 1$, the mean minimum distance is $d_S = 16.65$. This value decreased to $d_S = 8.92$ for $V = 3$, indicating a reduction in the influence of demographic noise for population density in the larger habitat. The d_S values for the Poisson-binomial LPA model (Figure 10.4) underestimate the observed deviations from the attractor, suggesting that other mechanisms of demographic stochasticity are needed [12]. Nevertheless, in these data the ratio of the two d_S values is $8.92/16.65 = 0.534$, which is not far from the value $1/\sqrt{3} = 0.577$ predicted by the inverse square root rule.

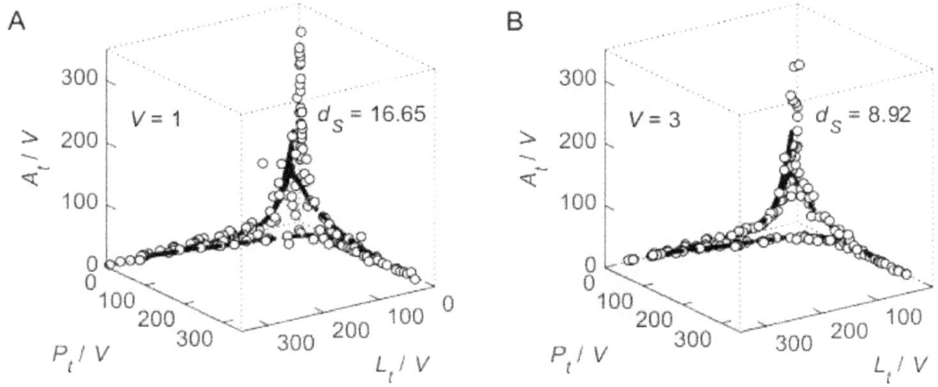

Figure 10.7 Observed life-stage densities (open circles) and model-predicted chaotic attractor (filled points) in state space. (A) 20 g of flour medium ($V = 1$). (B) 60 g of flour medium ($V = 3$). The mean minimum distance d_S of the state variables from the chaotic attractor is listed in each panel.

10.5 CONCLUSIONS

The influence of demographic stochasticity on population densities decreases inversely with the square root of the overall population numbers as determined by habitat size. For populations with a single age class fluctuating around a stable equilibrium, the influence of demographic stochasticity can be measured using the coefficient of variation in population numbers. For populations with several age/stage groups and complex dynamics, the influence of demographic stochasticity can be measured using the mean minimum distance of population densities from the predicted deterministic attractor in state space. This prediction is robust to the type of mechanisms used to produce demographic fluctuations. Data from a laboratory experiment using flour beetles at two different habitat sizes fluctuating around a chaotic attractor support the inverse square root rule. The results describe how the noise to signal ratio for deterministic dynamics decreases as habitat size increases.

Bibliography

[1] Anderson R, Gordon DM, Crawley M, Hassell, MP (1982) Variability in the abundance of animal and plant species. Nature 296:245–248. https://doi.org/10.1038/296245a0

[2] Bartlett MS (1960) Stochastic Population Models. London, Methuen.

[3] Costantino RF, Desharnais, RA (1991) Population Dynamics and the *Tribolium* Model: Genetics and Demography. Monographs on Theoretical and Applied Genetics 13. New York, Springer-Verlag.

[4] Costantino RF, Desharnais RA, Cushing JM, Dennis B (1997) Chaotic dynamics in an insect population. Science 275:389–391. https://doi.org/10.1126/science.275.5298.389

[5] Dennis B, Desharnais RA, Cushing JM, Henson SM, Costantino RF (2001) Estimating chaos and complex dynamics in an insect population. Ecological Monographs 71:277–303. https://doi.org/10.1890/0012-9615(2001)071[0277:ECACDI]2.0.CO;2

[6] Desharnais RA, Costantino RF (1982) The approach to equilibrium and the steady-state probability distribution of adult numbers in *Tribolium brevicornis*. American Naturalist 119:102–111. https://doi.org/10.1086/283893

[7] Desharnais RA, Costantino RF, Cushing JM, Henson SM, Dennis B, King AA (2006) Experimental support of the scaling rule for demographic stochasticity. Ecology Letters 9:537–547. https://doi.org/10.1111/j.1461-0248.2006.00903.x

[8] Desharnais RA, Henson SM, Costantino RF, Dennis B (2023) Capturing chaos: A multidisciplinary approach to nonlinear population dynamics. Journal of Difference Equations and Applications 1–38. https://doi.org/10.1080/10236198.2023.2260013

[9] Johnson NL, Kemp WK, Kotz S (2005) Univariate Discrete Distributions. Third Edition. Hoboken, New Jersey, John Wiley & Sons.

[10] Leslie PH (1958) A stochastic model for studying the properties of certain biological systems by numerical methods. Biometrika 45:16–31. https://doi.org/10.2307/2333042

[11] May RM (1973) Stability and Complexity in Model Ecosystems. Princeton, New Jersey, Princeton University Press.

[12] Melero AI (2008) Explicit Stochastic Models for the Population Dynamics of *Tribolium*. MS Thesis, California State University at Los Angeles.

[13] Nisbet RM, Gurney WSC (1982) Modelling Fluctuating Populations. Chichester, England, Wiley.

[14] Watt KEF (1968) Ecology and Resource Management. New York, McGraw Hill.

CHAPTER 11

Controlling Chaos

A key feature of chaotic systems is their sensitivity to initial conditions. A small perturbation will have a large impact on the long-term trajectory of the system. Can one use this sensitivity to bring about desired changes in the dynamics of a population? How and when should controlling perturbations be applied? In this chapter, we show that the sensitivity of a chaotic system varies in state space and this can be measured using the *Local Lyapunov Exponent* (LLE). We show how a model-guided understanding of state-space sensitivity leads to a protocol where small perturbations are used to bring about large changes in a population's dynamics and provide an experimental demonstration using the *Tribolium* laboratory system.

11.1 SENSITIVITY TO INITIAL CONDITIONS

An intriguing attribute of chaotic systems is their sensitivity to initial conditions. For a chaotic attractor, two trajectories, initiated on the attractor but with slightly different initial conditions, will diverge from one another at an exponential rate. This means that any error in the estimation of the state of a chaotic system will be magnified with time, making long-term predictions difficult, even in the absence of stochastic noise. Edward Lorenz, one of the pioneers of chaos theory, posed the following question [10]: "Does the flap of a butterfly's wings in Brazil set off a tornado in Texas?" Based on this analogy and its popularization by James Gleick [7], the sensitivity of chaos to perturbations is often referred to as the "butterfly effect."

One way to measure the rate of divergence (or convergence) of two nearby trajectories in a dynamic system is by means of the *Lyapunov exponent* λ. The Lyapunov exponent (LE) measures the rate of divergence of two trajectories with infinitesimally small differences in their initial conditions averaged over the attractor. Consider a system of difference equations of the form

$$\begin{aligned} x_{1,t+1} &= f_1\left(x_{1,t}, x_{2,t}, \ldots, x_{s,t}\right), \\ &\vdots \\ x_{s,t+1} &= f_s\left(x_{1,t}, x_{2,t}, \ldots, x_{s,t}\right). \end{aligned} \tag{11.1}$$

Let $\mathbf{x}_t = [x_{1,t}, \ldots, x_{s,t}]$ be the vector[1] of state variables at time t. Define the matrix J_t to

[1] We use the square-bracket notation $[x_{1,t}, \ldots, x_{s,t}]$ to mean the column vector $(x_{1,t}, \ldots, x_{s,t})^T$, which is the transpose of the row vector $(x_{1,t}, \ldots, x_{s,t})$.

DOI: 10.1201/9781003467120-11

be the Jacobian of the system (11.1) at time t, i.e.,

$$J_t = \begin{pmatrix} \frac{\partial f_1}{\partial x_1} & \cdots & \frac{\partial f_1}{\partial x_s} \\ \vdots & \ddots & \vdots \\ \frac{\partial f_s}{\partial x_1} & \cdots & \frac{\partial f_s}{\partial x_s} \end{pmatrix}_{\mathbf{x}=\mathbf{x}_t} . \tag{11.2}$$

The LE can be defined as

$$\lambda = \lim_{t\to\infty} \frac{1}{t} \ln \|J_t J_{t-1} \cdots J_2 J_1\| \tag{11.3}$$

for any initial condition $\mathbf{x}_0$ that converges onto the attractor [3]. Here "$\|\cdot\|$" denotes any norm of the matrix (for example, the square root of the sum of the squares of the elements of the matrix or the sum of the absolute values of the components). A positive LE ($\lambda > 0$) indicates an overall tendency for nearby trajectories to diverge on the attractor; this is often used as an indicator of possible chaotic dynamics. A negative LE ($\lambda < 0$) means that nearby trajectories converge, as is the case for stable equilibria and periodic cycles. A zero LE ($\lambda = 0$) occurs for aperiodic trajectories on an invariant loop.

11.1.1 Ricker Model

As an example, consider the deterministic discrete-time Ricker model

$$x_{t+1} = bx_t \exp(-cx_t) . \tag{11.4}$$

For this system with only one state variable, the Jacobian at time t is simply the derivative of the right side of (11.4), $bx \exp(-cx)$, with respect to x evaluated at $x = x_t$, namely

$$J_t = \left. \frac{d(bx\exp(-cx))}{dx} \right|_{x=x_t} = b\exp(-cx_t)(1 - cx_t) . \tag{11.5}$$

A substitution of expression (11.5) into equation (11.3) gives

$$\lambda = \ln b + \lim_{t\to\infty} \frac{1}{t} \sum_{i=1}^{t} (\ln|1 - cx_i| - cx_i). \tag{11.6}$$

For $b = 30$, the attractor is chaotic with a positive Lyapunov exponent $\lambda = 0.470$.

Figure 11.1A shows two trajectories for the chaotic Ricker model with different initial densities. The filled circles are the simulation results for an initial condition of $x_0 = 150$, whereas the open circles are for an initial population density of $\tilde{x}_0 = 150.1$. The two time series remain close until $t = 11$ when their divergence becomes more obvious. For $t \geq 15$, there is little synchrony between the densities of the two populations.

11.1.2 LPA Model

The LPA model also exhibits sensitivity to initial conditions in the chaotic region of parameter space studied in the route-to-chaos experiment [2, 3] discussed in Chapter 7. For a habitat size of $V = 1$, we have

$$\begin{aligned} L_{t+1} &= bA_t \exp(-c_{el}L_t - c_{ea}A_t) \\ P_{t+1} &= L_t(1 - \mu_l) \\ A_{t+1} &= P_t \exp(-c_{pa}A_t) + A_t(1 - \mu_a) . \end{aligned} \tag{11.7}$$

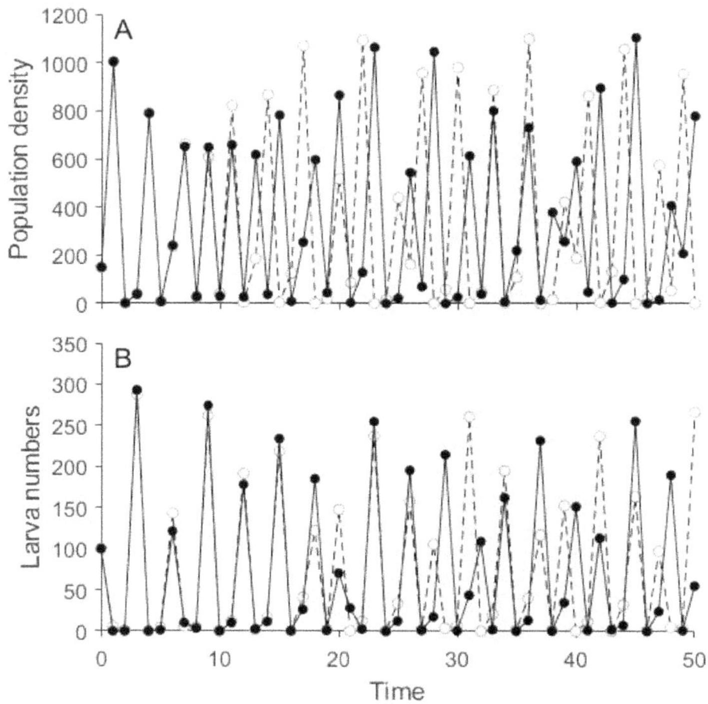

Figure 11.1 Demonstration of sensitivity to initial conditions for two chaotic population models. A. The Ricker model (11.4) with $b = 30$, $c = 0.01$, $x_0 = 150$ (solid circles), $\tilde{x}_0 = 150.1$ (open circles). B. The LPA model (11.7) with $b = 10.45$, $c_{el} = 0.01731$, $c_{ea} = 0.01310$, $c_{pa} = 0.35$, $\mu_l = 0.2000$, $\mu_a = 0.96$, $L_0 = \tilde{L}_0 = 100$, $P_0 = \tilde{P}_0 = 0$, $A_0 = 0$ (solid circles), $\tilde{A}_0 = 3$ (open circles).

The Jacobian matrix for this model is

$$J_t = \begin{pmatrix} -c_{el} b A_t g_{1t} & 0 & b\,(1 - c_{ea} A_t)\, g_{1t} \\ 1 - \mu_l & 0 & 0 \\ 0 & g_{2t} & -c_{pa} P_t g_{2t} + (1 - \mu_a) \end{pmatrix}, \tag{11.8}$$

where $g_{1t} = \exp\left(-c_{el} L_t - c_{ea} A_t\right)$ and $g_{2t} = \exp\left(-c_{pa} A_t\right)$. In the route-to-chaos experiment, the following parameter estimates were obtained: $b = 10.45$, $c_{el} = 0.01731$, $c_{ea} = 0.01310$, and $\mu_l = 0.2000$. For the remaining two parameters, the experimental protocol included a treatment group with $c_{pa} = 0.35$ and $\mu_a = 0.96$, yielding chaotic dynamics. Using these values in equations (11.8) and (11.3) we find that $\lambda = 0.096$.

Figure 11.1B shows two trajectories for the chaotic LPA model (11.7) with different initial densities. The filled circles are the simulation results for an initial condition of $[L_0, P_0, A_0] = [100, 0, 0]$, whereas the open circles are for an initial population density of $[\tilde{L}_0, \tilde{P}_0, \tilde{A}_0] = [100, 0, 3]$. The two time series remain close until $t = 18$ when their divergence becomes more obvious. For $t \geq 28$, there is little synchrony between the densities of the two populations.

11.2 USING SENSITIVITY TO CONTROL CHAOS

Since chaotic systems are sensitive to initial conditions, it was been suggested that small perturbations might be used to influence their dynamics [9, 11, 14]. The idea is to "nudge" the parameters or state variables at points in the trajectory where the system is sensitive to changes, producing a large desired effect from small perturbations. Ecologists have suggested that this concept could be applied to populations with chaotic dynamics [5, 8, 15, 16].

11.2.1 Local Lyapunov Exponent

The LE (11.3) measures the overall rate of divergence (or convergence) of two nearby trajectories as they move through state space on their approach to the attractor. If the attractor is a stable equilibrium point, $\mathbf{x}^* = [x_1^*, x_2^*, \ldots, x_s^*]$, of the system (11.1), then the LE is equal to the logarithm of the modulus of the largest eigenvalue of the Jacobian matrix (11.2), evaluated at $\mathbf{x}_t = \mathbf{x}^*$. In this case, λ is the rate of convergence of $\mathbf{x}_t$ to $\mathbf{x}^*$ in the neighborhood of the equilibrium. The *local Lyapunov exponent* (LLE) extends this idea to other points in state space. For any point $\mathbf{x}$ in state space, the LLE $\lambda_{\mathbf{x}}$ is defined as the logarithm of the modulus of the largest eigenvalue of the Jacobian matrix (11.2), evaluated at $\mathbf{x}_t = \mathbf{x}$. The LLE measures the effect of a small perturbation on the system trajectory at the point $\mathbf{x}$.

In chaotic systems, a subset of points on the attractor will have positive LLEs ($\lambda_{\mathbf{x}} > 0$), accounting for the overall sensitivity to initial conditions.[2] The idea behind controlling chaos is to apply small perturbations to the system whenever it enters a region of state space where the LLE is positive. Small perturbations in this region can have a large effect on the system.

11.2.2 Ricker Model

The LLE for the Ricker model (11.4) is

$$\lambda_x = \ln(b) - cx + \ln|1 - cx|, \tag{11.9}$$

which was obtained by taking the logarithm of the absolute value of the Jacobian (11.5) with x in place of x_t. The LLE for the Ricker model is plotted in Figure 11.2 as a function of population density x, for $b = 30$ and $c = 0.01$. The LLE starts out with large positive values, but decreases to $-\infty$ as x approaches $c^{-1} = 100$. The LLE increases to positive values again, peaks at $x = 200$, and decreases again, becoming negative for large population densities. Small perturbations applied when x is near zero should have a large effect on population trajectories.

To demonstrate the use of small perturbations, denoted by Δ_t, to change the dynamics of a chaotic system, we implemented the following version of the Ricker model:

$$\begin{aligned} X_{t+1} &= bx_t \exp(-cx_t) \\ \Delta_{t+1} &= \begin{cases} \delta, & \text{if } X_{t+1} < x_c \\ 0, & \text{if } X_{t+1} \geq x_c \end{cases} \\ x_{t+1} &= X_{t+1} + \Delta_{t+1}. \end{aligned} \tag{11.10}$$

Whenever population density falls below a critical value, x_c, a small perturbation δ is added to the population density. Figure 11.3A shows a simulation of the unperturbed Ricker model (11.4) for $b = 30$, $c = 0.01$, and $x_0 = 150$. Figure 11.3B shows a simulation of the model (11.10) using the same parameter values and initial condition as in Figure 11.3A, with

[2]If all the points on the attractor had negative LLEs, then the iterates of a set of points arbitrarily near a point on the attractor would eventually occupy an increasingly smaller volume, since nowhere on the attractor can the points diverge. Thus a positive dominant LE implies that least some subset of points on the attractor have positive LLEs. See Section 7.2.

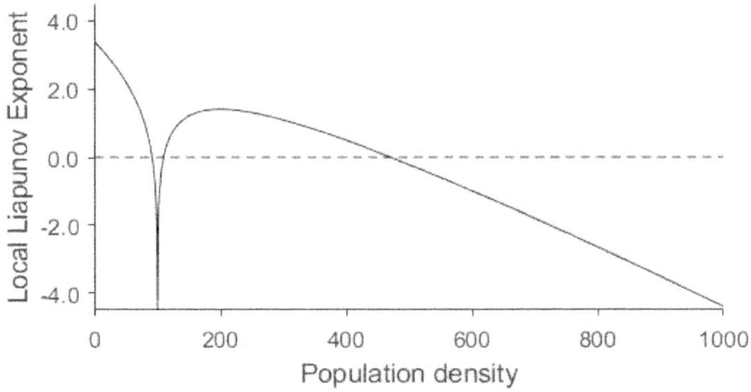

Figure 11.2 The local Lyapunov exponent (11.9) for the Ricker model (11.4) with $b = 30$ and $c = 0.01$ as a function of population density x.

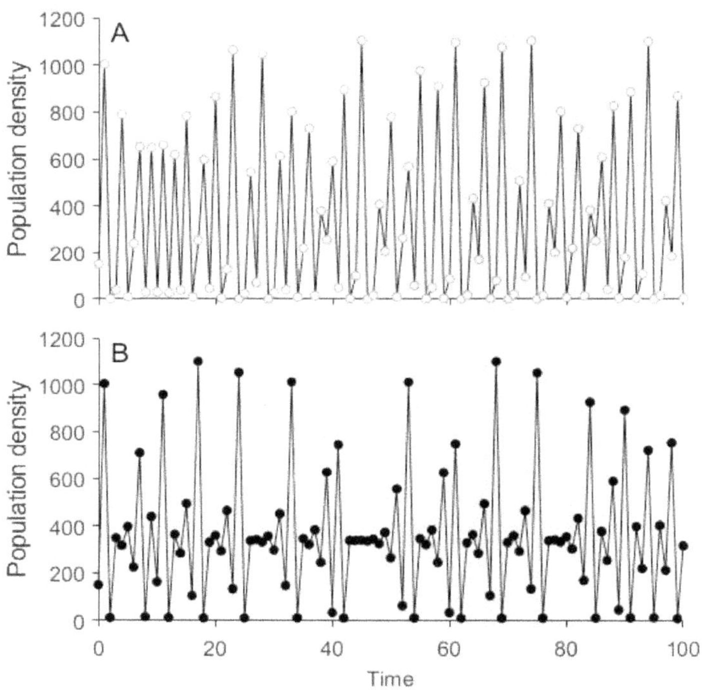

Figure 11.3 Simulations of the Ricker model demonstrating how small perturbations can alter the trajectory of a chaotic system. A. Simulation of the Ricker model (11.4) with $b = 30$, $c = 0.01$, and $x_0 = 150$. B. Simulation of the perturbed Ricker model (11.10) using the same parameter values and initial condition as in panel A, but with $x_c = 10$ and $\delta = 12$.

$x_c = 10$ and $\delta = 12$. The perturbed system spends more time near the unstable equilibrium $x^* = (\ln b)/c \approx 340.1$ with some widely-spaced outbreaks.

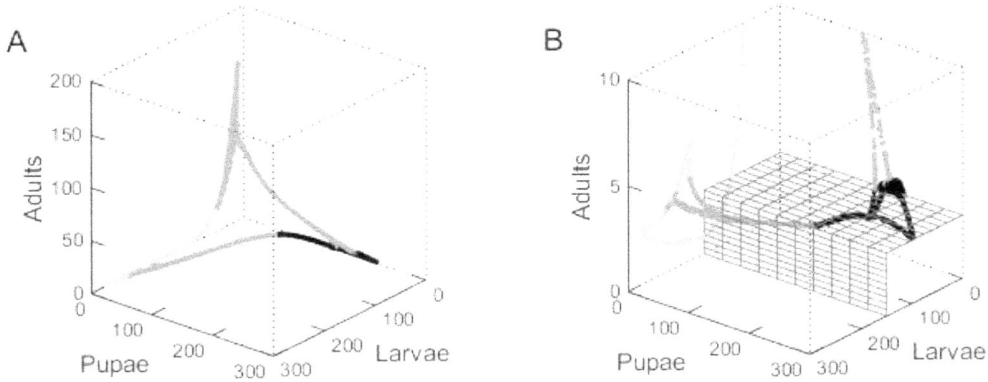

Figure 11.4 A. Distribution of local Lyapunov exponents (LLEs) for 2000 points on the chaotic attractor for the LPA model (11.7) with parameter values given in Figure 11.1B. Points on the attractor are colored light gray for negative LLEs ($\lambda_{\mathbf{x}} < 0$), dark gray for small to moderate positive LLEs ($0 \leq \lambda_{\mathbf{x}} \leq 3$), and black for large LLEs ($\lambda_{\mathbf{x}} > 3$). B. Magnified region near the base of the attractor. The grid lines circumscribe a "hot region" ($L \leq 150$ and $A \leq 3$) where the largest positive LLEs are concentrated.

11.2.3 LPA Model

To obtain the LLE for the LPA model (11.7) for any point $[L, P, A]$ in state space, one substitutes $L = L_t$, $P = P_t$, and $A = A_t$ into the Jacobian (11.8) and computes the three eigenvalues ϕ_1, ϕ_2, and ϕ_3 for the matrix. The LLE is given by

$$\lambda_{\mathbf{x}} = \ln \max \left(|\phi_1|, |\phi_2|, |\phi_3| \right), \tag{11.11}$$

for $\mathbf{x} = [L, P, A]$. The modulus $|\phi|$ equals the absolute value of the eigenvalue if it is real, and, if ϕ is complex, $|\phi| = \sqrt{\alpha^2 + \beta^2}$ for $\phi = \alpha \pm \beta i$.

Using the same parameter values as in Figure 11.1B, we computed the LLE for 2000 points on the chaotic attractor. The values of $\lambda_{\mathbf{x}}$ ranged from -1.03 to 3.92. Figure 11.4A shows these values coded by color: light gray for $\lambda_{\mathbf{x}} < 0$, dark gray for $0 \leq \lambda_{\mathbf{x}} \leq 3$, and black for $\lambda_{\mathbf{x}} > 3$. The largest LLEs occur where larval and adult numbers are small and numbers of pupae are large. In Figure 11.4B, the scale for adult numbers is modified to magnify the base of the attractor. The grid represents a region of state space where $L \leq 150$ and $A \leq 3$. This "hot region" is where the largest positive LLEs are concentrated.

Figure 11.5 demonstrates how small perturbations can drastically alter the dynamics of the chaotic LPA model. Figure 11.5A shows the larval time series for the unperturbed model with an initial condition $[L_0, P_0, A_0] = [250, 5, 100]$. Figure 11.5B shows the simulation points in state space for $t = 9$ to $t = 200$ using the same parameter values and initial condition. The initial time series for $t \leq 8$ was omitted to remove transients; the remaining points outline the deterministic chaotic attractor. We then implemented a simulation using the same parameter values and initial condition, but with the following perturbation scheme:

$$\begin{aligned} Y_{t+1} &= P_t \exp\left(-c_{pa} A_t\right) + A_t \left(1 - \mu_a\right) \\ \Delta_{t+1} &= \begin{cases} 3 & \text{if} \quad Y_{t+1} \leq 3 \quad \text{and} \quad L_{t+1} \leq 150 \\ 0 & \text{otherwise} \end{cases} \\ A_{t+1} &= Y_{t+1} + \Delta_{t+1}. \end{aligned} \tag{11.12}$$

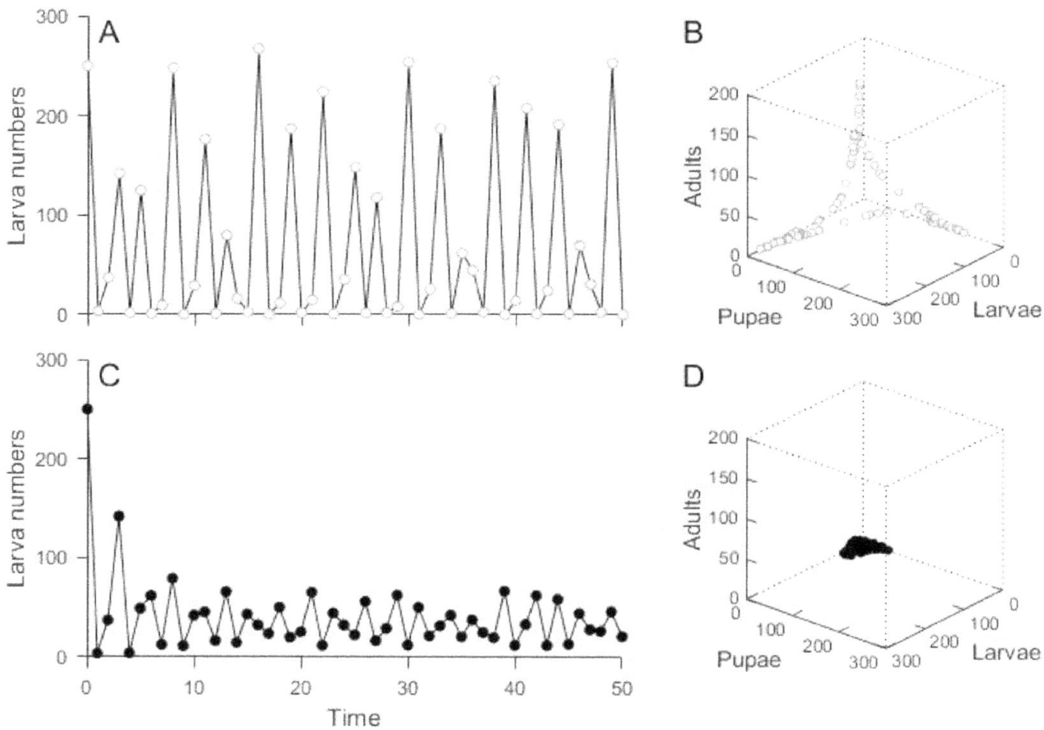

Figure 11.5 Simulations of the LPA models demonstrating how small perturbations can alter the trajectory of a chaotic system. A. Larval numbers from a simulation of the autonomous LPA model (11.7) with parameter values given in Figure 11.1B. B. Simulation results from the autonomous LPA model (11.7) plotted in state space. C. Simulation of the perturbed LPA model (11.12) using the same parameter values and initial condition as in panel A, but with three adults added whenever the system enters the hot region of state space (Figure 11.4B). D. Simulation results from the perturbed LPA model (11.12) plotted in state space.

with L_{t+1} and P_{t+1} computed using equations (11.7). In other words, if the population fell into the hot region of the attractor (Figure 11.4B), then three adults were added, otherwise, no perturbation was applied. Figure 11.5C shows the larval time series for the perturbed model (11.12). The magnitudes of the fluctuations were greatly reduced. Figure 11.5D shows the simulation points for the perturbation model in state space for $t = 9$ to $t = 200$. The simulation points are clustered around the unstable equilibrium $[L^*, P^*, A^*] = [28.0, 22.4, 4.6]$ of the unperturbed model (11.7). Small perturbations applied in the hot region of the attractor had a large impact on the trajectory of the population model.

11.3 EXPERIMENTALLY CONTROLLING CHAOS

The idea of controlling chaos has been tested experimentally in a variety of physical, electrical, and chemical systems [1, 13], but there are only a few cases of its application to biology [6, 12]. We are aware of only one example in ecology: our experimental test using *Tribolium* and the LPA model [4], the results of which are summarized below.

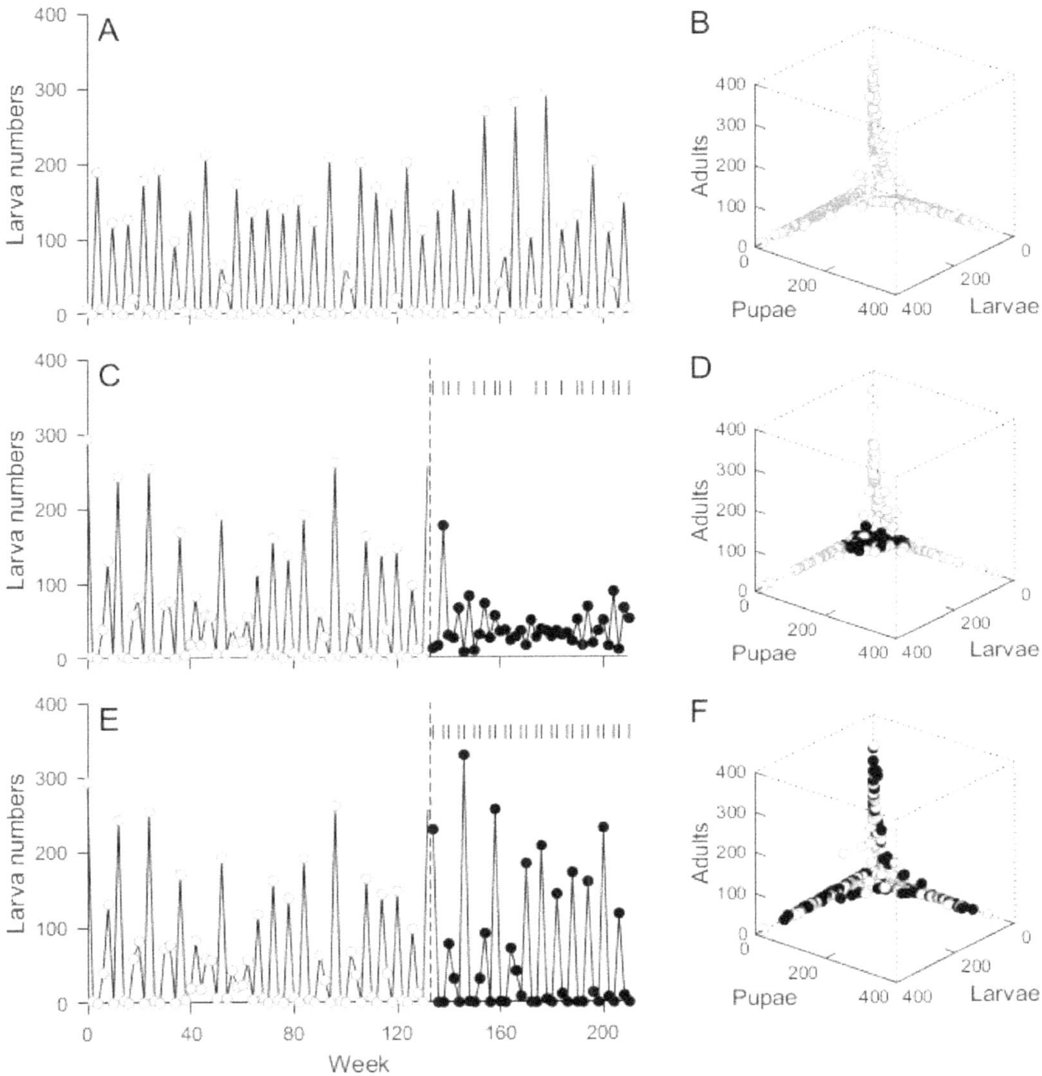

Figure 11.6 Results of the control-of-chaos experiment. A-B. Control. C-D. In-box treatment. E-F. Out-box treatment. The panels on the left show the larval numbers for one replicate from each treatment; the panels on the right show the data for all the replicates in state space. The open circles are for unperturbed populations. The solid circles are for the populations subjected to the perturbations starting at week 134 (dashed lines). The vertical ticks marks on panels C and E indicate when the perturbations were applied.

11.3.1 Experimental Design

Nine flour beetle populations were cultured using the protocol from the route-to-chaos experiment with $c_{pa} = 0.35$ and $\mu_a = 0.96$ (see Chapter 7) that placed them in the regime of chaotic dynamics [2, 3]. After 132 weeks, the nine populations were randomly divided into three treatments, each with three replicates: (1) an "in-box" treatment where three adults were added whenever the populations fell into the hot region of state space (Figure 11.4); (2) an "out-box" treatment where three adults were added whenever the populations fell outside the hot region of state space; and (3) and unperturbed control. The experiment was terminated at week 210.

The in-box treatment mirrors the protocol (11.12). The out-box treatment used following perturbation scheme:

$$\begin{aligned} Y_{t+1} &= P_t \exp\left(-c_{pa} A_t\right) + A_t \left(1 - \mu_a\right) \\ \Delta_{t+1} &= \begin{cases} 3 & \text{if} \quad Y_{t+1} > 3 \quad \text{or} \quad L_{t+1} > 150 \\ 0 & \text{otherwise} \end{cases} \\ A_{t+1} &= Y_{t+1} + \Delta_{t-1}, \end{aligned} \tag{11.13}$$

with L_{t+1} and P_{t+1} computed using equations (11.7). The out-box treatment serves as a negative control; perturbations only occur when the system is outside the most sensitive region of state space.

11.3.2 Experimental Results

The results appear in Figure 11.6. The panels on the left side show the time series for larval numbers of one replicate from each treatment. The panels on the right side show state space plots for all the data from each treatment. The open and closed circles are for unperturbed and perturbed populations, respectively. Once the in-box protocol (11.12) was initiated at week 132, it had the predicted effect of reducing the magnitude of the life-stage fluctuations (Figure 11.6B,D). The out-box protocol (11.13) had no obvious effect on the populations (Figure 11.6E,F), despite the fact that perturbations were applied more frequently (vertical ticks in Figure 11.6). This experiment provides concrete support for the idea that small carefully chosen perturbations can elicit large desirable changes in chaotic population dynamics.

11.4 CONCLUSIONS

Populations with chaotic dynamics are sensitive to initial conditions. This means that a small nudge to the system can have a large impact on the future trajectory of the population. This sensitivity may vary over state space, with "hot" and "cold" regions where interventions have bigger or smaller impacts. This phenomenon can be exploited by using small well-timed perturbations to alter a chaotic population's dynamics toward a desirable outcome. This idea was demonstrated using simulations of the Ricker and LPA models and verified experimentally using laboratory populations of the flour beetle *Tribolium.*

Bibliography

[1] Boccaletti S, Grebogi C, Lai YC, Mancini H, Maza D (2000) The control of chaos: Theory and applications. Physics Reports 329:103–197. https://doi.org/10.1016/S0370-1573(99)00096-4

[2] Costantino RF, Desharnais RA, Cushing JM, Dennis B (1997) Chaotic dynamics in an insect population. Science 275:389–391. https://doi.org/10.1126/science.275.5298.389

[3] Dennis B, Desharnais RA, Cushing JM, Henson SM, Costantino RF (2001) Estimating chaos and complex dynamics in an insect population. Ecological Monographs 71:277–303. https://doi.org/10.1890/0012-9615(2001)071[0277:ECACDI]2.0.CO;2

[4] Desharnais, RA, Costantino RF, Cushing JM, Henson SM, Dennis B (2001) Chaos and population control of insect outbreaks. Ecology Letters 4:229–235. https://doi.org/10.1046/j.1461-0248.2001.00223.x

[5] Doebeli M (1993) The evolutionary advantage of controlled chaos. Proceedings of the Royal Society of London 254:281–285. https://doi.org/10.1098/rspb.1993.0157

[6] Garfinkel A, Spano ML, Ditto WL, Weiss JN (1992) Controlling cardiac chaos. Science 257:1230–1235. https://doi.org/10.1126/science.1519060

[7] Gleick, J (1987) Chaos: Making a New Science. New York, Viking.

[8] Hilker FM, Westerhoff FH (2007) Preventing extinction and outbreaks in chaotic populations. American Naturalist 170:232–241. https://www.journals.uchicago.edu/doi/10.1086/518949

[9] Kapitaniak, T (1996) Controlling Chaos. San Diego, Academic Press.

[10] Lorenz EN (1985) The butterfly effect. In Abraham R, Yoshisuke U (eds.), The Chaos Avant-garde: Memoirs of the Early Days of Chaos Theory, First Edition, pp 91–94. Singapore, World Scientific Publishing Company.

[11] Ott E, Sauer T, Yorke JA (1994). Coping with Chaos. New York, John Wiley.

[12] Schiff, SJ, Jerger K, Duong DH, Chang T, Spano ML, Ditto WL (1994) Controlling chaos in the brain. Nature 370:615–620. https://doi.org/10.1038/370615a0

[13] Schöll E, Schuster HG, eds. (2007) Handbook of Chaos Control. First Edition. Weinheim, Wiley-VCH.

[14] Shinbrot T, Grebogi C, Ott E, Yorke, JA (1993). Using small perturbations to control chaos. Nature 363:411–417. https://doi.org/10.1038/363411a0

[15] Shulenburger L, Lai YC, Yalçinkaya T, Holt RD (1999) Controlling transient chaos to prevent species extinction. Physics Letters A 260:156–161. https://doi.org/10.1016/S0375-9601(99)00483-1

[16] Solé RV, Gamarra JG, Ginovart M, López D (1999) Controlling chaos in ecology: From deterministic to individual-based models. Bulletin of Mathematical Biology 61:1187–1207. https://doi.org/10.1006/bulm.1999.0141

IV

Recognizing Nonlinear Phenomena in Ecology

CHAPTER 12

Saddles and Manifolds

In this part of the book, Part IV, we consider whether nonlinear phenomena leave recognizable signatures in experimental time-series data. If one is presented with a time-series graph of data, can one detect patterns that suggest the possibility of certain types of nonlinear mechanisms, even if one does not yet have a mathematical model in hand? Of course, the eye often sees "patterns" in random sequences where no mechanistic patterns actually exist. Potential patterns and nonlinear phenomena must therefore be verified, either by statistical methods or by means of a rigorously validated mathematical model. But do we know how to spot potential nonlinear signatures in order to test them? In this chapter we demonstrate how one type of nonlinear structure in state space can be manifest in noisy population time series.

First let's examine the experimental time-series data for *Tribolium* shown in Figure 12.1. Here, we see animal numbers from two of four control replicates of an experiment conducted in 1978 by Jillson [9]. Replicate 13 features the noisy two-cycle-like fluctuations seen in three of the four replicates. In contrast, replicate 20 shows an early period of "flat," equilibrium-like dynamics, indicated by arrows in each life stage. Given that this occurs near the beginning of the experiment, one might think this culture is simply "settling in," becoming acclimatized to the experimental conditions. Or is this unique episode the signature of some nonlinear mechanism, and if so, why doesn't it appear in the other three control cultures?

Next consider a time series for *Tribolium* from an experiment conducted 20 years after the Jillson experiment, in a different laboratory. These data, shown in Figure 12.2, are from replicate 7 of the $c_{pa} = 0.05$ with $\mu_a = 0.96$ treatment of the route-to-chaos experiment described in Chapter 7 [1], where the LPA model is given by

$$\begin{aligned}
L_{t+1} &= bA_t \exp\left(-c_{el}L_t - c_{ea}A_t\right)\\
P_{t+1} &= L_t(1-\mu_l)\\
A_{t+1} &= P_t \exp\left(-c_{pa}A_t\right) + A_t(1-\mu_a).
\end{aligned}$$

At this value of c_{pa}, the LPA-predicted deterministic attractor is chaotic. The arrow in Figure 12.2 indicates a strange interlude of equilibrium-like dynamics that occurs in all three life stages. Clearly this occurrence is not due to the culture "settling in," given that it suddenly appears about 16 weeks into the experiment. The episode lasts for 10 time steps, which is 20 weeks, or five months! Imagine how Bob Costantino, who was counting beetles every two weeks, felt when he saw this replicate culture remaining flat at this low level, "doing the wrong thing," fortnight after fortnight, while the other replicates in the treatment flourished. Was the culture diseased or perhaps fundamentally changed in some

DOI: 10.1201/9781003467120-12

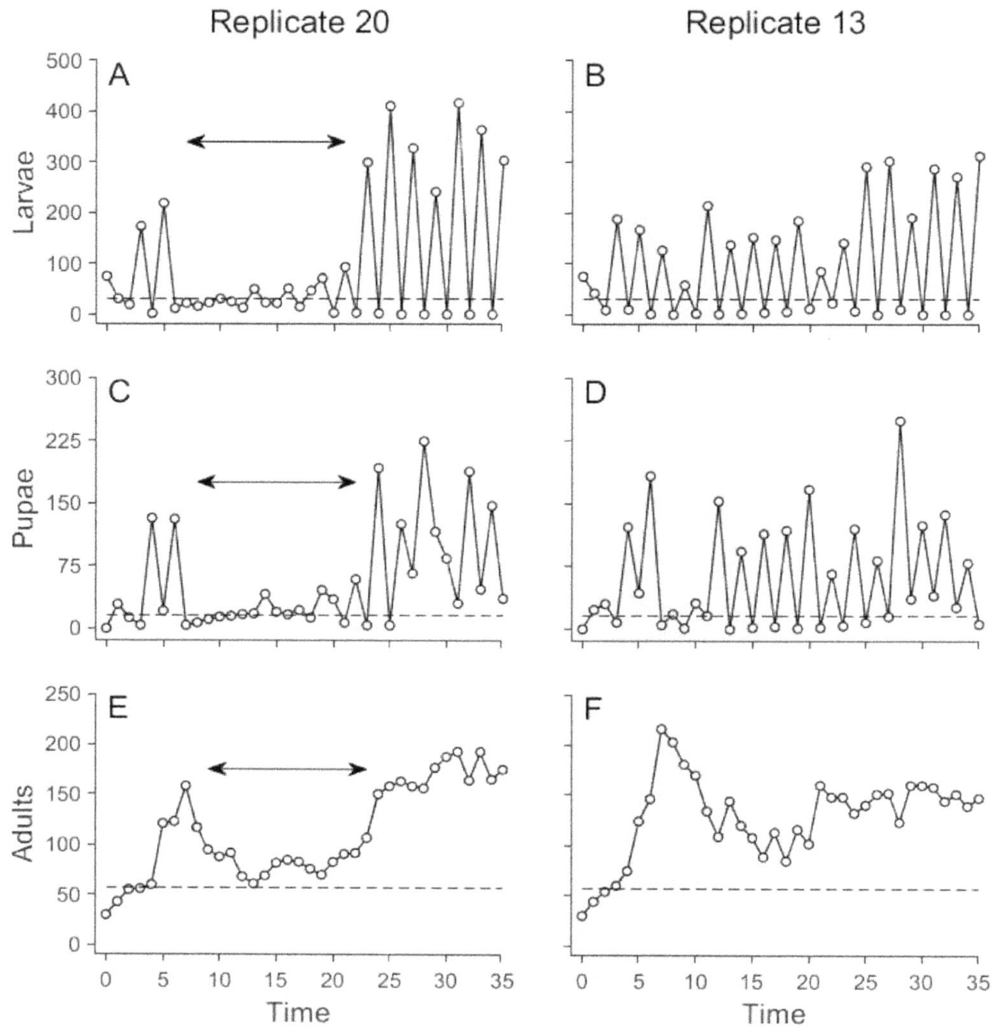

Figure 12.1 Experimental time series from two of four control replicates in the experiment conducted by Jillson [9]. Three of the cultures, including replicate 13, featured only noisy two-cycle fluctuations. Replicate 20, however, shows an early period of "flat," equilibrium-like dynamics (arrows).

way? Would you give up on this replicate? But after 10 censuses, the dynamics returned to their previous fluctuating vigor, leaving the question, "What just happened?"

For a third example, consider Figure 12.3. This also is from the route-to-chaos experiment, but from the $c_{pa} = 0.35$ with $\mu_a = 0.96$ treatment, replicate 13 [1]. At this value of c_{pa}, the LPA-predicted deterministic attractor is chaotic. The arrow in Figure 12.3 indicates an episode of equilibrium-like dynamics. Notably, this interlude occurs at about 354 weeks (nearly seven years) into the experiment and lasts about 30 weeks (over half a year)!

In this chapter we consider the mystery of such episodes of apparent equilibration in data time series. The key to understanding this phenomenon can be found by visualizing

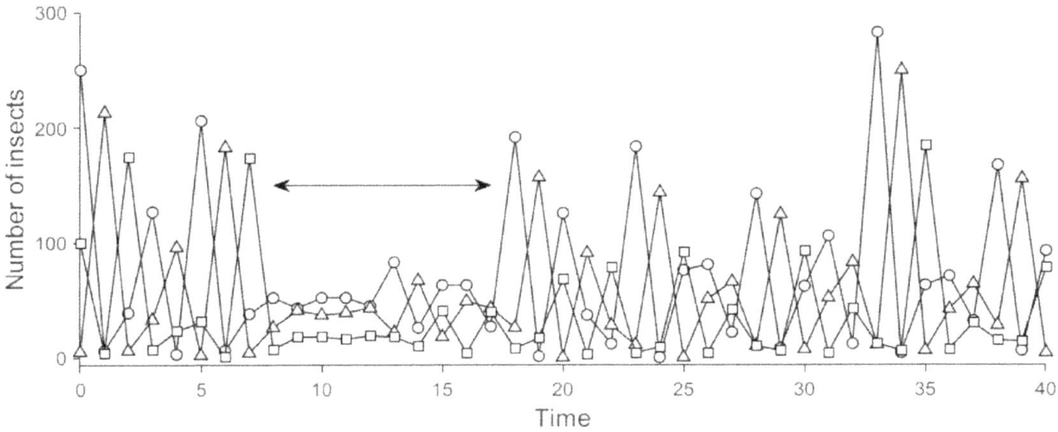

Figure 12.2 Time series of L-stage (circles), P-stage (triangles), and A-stage (squares) numbers for replicate 7 from the $c_{pa} = 0.05$ treatment of the route-to-chaos experiment described in Chapter 7 [1]. The arrow indicates an episode of "flat," equilibrium-like dynamics.

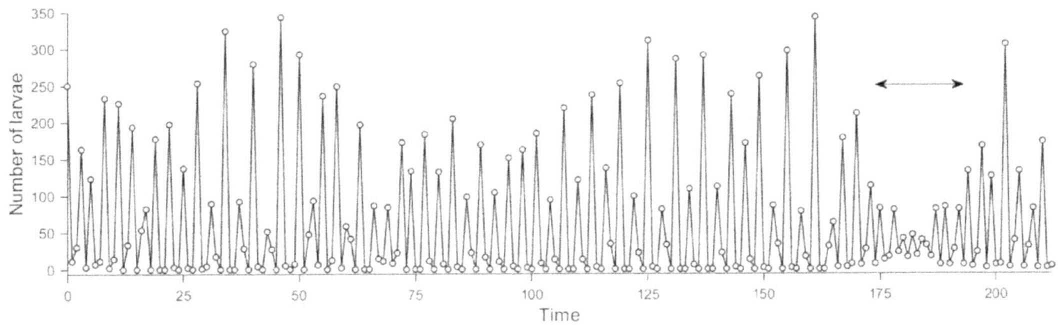

Figure 12.3 L-stage time series data for replicate 13 from the $c_{pa} = 0.35$ treatment of the route-to-chaos experiment described in Chapter 7 [1]. The arrow marks an episode of equilibrium-like dynamics.

invariant sets in state space.[1] As discussed in Chapter 3, deterministic nonlinear systems can have asymptotically stable, neutrally stable, and unstable invariant sets in state space. Some unstable invariant sets are repelling, but others are saddles with stable and unstable manifolds. Stochastic nonlinear systems may repeatedly fall under the influence of an unstable invariant set, especially if that invariant set is a saddle. For example, visualize in state space a trajectory being stochastically bumped away from an attractor and landing near the stable manifold of a saddle equilibrium. The trajectory may then move in the direction of the stable manifold toward the saddle and linger in its vicinity before moving back, under the influence of the unstable manifold, toward the attractor. The transitory approach toward and divergence from the saddle is called a *saddle flyby*. Flybys can happen repeatedly in a stochastic system, and such events are manifest as episodes of apparent equilibration

[1]An invariant set is a set of points in state space that is invariant under the forward evolution of the dynamical system.

in the time series. *Saddle flybys are an important way in which unstable invariant sets can leave their signature in stochastic time-series data.*

12.1 A TWO-DIMENSIONAL RICKER SADDLE

We begin with an example of a saddle in the two-dimensional Ricker-type model that we briefly considered in Chapter 3: the "Ricker model with senescence," in which individuals enter a second, senescent (post-reproductive), life stage after one time step:

$$\begin{aligned} x_{t+1} &= bx_t e^{-cx_t} \\ y_{t+1} &= x_t + sy_t, \end{aligned} \tag{12.1}$$

with $b, c > 0$ and $0 < s < 1$.

Because the first equation does not involve y, the dynamics for x are identical to those of the Ricker model (3.3) in Chapter 3, with two equilibrium branches $x_e = 0$ and $x_e = (\ln b)/c$ in the b-x_e plane, the latter of which is positive if $b > 1$. As illustrated in Figure 3.3 in Chapter 3, the extinction equilibrium destabilizes and the positive equilibrium becomes stable as b increases through $b = 1$, in a typical exchange of stability. The positive equilibrium itself goes on to destabilize in a two-cycle bifurcation as b increases through $b = e^2$, as discussed in Chapter 3.

What about the dynamics for y in model (12.1)? From the fixed point equation $y = x + sy$ for the senescent stage, we obtain $y_e = x_e/(1 - s)$, and so model (12.1) has, as a function of b, two equilibrium branches

$$[0, 0] \text{ and } \left[\frac{1}{c} \ln b, \frac{1}{c(1-s)} \ln b\right], \tag{12.2}$$

the second of which is positive when $b > 1$.[2]

In Chapter 3, we asked what happens to the positive equilibrium of model (12.1) when it gives up its stability to the new-born two-cycle at $b = e^2$ (Figure 3.3 in Chapter 3). The equilibrium still exists, now unstable, but does it cease to influence the dynamics? As we noted in Chapter 3, the answer is a resounding "no!" *Unstable invariant sets continue to influence the trajectories of populations in state space.*

We noted in Chapter 3 that the positive equilibrium (12.2) of model (12.1) has eigenvalues $\lambda = s$ and $\lambda = 1 - \ln b$. As b increases through the critical value $b = e^2$, one eigenvalue ($\lambda = s$) remains between 0 and 1, with eigenvector $[0, 1]$, whereas the other ($\lambda = 1 - \ln b$) decreases through the value -1, with eigenvector $[1 - s - \ln b, 1]$. An equilibrium with one eigenvalue satisfying $|\lambda| < 1$ and another satisfying $|\lambda| > 1$ is called a *saddle*. The line in state space through the saddle equilibrium that is determined by the eigenvector $[0, 1]$, which is in this case a vertical line, is tangent to the *stable manifold*, the invariant set along which trajectories in state space approach the saddle equilibrium. The line through the saddle that is determined by the vector $[1 - s - \ln b, 1]$, which is a line with slope $1/(1 - s - \ln b)$, is tangent to the *unstable manifold*, along which trajectories move away from the saddle equilibrium in an oscillatory fashion and approach the two-cycle attractor. Trajectories that land close to the stable manifold move toward the saddle equilibrium before moving away in the direction of the unstable manifold; such a trajectory is called a *saddle flyby*.

Figure 12.4 shows a saddle flyby, both in state space and as a time series, for model (12.1) with $b = 8$, $c = 0.01$, $s = 0.7$, and initial condition $[x_0, y_0] = [190, 44]$. The saddle, at the intersection of the two manifolds, has coordinates $[207.94, 693.15]$. The coordinates of the

[2]We use square-bracket notation $[x, y]$ to denote the column vector $(x, y)^T$, which is the transpose of the row vector (x, y).

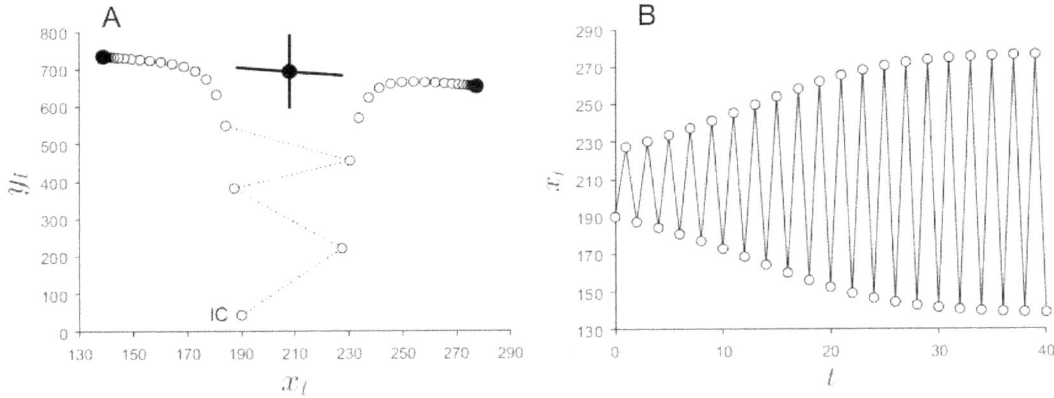

Figure 12.4 Deterministic saddle flyby for model (12.1) with $b = 8$, $c = 0.01$, $s = 0.7$, and initial condition $[x_0, y_0] = [190, 44]$. A. Shown in state space, the trajectory approaches the saddle (intersection of cross) along the stable manifold (vertical line), and exits along the unstable manifold as it moves toward the two-cycle attractor (solid circles). The first few steps are connected with dotted lines. B. Corresponding time series for x.

attracting two-cycle are $[277.26, 652.37]$ and $[138.63, 733.92]$. The eigenvalues are $\lambda = 0.7$ with eigenvector $[0, 1]$ and $\lambda = -1.08$ with eigenvector $[-1.78, 1]$. The linearized stable manifold is a vertical line, and the linearized unstable manifold is a line with slope $-1/1.78 = -0.561\,80$. The trajectory, which begins close to the stable manifold, first approaches the saddle in the direction of the stable manifold and then departs, oscillating back and forth in the direction of the unstable manifold toward the stable two-cycle. The distance between the points on the trajectory indicates the speed. Locally (close to the saddle), along the stable manifold, the trajectory's displacement from the saddle decays like $\lambda^t = (0.7)^t$, slowing down as it approaches the saddle. Locally, along the unstable manifold, the displacement grows like $\lambda^t = (-1.80)^t$, speeding up as it leaves the saddle (but slowing again as it approaches the stable two-cycle).

The saddle flyby in Figure 12.4 is a deterministic trajectory whose initial condition was deliberately placed close to the stable manifold. In a noisy system, stochasticity may bump the trajectory close to the saddle or its stable manifold, initiating a saddle flyby, and this can happen repeatedly. To illustrate, let us now consider a stochastic version of model (12.1) with demographic noise,

$$x_{t+1} = \left(\sqrt{bx_t e^{-cx_t}} + \sigma_1 \varepsilon_{1t}\right)^2 \tag{12.3}$$
$$y_{t+1} = \left(\sqrt{x_t + sy_t} + \sigma_2 \varepsilon_{2t}\right)^2,$$

in which the ε_{it} are standard normal random variables (mean zero and standard deviation one) uncorrelated in time and with each other, and $\sigma_i > 0$.

Figure 12.5 shows a stochastic simulation of model (12.3) as a time series and in state space. In Figure 12.5B-E we see that, for $t = 0 - 21$, the trajectory bounces back and forth near the stable two-cycle.[3] At $t = 22$, the trajectory is stochastically knocked near

[3]The notation "for $t = n_1 - n_2$", where n_1 and n_2 are two integers $n_1 < n_2$, means "for t equal to all sequential integer values beginning with n_1 and ending with n_2."

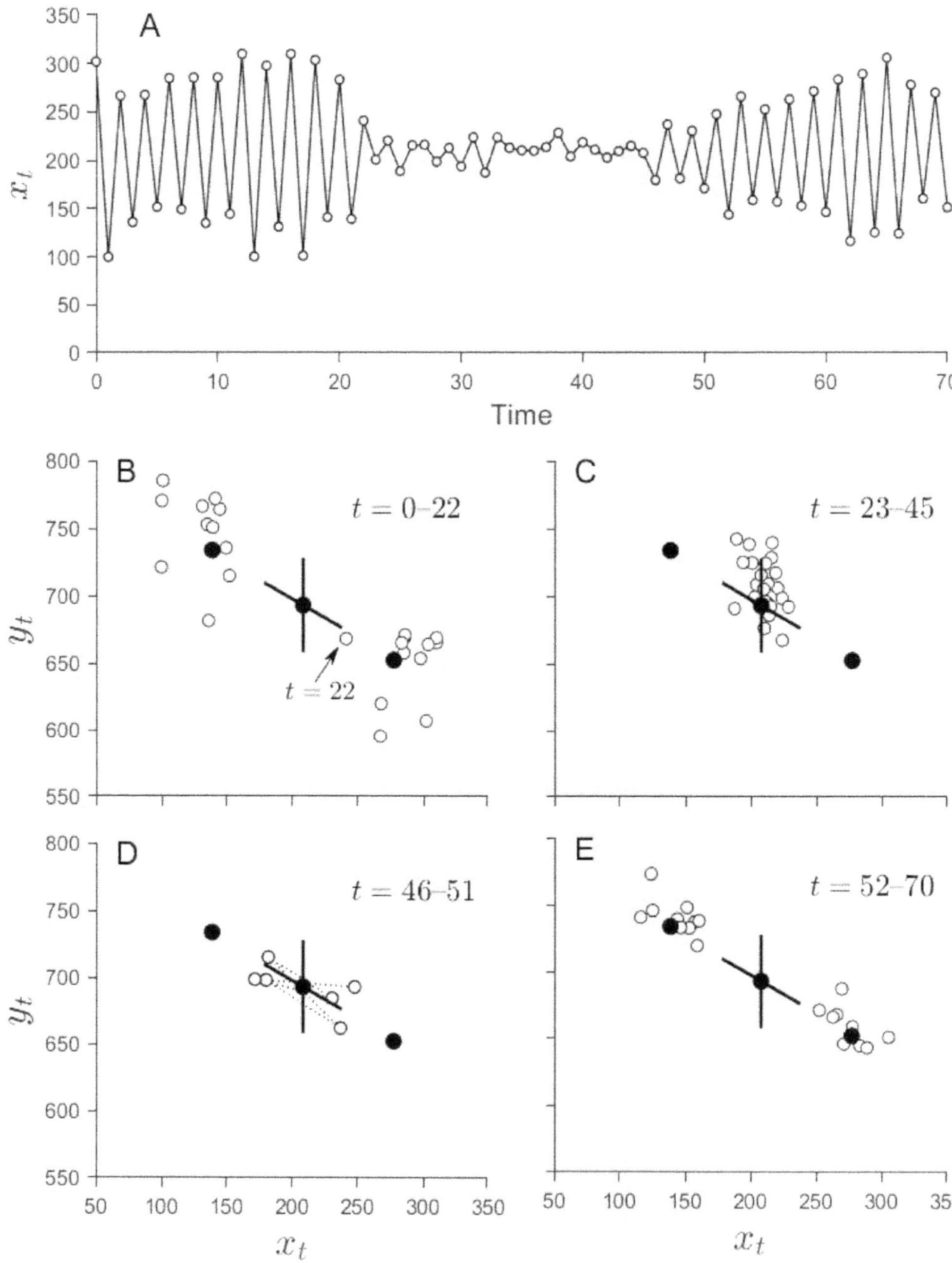

Figure 12.5 Stochastic simulation of model (12.3) with $b = 8$, $c = 0.01$, $s = 0.7$, $\sigma_1 = 0.5$, and $\sigma_2 = 0.2$. A. Time series with episode of apparent equilibration for $t = 23$–45. B-E. Trajectory in state space. For $t = 0$–21, trajectory bounces back and forth near the stable two-cycle. At $t = 22$, trajectory is stochastically knocked near the saddle equilibrium, around which it hovers for $t = 23$–45. During $t = 46$–51, trajectory leaves saddle, oscillating along the unstable manifold, and returns to the stable two-cycle ($t = 52$–70).

the saddle equilibrium, around which it hovers for $t = 23 - 45$. During $t = 46 - 51$, the trajectory leaves the saddle, oscillating along the unstable manifold, and returns to the stable two-cycle ($t = 52 - 70$). The saddle flyby during $t = 23 - 45$ appears as an episode of apparent equilibration in the corresponding time series (Figure 12.5A).

12.2 SADDLES IN THREE DIMENSIONS

In two dimensions, there are (only) two eigenvalues. They must be both real-valued or else they must be a complex conjugate pair. Complex conjugates have the same modulus[4], and so, in the generic situation considered here[5], they are either both inside or both outside the unit circle in the complex plane. Because a discrete-time saddle requires that one eigenvalue be inside the unit circle in the complex plane and one be outside, that is, $|\lambda_1| < 1$ and $|\lambda_2| > 1$, we can see that for a two-dimensional saddle both eigenvalues must be real. The generic topological configuration for a two-dimensional saddle in state space therefore consists of two one-dimensional (curvilinear) manifolds. One manifold is stable, and is tangent to the line determined by the eigenvector v_1 associated with eigenvalue λ_1, and the other is unstable and tangent to the line determined by v_2. Depending on the sign of the eigenvalues, the trajectories on the manifolds may approach or leave the saddle either by jumping back and forth in an oscillatory fashion (like λ^t where $\lambda < 0$) or by proceeding in a monotonic fashion (like λ^t where $\lambda > 0$).

In three dimensions, either all three eigenvalues are real, or else one eigenvalue is real and the other two are a complex conjugate pair (of equal modulus). Generically, a three-dimensional saddle therefore falls into one of two cases:

1. $|\lambda_1| > 1, \quad |\lambda_2| < 1, \quad |\lambda_3| < 1$
2. $|\lambda_1| < 1, \quad |\lambda_2| > 1, \quad |\lambda_3| > 1,$

where λ_1 is real and $\lambda_{2,3}$ are either real or a complex conjugate pair.

A three-dimensional saddle's topological configuration thus has a one-dimensional curvilinear manifold and a two-dimensional surface manifold. In case 1 above, the curvilinear manifold is unstable and tangent to the line determined by eigenvector v_1; and the surface manifold is stable and tangent to a plane which is determined by the eigenvectors v_2 and v_3. Along the one-dimensional unstable manifold associated with λ_1, trajectories leave the saddle in either a back-and-forth oscillatory fashion or a monotone fashion (depending on whether $\lambda_1 < 0$ or $\lambda_1 > 0$). If the two eigenvalues associated with the stable surface manifold are real, then their (real) eigenvectors span a plane tangent to the surface. If the two eigenvalues associated with the surface manifold are complex conjugates, then their complex conjugate eigenvectors $v \pm wi$ determine the linearized surface manifold in the following way: The real and imaginary parts v and w (which are themselves real vectors) span a plane tangent to the surface. In the case of complex eigenvalues, the local dynamics on the surface manifold consist of rotations which approach the saddle. Case 2 above is similar but with reversed stabilities.

The dynamics of the LPA model include many examples of saddles. We now return to the three *Tribolium* data time series shown at the beginning of this chapter and consider whether the LPA model can shed light on their dynamics.

---

[4]The modulus of a complex conjugate pair is $|\alpha \pm \beta i| = \sqrt{\alpha^2 + (\pm\beta)^2} = \sqrt{\alpha^2 + \beta^2}$.

[5]By *generic* we mean situations in which the eigenvalues are distinct (not equal) and do not lie on the boundary of the unit circle in the complex plane (do not have modulus of one).

12.3 JILLSON SADDLE

Let's consider again the *Tribolium* data time series in Figure 12.1, which show animal numbers from two of the four control replicates of an experiment by Jillson [9].

In order to explore the configuration of state space, we must know the parameter values. Cushing et al. [4] obtained from the four control cultures the conditional least squares (CLS) parameter estimates $b = 4.445$, $c_{el} = 0.05841, c_{ea} = 0.005784$, $\mu_l = 0.4794$, and $c_{pa} = 0.01053$ using the environmental stochastic LPA model, where the value of $\mu_a = 0.1542$ was computed directly from the observed numbers of dead adults. At these parameter values, the LPA model predicts a two-cycle attractor with values

$$[0, 138.1, 118.2]\,, \qquad [265.2, 0, 139.7]$$

and an unstable equilibrium

$$[L^*, P^*, A^*] = [30.5, 15.9, 56.7]\,.$$

The Jacobian

$$J = \begin{pmatrix} -c_{el}bAe^{-c_{el}L-c_{ea}A} & 0 & be^{-c_{el}L-c_{ea}A}(1-c_{ea}A) \\ 1-\mu_l & 0 & 0 \\ 0 & e^{-c_{pa}A} & -c_{pa}Pe^{-c_{pa}A}+(1-\mu_a) \end{pmatrix}, \tag{12.4}$$

evaluated at the equilibrium, is

$$\begin{pmatrix} -1.78 & 0 & 0.362 \\ 0.521 & 0 & 0 \\ 0 & 0.550 & 0.754 \end{pmatrix}.$$

This matrix has three real eigenvalues:

$$\lambda_1 = 0.804,\ \ \lambda_2 = -0.0733,\ \ \lambda_3 = -1.76,$$

and hence the equilibrium is a saddle with a one-dimensional unstable manifold and a two-dimensional stable manifold.

Corresponding eigenvectors are

$$\begin{aligned} v_1 &= [0.138, 0.0893, 0.986] \\ v_2 &= [0.116, -0.827, 0.550] \\ v_3 &= [0.957, -0.283, 0.0620]. \end{aligned}$$

The linearized unstable manifold is therefore a line in the direction of vector v_3, and the nonlinear unstable manifold is curvilinear and tangent to this line at the saddle point. The equation for the line in state space is

$$\frac{L-30.5}{0.957} = \frac{P-15.9}{-0.283} = \frac{A-56.7}{0.0620},$$

or, using a parametric vector equation,

$$\begin{pmatrix} L \\ P \\ A \end{pmatrix} = \begin{pmatrix} 30.5 \\ 15.9 \\ 56.7 \end{pmatrix} + \alpha \begin{pmatrix} 0.957 \\ -0.283 \\ 0.0620 \end{pmatrix}, \qquad \alpha \in R.$$

The linearized stable manifold is a plane spanned by the two eigenvectors v_1 and v_2. To find an equation of this plane, we note that a vector orthogonal to v_1 and v_2 (and hence to the plane) is the vector cross product

$$v_1 \times v_2 = [0.865, 0.0385, -0.124] .$$

Given any non-equilibrium point $[L, P, A]$ on the surface, the vector $[L, P, A] - [L^*, P^*, A^*]$ is parallel to the surface. Since the orthogonal vector $v_1 \times v_2$ and the parallel vector $[L, P, A] - [L^*, P^*, A^*]$ are orthogonal, their dot product must be zero:

$$(v_1 \times v_2) \circ ([L, P, A] - [L^*, P^*, A^*]) = 0.$$

From this requirement we obtain the equation for the plane:

$$0.865\,(L - 30.5) + 0.0385\,(P - 15.9) - 0.124\,(A - 56.7) = 0,$$

that is,

$$0.865L + 0.0385P - 0.124A = 20.0.$$

How does the configuration of state space help us understand the dynamics of replicate 20 of the Jillson experiment (Figure 12.1)? The data are shown in state space in Figure 12.6. Here we see the saddle equilibrium (solid circle lying on the planar surface), its two-dimensional stable manifold (planar surface), its one-dimensional unstable manifold (line), and the stable two-cycle (two solid circles lying off the surface). The culture begins ($t = 0$) near the stable manifold (Figure 12.6A), then lands near the saddle, quickly exits down the unstable manifold, and approaches the two-cycle (Figure 12.6A). At $t = 7$ (Figure 12.6B) the culture is knocked back to the stable manifold, which it follows to the saddle, hovering around the saddle until $t = 16$, at which point it leaves the saddle, bouncing back and forth over and under the plane, under the influence of the unstable manifold (Figure 12.6C), and finally returning to the two-cycle (Figure 12.6D).

12.4 ROUTE-TO-CHAOS DATA: $C_{PA} = 0.05$ SADDLE

Next, let's return to the time series for *Tribolium* in Figure 12.2. These data are from an experiment conducted in a different laboratory 20 years after the Jillson experiment. They are from replicate 7 of the $c_{pa} = 0.05$ with $\mu_a = 0.96$ treatment of the route-to-chaos experiment described in Chapter 7 [1].

Let's consider the architecture of state space in which these data occur. For the parameters estimated from the entire route-to-chaos data set ($b = 10.45$, $c_{el} = 0.01731$, $c_{ea} = 0.01310$, $\mu_l = 0.2000$) with manipulated parameter values $c_{pa} = 0.05$ and $\mu_a = 0.96$, the deterministic attractor of the LPA model is chaotic [5]. The equilibrium, unstable at these parameter values, is located approximately at

$$[L^*, P^*, A^*] = [57.0, 45.6, 18.7] .$$

The Jacobian (12.4), evaluated at the equilibrium, has a real eigenvalue of approximately

$$\lambda_1 = 0.409$$

and a complex conjugate pair of eigenvalues, approximately

$$\lambda_{2,3} = -1.13 \pm 0.706i,$$

with corresponding eigenvectors

$$v_1 = [0.439, 0.858, 0.266]$$
$$v_{2,3} = [0.828, -0.422, -0.0503] \pm i[0, -0.265, 0.254].$$

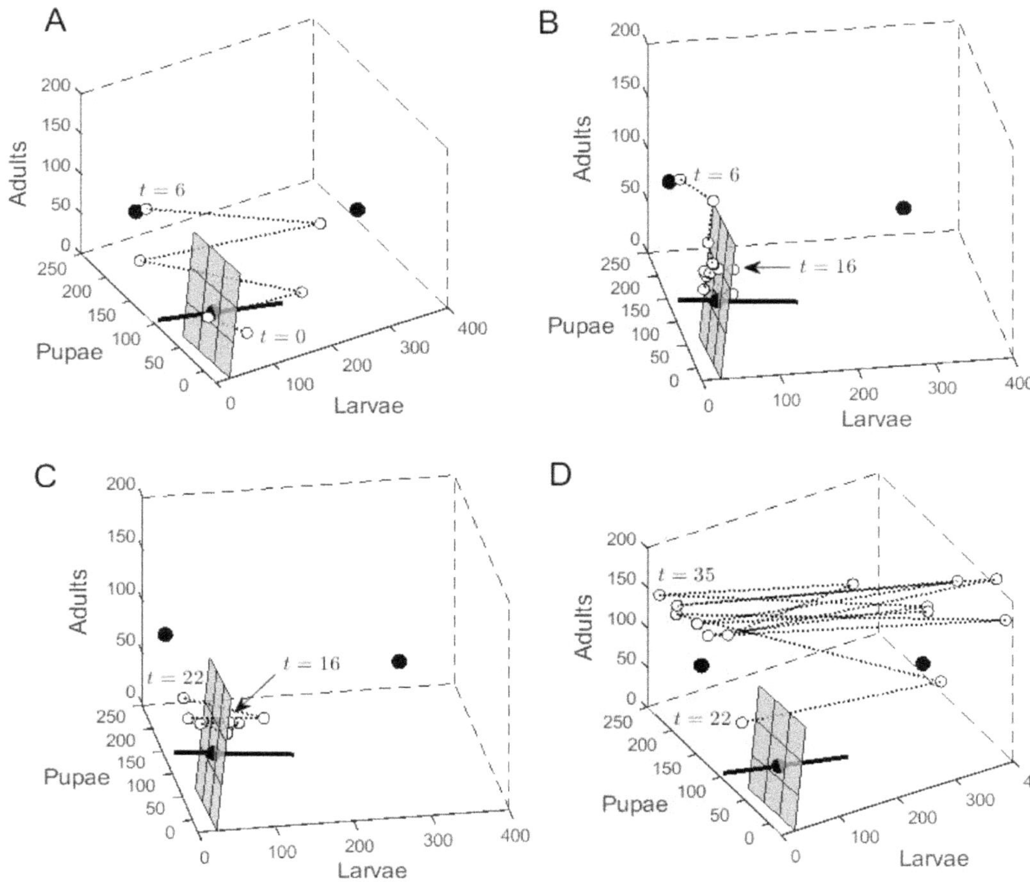

Figure 12.6 Replicate 20 data from the Jillson experiment [9] (open circles) in state space, along with the saddle equilibrium (solid circle lying on the surface), its two-dimensional stable manifold (planar surface), its one-dimensional unstable manifold (line), and the stable two-cycle (two solid circles lying off the surface). A. The culture begins near the stable manifold, then lands near the saddle, exits down the unstable manifold, and approaches the stable two-cycle. B. At $t = 7$, the culture is knocked back to the stable manifold, which it follows to the saddle, hovering around the saddle until $t = 16$. C. The culture then leaves the saddle, bouncing back and forth over and under the plane, under the influence of the unstable manifold, until $t = 22$. D. The culture then returns to the stable two-cycle.

Note that $|\lambda_1| < 1$ and $|\lambda_{2,3}| \simeq \sqrt{1.13^2 + 0.71^2} \simeq 1.33 > 1$, and hence the equilibrium is a saddle with one-dimensional stable manifold and two-dimensional unstable manifold.

The equation of the linearized stable manifold is

$$\frac{L - 57.0}{0.439} = \frac{P - 45.6}{0.858} = \frac{A - 18.7}{0.266}.$$

To obtain information about the properties of the unstable surface manifold, we can write the complex conjugate pair of eigenvalues as $\lambda_{2,3} = re^{\pm i\varphi}$, with $r = |\lambda_{2,3}| \simeq 1.33$

and $\tan\varphi = 0.706/1.13$, that is, $\varphi \simeq 2.58$. This implies a rotational motion away from the saddle equilibrium with an angle of $2\pi/\varphi \simeq 2.44$ radians or about 140°. The rotational movement away from the saddle is predicted to lie on a two-dimensional surface tangent to the plane determined by the eigenvectors v_2 and v_3. In particular, the real and imaginary parts of these complex eigenvectors span a plane parallel to the linearized plane of rotation. A vector orthogonal to this plane is the vector cross product

$$\begin{aligned}\mathrm{Re}(v_{2,3}) \times \mathrm{Im}(v_{2,3}) &= [0.828, -0.422, -0.0503] \times [0, -0.265, 0.254] \\ &= [-0.120, -0.210, -0.219].\end{aligned}$$

Given any (nonequilibrium) point $[L, P, A]$ on the plane, the equation of the plane is obtained by ensuring that

$$(\mathrm{Re}(v_{2,3}) \times \mathrm{Im}(v_{2,3})) \circ ([L, P, A] - [L^*, P^*, A^*]) = 0,$$

which leads to the following scalar equation for the (linearized) plane of rotation:

$$-0.120\,(L - 57.0) - 0.210\,(P - 45.6) - 0.219\,(A - 18.7) = 0,$$

that is,

$$0.120L + 0.210P + 0.219A = 20.5.$$

Given this understanding of the configuration of state space, we can now place the data time series (t = 0–40 in Figure 12.2) within state space, breaking it down into four time intervals:

- For t = 0–7, the population appears to be fluctuating around the attractor (Figure 12.7A).
- At $t = 8$, the population moves to the vicinity of the unstable equilibrium and lingers there from t = 8–12 (Figure 12.7B). During this period, the means and standard errors of the L_t, P_t, and A_t life stages are 48.8 ± 2.0, 37.2 ± 3.0, and 15.6 ± 2.2, respectively, which are near the predicted saddle.
- From t = 13–17, the population spirals away from the equilibrium (Figure 12.7C), during which time the one-step transitions formed angles with the unstable equilibrium with mean and standard error of $144.2° \pm 7.5°$. If the life stage values during the spiral away from the saddle are near the predicted plane of rotation, the orthogonal state space distances from the data points to the plane should be relatively small. From t = 13–17, the mean and standard error of these distances is 3.41 ± 1.06. By contrast, the mean and standard errors of the distances for the time periods t = 1–7, t = 8–12, t = 18–40 are 48.66 ± 10.75, 10.46 ± 2.98, and 29.79 ± 5.32, respectively. Thus, life-stage values are closest to the plane of rotation during the rotational period.
- For t = 18–40, the population moves back to the attractor (Figure 12.7D).

12.5 ROUTE-TO-CHAOS DATA: $C_{PA} = 0.35$ SADDLE

For the final example, we return to Figure 12.3. This data time series is also from the route-to-chaos experiment described in Chapter 7 [1], but this time from replicate 13 of the $c_{pa} = 0.35$ with $\mu_a = 0.96$ treatment. For the parameters estimated from the entire route-to-chaos data set ($b = 10.45$, $c_{el} = 0.01731$, $c_{ea} = 0.01310$, $\mu_l = 0.2000$) with manipulated parameter values $c_{pa} = 0.35$ and $\mu_a = 0.96$, the deterministic attractor of the LPA model

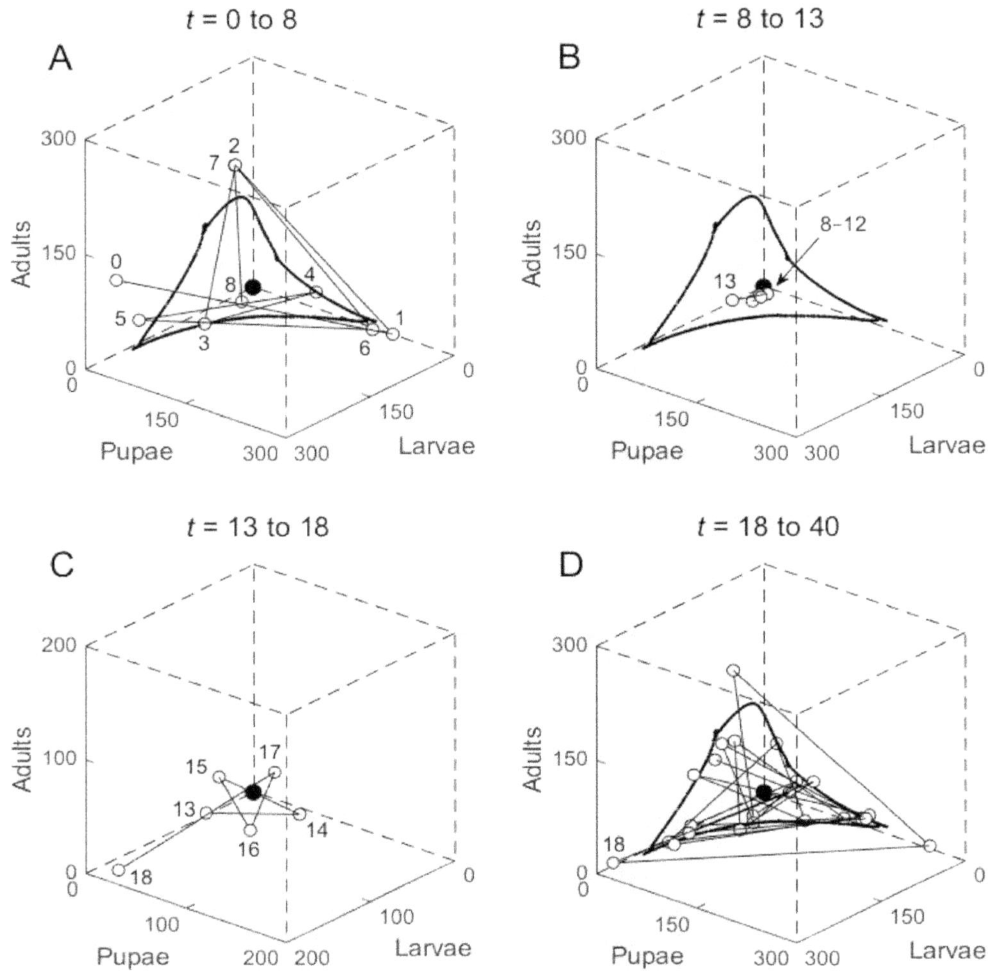

Figure 12.7 Saddle flyby in replicate 7 of the $c_{pa} = 0.05$ treatment of the route-to-chaos experiment described in Chapter 7, shown here in state space [1]. The corresponding time series is shown in Figure 12.2. The solid curve is the chaotic attractor and the solid circle is the unstable equilibrium. Open circles show the data trajectory connected by lines. A. During $t = 0$–8, the population moves around the chaotic attractor. B. At time $t = 8$, the population lands near the unstable equilibrium and hovers there until $t = 13$. C. During $t = 13$–18, the population spirals away from the unstable equilibrium. D. The population then resumes its movement around the chaotic attractor.

is again chaotic [5]. The arrow in Figure 12.3 indicates a saddle flyby occurring at about 354 weeks (nearly seven years) into the experiment and lasting about 30 weeks (over half a year). The state space architecture in this example is similar to the previous example from the $c_{pa} = 0.05$ treatment, but we include it to emphasize four important points about saddle flybys:

1. Saddle flybys are not uncommon events;
2. Saddle flybys can last a long time;

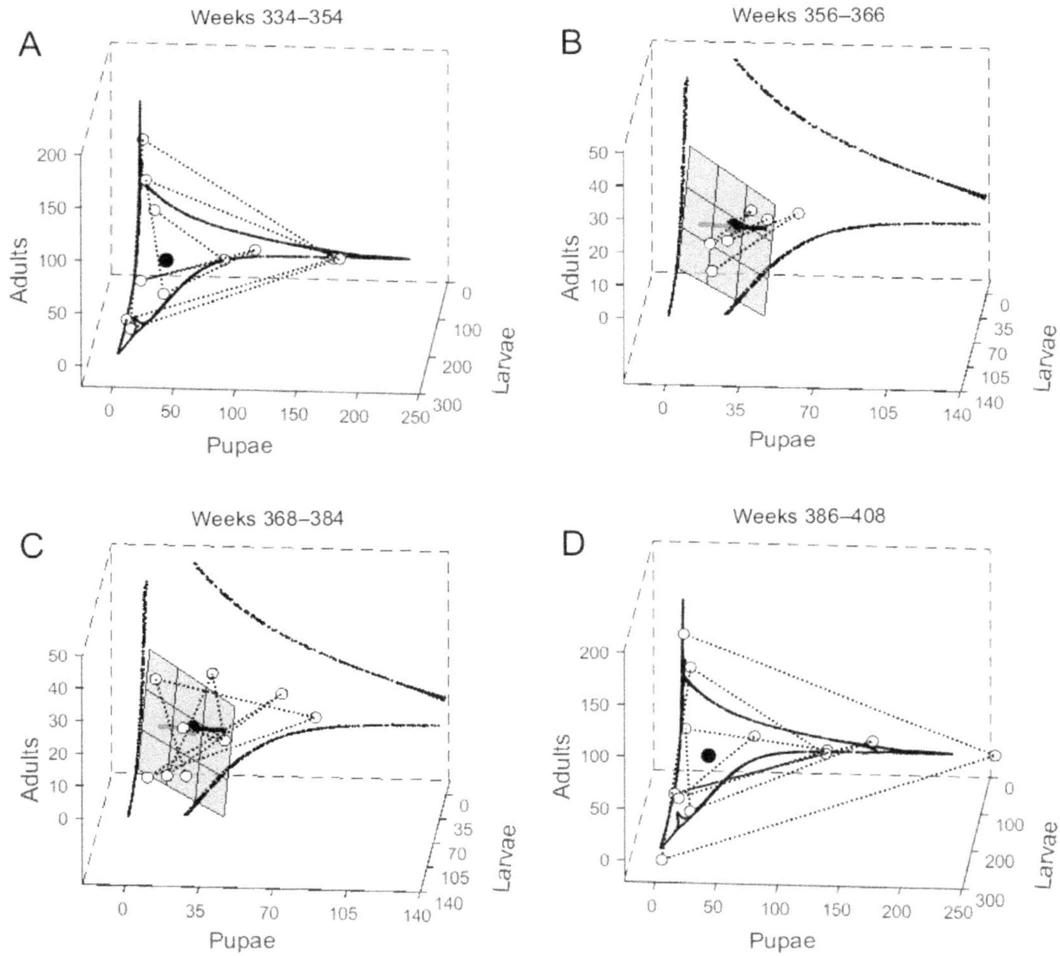

Figure 12.8 Data for weeks 334–408 (open circles), shown in state space, of replicate 13 from the $c_{pa} = 0.35$ treatment of the route-to-chaos experiment described in Chapter 7 [1]. A. For weeks 334–354, the population is on the chaotic attractor (solid curve). B. For weeks 356–366, the population lingers around the saddle (solid circle). C. For weeks 368–384, the population spirals away from the saddle along the unstable rotational planar manifold. D. For weeks 386–408, the population returns to the chaotic attractor.

3. Saddle flybys can occur late in a time series; and
4. Saddle flybys can occur suddenly and without warning, thanks to stochasticity, which can bump a population into a region of state space that is under influence of a saddle's stable manifold.

In this example the saddle equilibrium is $[28.0, 22.4, 4.63]$, with eigenvalues $\lambda_1 = 0.473$ and $\lambda_{2,3} = -1.24 \pm 0.614$, and corresponding eigenvectors $v_1 = [0.507, 0.857, 0.0855]$ and $v_{2,3} = [0.856, -0.444, -0.113] \pm i[0, -0.221, 0.0923]$. Note that $|\lambda_1| < 1$ and $|\lambda_{2,3}| = 1.38 > 1$. Thus, this saddle has a one-dimensional stable manifold and a two-dimensional rotational unstable manifold. The equation of the linearized stable manifold is

$$\frac{L - 28.0}{0.705} = \frac{P - 22.4}{0.857} = \frac{A - 4.63}{0.0855}$$

and the equation of the linearized unstable manifold is

$$6.60\,(L - 28.0) + 7.90\,(P - 22.4) + 18.9\,(A - 4.63) = 0,$$

that is,

$$6.60L + 7.90P + 18.9A = 449.$$

Figure 12.8 places the data (open circles) for weeks 334–408 within state space. During weeks 334–354, the population is on the chaotic attractor (Figure 12.8A; solid curve). At week 356, the population is stochastically bumped near the saddle (solid circle), around which it lingers until week 366 (Figure 12.8B). During weeks 368–384, the population spirals away from the saddle along the unstable rotational planar manifold (Figure 12.8C). During weeks 386–408, the population returns to the chaotic attractor (Figure 12.8D).

12.6 CONCLUSIONS

In this chapter, we have considered the mystery of episodes of apparent equilibration in data time series. There are two keys for unlocking this mystery:

- Visualization of the configuration of state space with its stable and unstable invariant sets, including saddles and their manifolds; and
- Awareness of the *invisible hand of stochasticity* that can cause populations to visit and revisit these entities in state space.

These two keys allow us to understand *saddle flybys*, which occur when a population is stochastically bumped away from an attractor, lands near the stable manifold of a saddle, moves along the direction of the stable manifold toward the saddle, lingers in the saddle's vicinity, and finally moves back, along the direction of the unstable manifold, toward the attractor.

Saddle flybys are one way in which unstable invariant sets can leave their signature in stochastic time series data. In this chapter, we have focused on saddle equilibria, but other types of invariant sets, such as cycles, can be saddles, as well. For example, flybys of saddle cycles in flour beetle populations were discussed in Chapter 9 and are again discussed in Chapter 14 [6, 7, 8].

Bibliography

[1] Costantino RF, Desharnais RA, Cushing JM, Dennis B (1997) Chaotic dynamics in an insect population. Science 275:389–391. https://doi.org/10.1126/science.275.5298.389

[2] Costantino RF, Cushing JM, Dennis B, Desharnais RA, Henson SM (1998) Resonant population cycles in temporally fluctuating habitats. Bulletin of Mathematical Biology 60:247–273. https://doi.org/10.1006/bulm.1997.0017

[3] Costantino RF, Desharnais RA, Cushing JM, Dennis B, Henson SM, King AA (2005) Nonlinear stochastic population dynamics: The flour beetle *Tribolium* as an effective tool of discovery. In Desharnais RA (ed.), Advances in Ecological Research, pp. 101–141. New York, Academic Press.

[4] Cushing JM, Dennis B, Desharnais RA, Costantino RF (1998) Moving toward an unstable equilibrium: Saddle nodes in population systems. Journal of Animal Ecology 67:298–306. https://doi.org/10.1046/j.1365-2656.1998.00194.x

[5] Dennis B, Desharnais RA, Cushing JM, Henson SM, Costantino RF (2001) Estimating chaos and complex dynamics in an insect population, Ecological Monographs 71:277–303. https://doi.org/10.1890/0012-9615(2001)071[0277:ecacdi]2.0.co;2

[6] Henson SM, Costantino RF, Cushing JM, Dennis B, Desharnais RA (1999) Multiple attractors, saddles, and population dynamics in periodic habitats. Bulletin of Mathematical Biology 61:1121–1149. https://doi.org/10.1006/bulm.1999.0136

[7] Henson SM (2000) Multiple attractors and resonance in periodically forced population models. Physica D: Nonlinear Phenomena 140:33–49. https://doi.org/10.1016/s0167-2789(99)00231-6

[8] Henson SM, Costantino RF, Desharnais RA, Cushing JM, Dennis B (2002) Basins of attraction: Population dynamics with two stable 4-cycles. Oikos 98:17–24. https://doi.org/10.1034/j.1600-0706.2002.980102.x

[9] Jillson DA (1980) Insect populations respond to fluctuating environments. Nature 288:699–700. https://doi.org/10.1038/288699a0

[10] King AA, Costantino RF, Cushing JM, Henson SM, Desharnais RA, Dennis B (2004) Anatomy of a chaotic attractor: Subtle model-predicted patterns revealed in population data. Proceedings of the National Academy of Sciences USA 101:408–413. https://doi.org/10.1073/pnas.2237266100

CHAPTER 13

Cycles and Phase Switching

Many studies have reported cycles in time series of biological populations, including paramecia [9], bean weevils [17], blow flies [15], and flour beetles [8, 10, 14]. Cycles in a population system can be caused by nonlinear mechanisms inherent in the population itself, or can be forced by a periodic environment. In this chapter, we consider noisy cycles in constant environments. Do the underlying deterministic mechanisms that cause cycling leave measurable signatures in experimental data? And if so, can the stochastic deviations from the deterministic signal be explained within the architecture of state space?

We begin with a brief model-free examination of two-cycle signatures in some short noisy time series, and then discuss the power of using a well-validated mathematical model to identify cycles in noisy data. Finally, we consider the stochastic phenomenon of phase switching in cycling populations and show how it can be explained within the geometry of state space.

13.1 RECOGNIZING CYCLES IN TIME SERIES AND STATE SPACE

Cover up the figure captions and consider the two short time series shown in Figure 13.1. Can you decide which of these time series is a noisy two-cycle and which is a purely random sequence with no underlying nonlinear mechanism? Distinguishing noisy cycles from purely random fluctuations by eye alone can be tricky depending on the level of noise; assertions of cycling require rigorous quantitative evidence.

Figure 13.1A is a noisy two-cycle. It was generated by adding Gaussian noise to the Ricker model, that is by the equation

$$x_{t+1} = bx_t e^{-cx_t} + \sigma\varepsilon_t, \tag{13.1}$$

with $b = 7.6$, $c = 0.01$, and $\sigma = 10$. The ε_t are standard normal random variables (mean zero and standard deviation one), uncorrelated in time. At these parameter values, the deterministic skeleton (that is, the model with $\sigma = 0$) has a stable two-cycle. By contrast, the time series in Figure 13.1B was generated by purely Gaussian fluctuations around a constant:

$$x_{t+1} = \overline{x} + s\varepsilon_t, \tag{13.2}$$

where $\overline{x}$ and s are the mean ($\overline{x} = 199$) and standard deviation ($s = 11$), respectively, of the values of the time series in Figure 13.1A.[1] Does the noisy cycle in Figure 13.1A really contain enough structure to be distinguishable from purely random fluctuations such as those in Figure 13.1B?

[1]The standard deviation s does not correspond to the σ value in model (13.1). Rather, s measures the variability of the time series, whereas σ measures the variability around the deterministic skeleton.

DOI: 10.1201/9781003467120-13

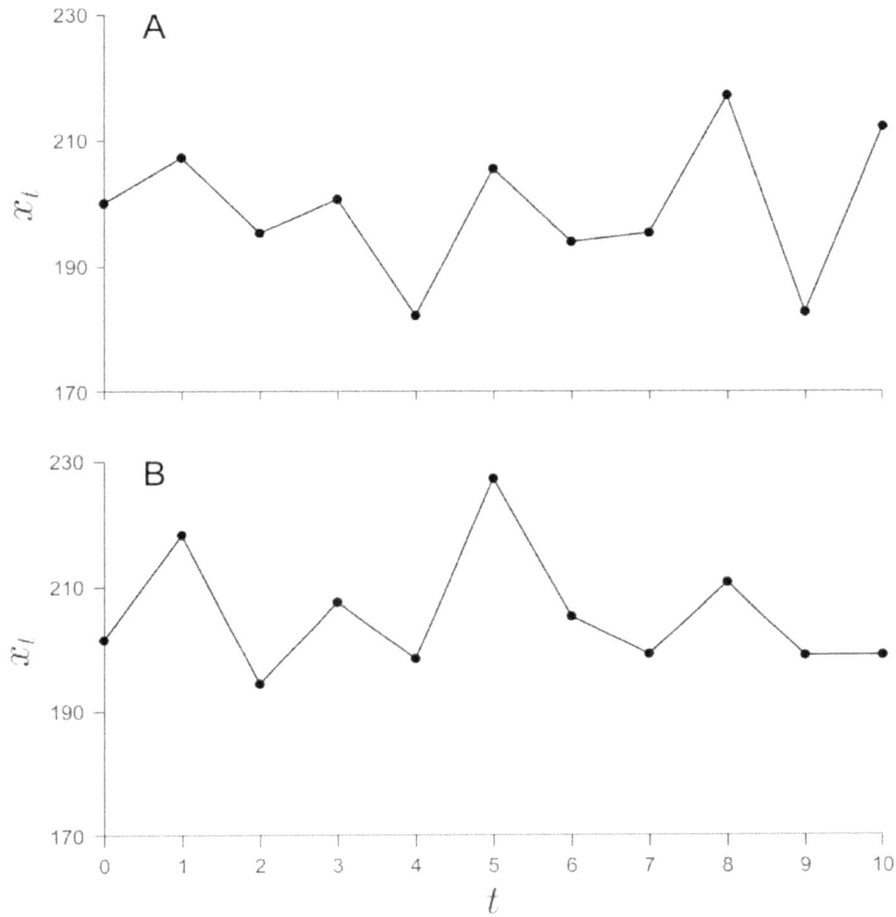

Figure 13.1 Stochastic time series. A. Noisy Ricker model (13.1) with $b = 7.5$, $c = 0.01$, and $\sigma = 10$; $TVV = 164.96$. B. Pure Gaussian noise generated by equation (13.2) with $\overline{x} = 199$ and $s = 11$; $TVV = 143.71$.

13.1.1 When the Underlying Deterministic Skeleton Is Unknown

There are a number of ways to detect cycles in noisy population data. Spectral methods can be used to estimate dominant periodicities of population time series in the frequency domain, as we will do in Chapter 16. In the time domain, the standard statistical approach is to calculate the autocorrelation function, in which peaks at different lags can indicate periodicities [2]. Both approaches require long time series of many observations. For short noisy time series such as the ones in Figure 13.1, we could use a simple approach developed in [1, 11, 16] to detect deterministic signatures of cycles. In this approach, one defines the "total vertical variation" of a time series $x_0, x_1, \cdots, x_n$ of length $n+1$ to be

$$TVV = \sum_{t=0}^{n-1} |x_{t+1} - x_t| . \tag{13.3}$$

In a graph of x_t vs. t, this is simply the total vertical "movement," up and down, along the time series. What if we were to compare TVV for a noisy two-cycle with TVV for a purely random sequence? We might suspect that, on average, a noisy two-cycle would generate a higher TVV due to the regular up-and-down influence imparted by the underlying oscillatory mechanism.

A comparison of TVV for the time series in Figure 13.1A ($TVV = 164.96$) to that in Figure 13.1B ($TVV = 143.71$) is consistent with this idea. But how likely is it that $TVV = 164.96$ is greater than what is expected due to chance alone? To answer this, we can compare the "observed" $TVV = 164.96$ value to the null distribution of TVV values estimated from a large number of purely random time series generated by equation (13.2) with $n = 10$, mean $\overline{x} = 199$, and standard deviation $s = 11$.[2] Figure 13.2A shows the null distribution generated from 10^6 random sequences. Because 6.6% of the null distribution lies to the right of our observed value of $TVV = 164.96$ (Figure 13.2A), we can say that the probability of obtaining, by chance alone, a TVV value greater than the observed is only 0.066.[3] Thus, we have quantified modest evidence that the time series in Figure 13.1A is influenced by a deterministic mechanism.

How easy would it be to obtain a false negative, that is, miss the underlying signal, with this method? As it turns out, very easy. To see why, consider the distribution of TVV (estimated by 10^6 simulations) of the noisy Ricker model (13.1) with $b = 7.6$, $c = 0.01$, and $\sigma = 10$, shown in Figure 13.2B. Although this distribution cannot be compared directly to the null distribution in Figure 13.2A, which is based on the standard deviation ($s = 11$) of the single realized Ricker time series in Figure 13.1A, we note that the Ricker distribution in Figure 13.2B is skewed heavily to the right, but also has a large overlap with the null distribution. This suggests that, for short time series, it is easy to obtain a false negative (Type II error; saying there is not an underlying deterministic two-cycle when actually there is one).

Now consider the *Tribolium* time series data in Figure 13.3A from an experiment conducted by Jillson [12]. From this time series of length 36 ($n = 35$), we obtain $TVV = 10,147$ by adding the TVV for the three life stages, and we compute standard deviations $s_L = 106.49$, $s_P = 64.7$, and $s_A = 41.64$ from the variability of each life stage without the assumption of a deterministic skeleton.[4] Purely Gaussian time series of length 36 with these standard deviations produce the null distribution of TVV values shown in Figure 13.3B. Of the 10^6 null simulations, 1.79% had TVV greater than the experimental value $TVV = 10,147$. Thus, the probability of obtaining, by chance alone, a TVV value greater than the observed is only 0.0179. This provides strong evidence that the time series in Figure 13.3A is indeed a noisy two-cycle, which is visually apparent in this case.

For two- and three-dimensional data, such as those in Figure 13.3A, one way to visualize noisy cycles is by graphing the data in state space, in which a noisy p-cycle should consist of p clusters of points. For example, Figure 13.4 shows the *Tribolium* time series data from Figure 13.3A in state space. The data form two loose clusters, which become especially visible when the first three transient data points (open circles) are removed. Measures of clustering or periodic return can be used to identify noisy cycles in state space. Other ways to study noisy cycles are also available. For example, in Chapter 9 we introduced a "lag metric comparison" (LMC) to quantify the existence, length, and frequency of cycle episodes

---

[2]Note that, when computing TVV for a purely Gaussian sequence generated by equation (13.2), the value of $\overline{x}$ is irrelevant because it cancels in each term $x_{t+1} - x_t$ of the TVV.

[3]In the parlance of introductory statistics, we could say that the observed TVV is *significant at the* $\alpha = 0.07$ *level* ($p = 0.066$). However, the cutoff value α for significance is arbitrary and artificial. It is better to simply state that the probability of obtaining a result at least as extreme as the observed, due to chance alone, is 0.066.

[4]These standard deviations do not correspond to the σ values in the stochastic LPA model. They measure the overall variability of each life stage, as opposed to the variability around a deterministic skeleton.

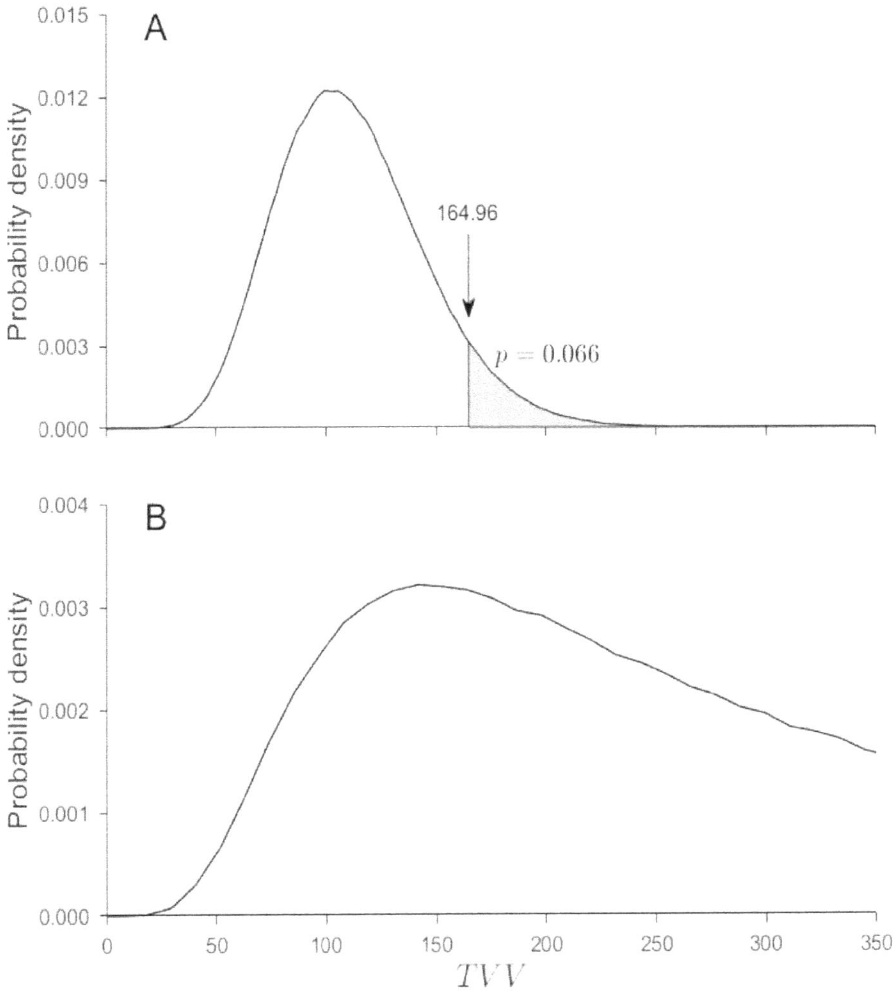

Figure 13.2 Distributions of TVV values. A. Null distribution of TVV estimated from 10^6 purely random time series with $n = 10$ and standard deviation $s = 11$. The value of $TVV = 164.96$ (arrow) computed for the time series in Figure 13.1A cuts off 6.6% of the right tail of the null distribution. B. Distribution of TVV for the noisy Ricker model (13.1) with $b = 7.5$, $c = 0.01$, and $\sigma = 10$ for time series with $n = 10$.

in stochastic data [13]. In Chapter 16, we will use spectral methods to identify periodicities in data.

13.1.2 Cycles Predicted by Well-Validated Mathematical Models

As we have seen above, signatures of cycles sometimes can be detected, even in very noisy and short time series, when the underlying deterministic skeleton is unknown. The best way, however, to identify and document cycles in population data is by means of a mechanistic mathematical model that has been validated on an independent data set. If the deterministic

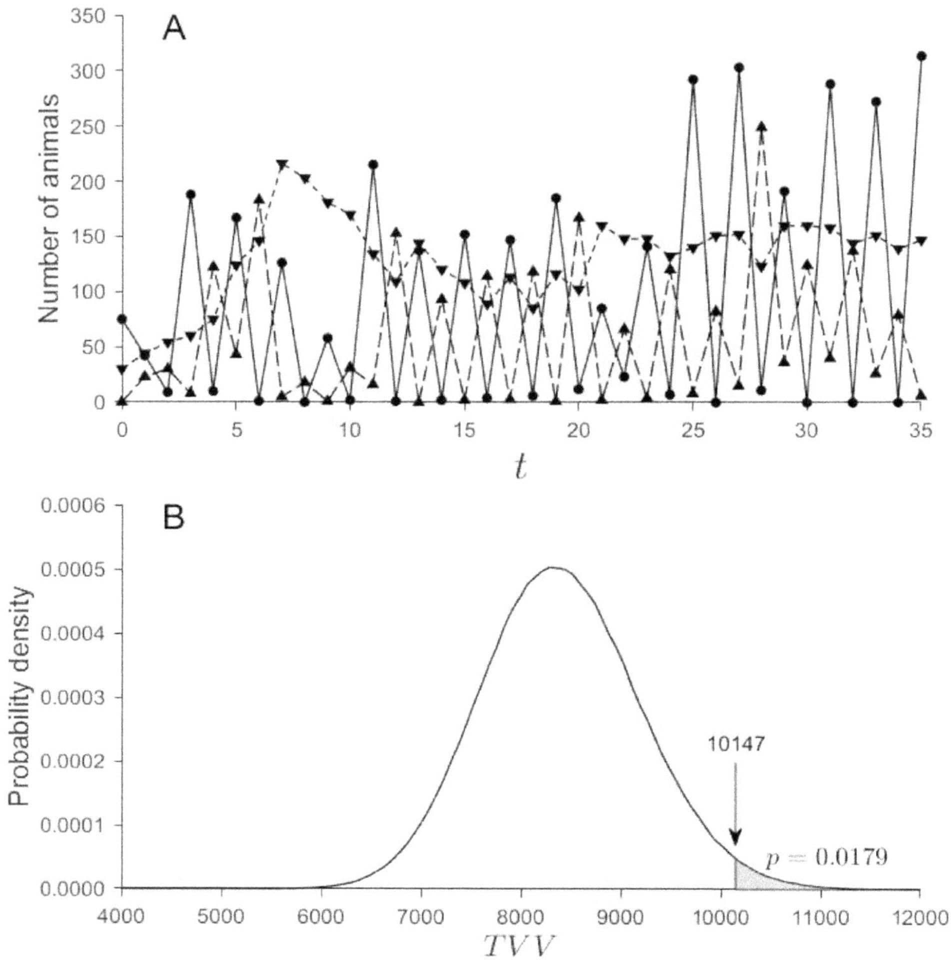

Figure 13.3 TVV for a *Tribolium* time series of length 36 from an experiment conducted by Jillson [12]. A. Time series for L-stage (solid), P-stage (short dashes), and A-stage (long dashes). $TVV = 10,147$ is the sum of the TVV for the three life stages. The standard deviations for each stage are $s_L = 106.49$, $s_P = 64.7$, and $s_A = 41.64$. B. Null distribution of TVV values from 10^6 Monte Carlo simulations of pure Gaussian noise time series of length 36 with these standard deviations. 1.79% of the null distribution lies to the right of the observed value $TVV = 10,147$.

skeleton of the validated model predicts cycles, this provides evidence that the observed fluctuations are noisy cycles.

Even more convincing evidence is obtained if a model can predict *transitions* between different types of attractors as a parameter is varied. For example, in Chapter 6, Figure 6.4 shows a bifurcation diagram with a "two-cycle bubble," in which the LPA model predicts a transition from an equilibrium to a two-cycle back to an equilibrium as one "tunes" the parameter μ_a from 0 to 0.73. Experimental treatments were placed at $\mu_a = 0.04$ (predicted equilibrium), $\mu_a = 0.27$ (predicted two-cycle), $\mu_a = 0.5$ (predicted two-cycle), and $\mu_a = 0.73$ (predicted equilibrium). In that same chapter, Figure 6.5A-D shows the experimental data

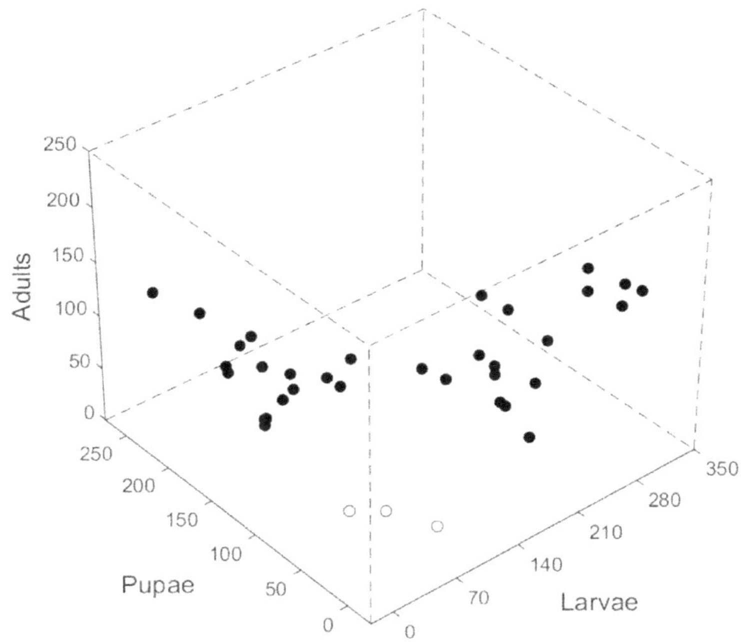

Figure 13.4 *Tribolium* time series from Figure 13.3A, shown here in state space. Data for times $t = 0$–2 are open circles; data for $t = 3$–35 are solid circles. The data form two loose clusters.

in state space. A single cluster of observed data in state space splits into two clusters which then merge back into a single cluster as the parameter μ_a is tuned experientially. Clearly this type of "bookend argument" provides strong evidence that the data in the treatments for $\mu_a = 0.27$ and $\mu_a = 0.5$ exhibit two-cycles.[5]

13.2 PHASE SWITCHING

Noise is ubiquitous in population time series, and cycling populations exhibit stochastic variations in amplitude, average, and phase. To what extent can these stochastic deviations be explained by the architecture of state space?

The phenomenon of *phase switching*, in particular, occurs when a periodic pattern "hiccups," then resumes cycling in an offbeat phase—for example, when a high-low pattern skips to a low-high pattern. In [10], we humorously christened these events as "chicken steps." Let's consider four examples of this phenomenon in population time series data (Figure 13.5).

The *T. castaneum* two-cycle data in Figure 13.5A are control replicates from [7], historically denoted "replicate A" (circles) and "replicate B" (squares), but renamed here and in the data archive as "replicate 1" and "replicate 2." Both replicates switch phase at week 8 (arrows). Replicate 2 (squares) switches phase again at week 28 (arrow), after which the replicates remain out of phase with each other.

The *T. castaneum* two-cycle data in Figure 13.5B are replicate 4 (squares) and replicate 22 (circles) from [3]. Replicate 4 does not switch phase. Replicate 22 switches phase at week 20, and is thereafter out of phase with replicate 4.

[5]Similarly, a bookend method was used in Chapter 7 to identify the influence of chaotic dynamics in the route-to-chaos experiments.

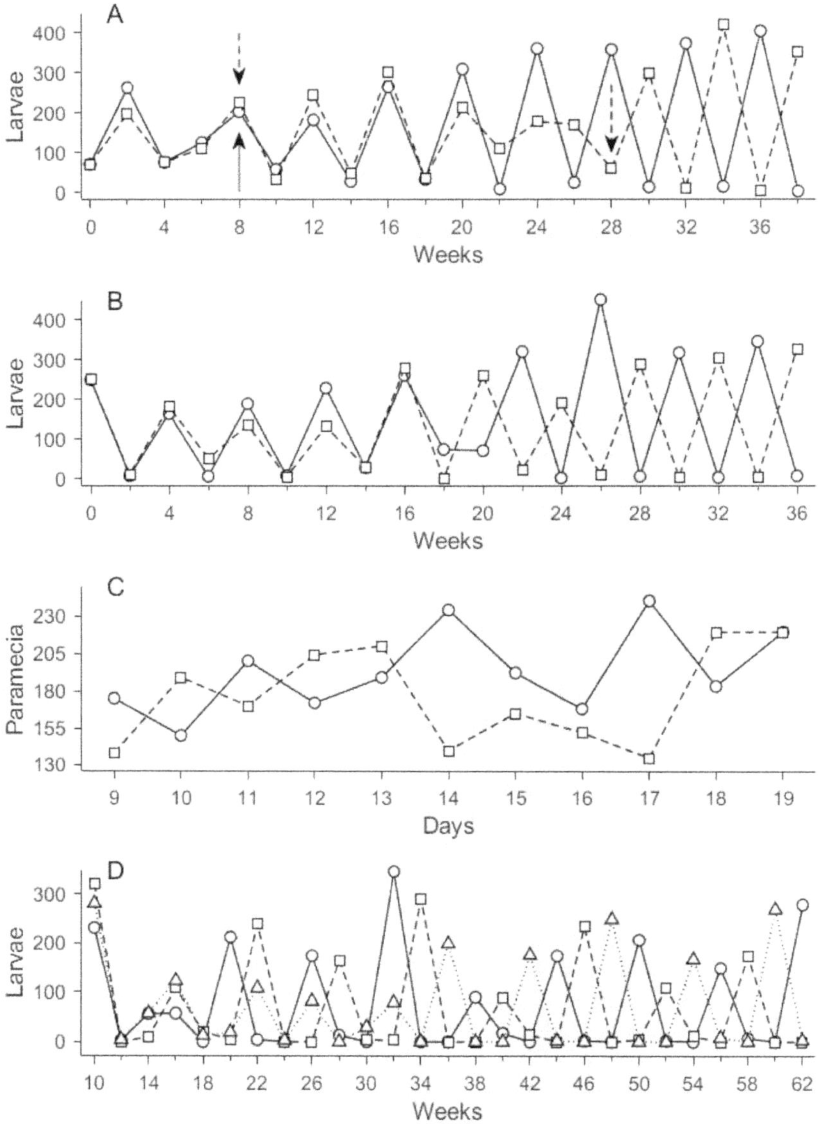

Figure 13.5 A. *Tribolium castaneum* two-cycle data from [7]. Replicates 1 (circles) and 2 (squares) both switch phase at week 8, and replicate 2 switches phase at week 28. B. *T. castaneum* two-cycle data from [3]. Replicate 4 (squares) does not change phase. Replicate 22 (circles) changes phase at week 20. C. *Paramecium caudatum* data from [9]. Transient data (days zero to eight) are omitted. Replicate 2 (circles) switches phase at days 14 and 16. Replicate 4 (squares) switches phase at days 13 and 17. D. *T. castaneum* three-cycle data from [4]. Transients (weeks zero to eight) are omitted. Replicate 23 (squares) does not shift phase. Replicate 3 (circles) switches phase once, at week 16. Replicate 18 (triangles) changes phase at week 26, when it synchronizes with replicate 3. It switches phase again at week 36, after which all three replicates are out of phase. This situation persisted until 80 weeks, the end of the experiment.

Two replicate *Paramecium caudatum* data time series from Gause [9] are shown in Figure 13.5C. Replicate 2 (circles) changes phase at days 14 and 16, whereas replicate 4 (squares) changes phase at days 13 and 17.

Three replicates of *T. castaneum* data from the route-to-chaos experiment [4] are shown in Figure 13.5D. The data show a three-cycle pattern of high-low-low. Replicate 23 (squares) does not shift phase. Replicate 3 (circles) changes phase once, at week 16. Replicate 18 (triangles) changes phase at week 26, at which time it synchronizes with replicate 3. At week 36, replicate 18 changes phase again. From week 36 on, until the end of the experiment at week 80, all three replicates remained out of phase with each other.

Phase switches typically lead to asynchronous experimental replicates. *From a practical point of view, phase switching means that averaging cycling replicates is, in general, a bad idea and can obscure the dynamics.* From a theoretical point of view, what is the explanation for stochastic phase switching, and does that explanation matter to our understanding of population dynamics?

13.2.1 Composite State Space

To gain insight into the phenomenon of phase switching, we need to think in terms of the *composite* model, that is, the model composed with itself which looks at every-other time step.[6]

If an *autonomous*[7] model has a stable two-cycle

$$x_0, x_1, x_0, x_1, \cdots,$$

then its out-of-phase sibling

$$x_1, x_0, x_1, x_0, \cdots$$

is also a stable two-cycle.[8] The composite model (which looks at every-other-step) thus has two stable fixed points (equilibria) of value x_0 and x_1. That is to say, each point on a two-cycle attractor of the original model is a stable fixed point for the composite. In addition, any fixed point of the original model is of course a fixed point of the composite.

For example, consider the two-dimensional "Ricker model with senescence" from Chapter 3, namely

$$\begin{aligned} x_{t+1} &= b x_t e^{-c x_t} \\ y_{t+1} &= x_t + s y_t, \end{aligned} \tag{13.4}$$

where $b, c > 0$ and $0 < s < 1$. As we showed in Chapter 3, this model has a non-trivial equilibrium[9]

$$\left[\frac{1}{c}\ln b, \frac{1}{c(1-s)}\ln b\right], \tag{13.5}$$

which is positive for $b > 1$ and has eigenvalues $\lambda_1 = 1 - \ln b$ and $\lambda_2 = s$. The equilibrium (13.5) is stable for $1 < b < e^2$ and unstable for $b > e^2$. In fact, as the parameter b is increased through the critical value $b_{cr} = e^2$, the eigenvalue λ_1 exits the unit circle in the complex plane at $\lambda_1 = -1$, and eigenvalue $\lambda_2 = s$ remains within the unit circle. Thus, as

---

[6] The composite of $x_{t+1} = f(x_t)$ is the model $x_{t+1} = f(f(x_t))$.

[7] An *autonomous* model is one which does not depend explicitly on time.

[8] Here, x is a real number for one-dimensional models, but is a vector for higher-dimensional systems. For example, in the three-dimensional LPA model, x is a triple $[L, P, A]$.

[9] We use square-bracket notation $[x, y]$ to denote the column vector $(x, y)^T$, which is the transpose of the row vector (x, y).

b increases through e^2, equilibrium (13.5) becomes a saddle and a stable two-cycle is born (Chapter 3).

At parameters $b = 8$, $c = 0.01$, and $s = 0.7$ for model (13.4), the values of the stable two-cycle are (rounded to the nearest individual)

$$[277, 652], \qquad [139, 734]$$

and the saddle equilibrium is

$$[208, 693].$$

Thus, the composite model has two stable fixed points [277, 652] and [139, 734], and one saddle fixed point [208, 693].

In Figure 13.6A, we see the x-component time series of the stable two-cycle (solid line), with even time steps marked by circles. The out-of-phase equivalent stable cycle (dashed line) is also shown, with even time steps marked by triangles. Figure 13.6B shows a stochastic simulation of the demographic noise version of model (13.5), namely

$$\begin{aligned} x_{t+1} &= \left(\sqrt{b x_t e^{-c x_t}} + \sigma_1 \varepsilon_{1t}\right)^2 \\ y_{t+1} &= \left(\sqrt{x_t + s y_t} + \sigma_2 \varepsilon_{2t}\right)^2, \end{aligned} \tag{13.6}$$

where $b = 8$, $c = 0.01$, $s = 0.7$, and $\sigma = 2$. Even time steps are shown with symbols. The stochastic time series switches phase at $t = 6$, at which point the every-other-step symbols change from circles to triangles.

Figure 13.6C shows the two stable fixed points of the *composite* model in composite state space (solid circle and solid triangle). The saddle fixed point (cross) lies on the basin boundary (curve) between the two stable fixed points. The basin boundary is the stable manifold of the saddle, and it separates the basins of attraction of the two stable fixed points. Deterministic trajectories starting to the left of the boundary approach the left stable fixed point (solid triangle), whereas trajectories starting to the right of the boundary approach the right stable fixed point (solid circle). *Stochastic trajectories, by contrast, can be stochastically knocked across the basin boundary one or more times.* The stochastic time series in Figure 13.6B, when plotted in composite state space in Figure 13.6C, illustrates such a basin jump, which occurs exactly when the every-other-step symbols switch from open circles to open triangles.

Thus, *each basin jump in the composite system's state space corresponds to a phase switch in the original system's time series.*

13.2.2 An Example From the LPA System

Let's return to the *T. castaneum* experimental data from [7], shown in Figure 13.5A.

The maximum likelihood parameters for the LPA model, based on the control replicates in [7], were given in [6] as $b = 11.6772$, $\mu_l = 0.5129$, $c_{pa} = 0.0178$, $c_{ea} = 0.0110$, $c_{el} = 0.0093$, and $\mu_a = 0.1108$ with variance-covariance matrix

$$\Sigma = \begin{pmatrix} 0.2771 & 0.0279 & 0.0098 \\ 0.0279 & 0.4284 & -0.0081 \\ 0.0098 & -0.0081 & 0.0111 \end{pmatrix}.$$

For these parameters, the LPA model predicts a saddle equilibrium of [124, 60, 97] and a two-cycle attractor consisting of the points [18, 158, 106] and [325, 9, 118] (rounded to the nearest individual).

Figure 13.7 presents the model predictions and experimental data both as time series and in composite state space, with one time step equal to two weeks.

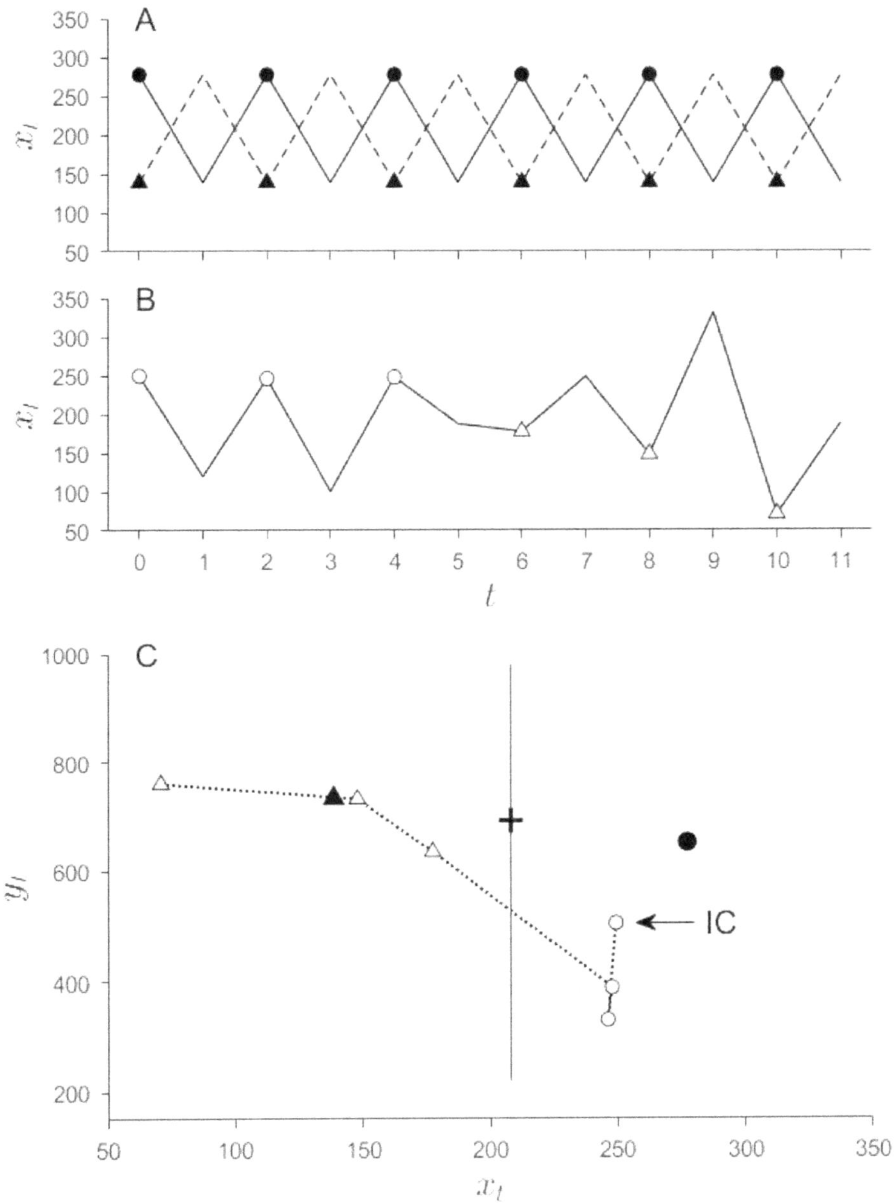

Figure 13.6 Predictions of the deterministic and stochastic Ricker models (13.4) and (13.6). A. Two stable out-of-phase deterministic two-cycle solutions "live" on the two-cycle attractor. The associated composite time series are indicated by every-other-step symbols. B. Stochastic time series showing a phase change at $t = 6$. C. Composite state space with two stable fixed points (large solid circle and triangle) and saddle (cross). The basins of attraction of the two stable fixed points are separated by the solid curve containing the saddle. The stochastic trajectory of the composite consists of every-other step of the stochastic time series in panel B. The stochastic jump across the basin boundary corresponds to the phase change in the time series in panel B. The initial condition is indicated by IC.

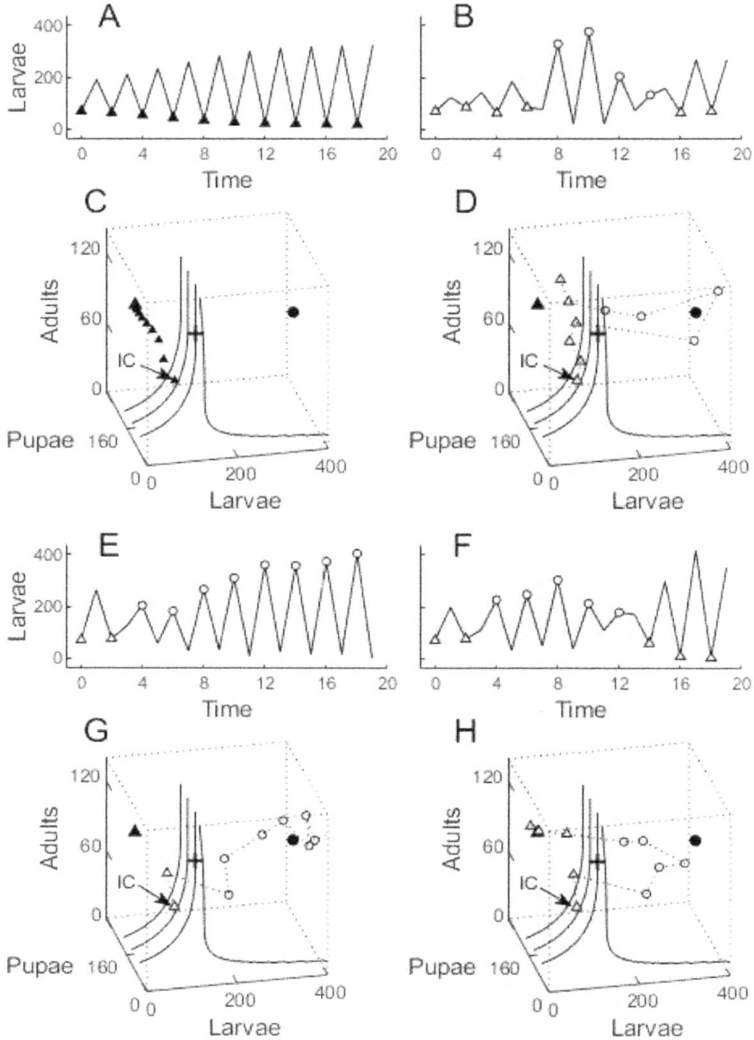

Figure 13.7 Stochastic simulations and experimental data. A. Deterministic model time series with initial condition [70, 35, 64] approaches a stable two-cycle. Composite (even) time steps are shown with solid triangles. B. Stochastic simulation with initial condition [70, 35, 64] has phase shifts at $t = 7$ and $t = 15$. C. Deterministic model time series from panel A, shown in composite state space (solid triangles), approaches fixed point [18, 158, 106], remaining to the left of the basin boundary (indicated by cross sections in the P = 0, 60, 120, and 180 planes). Also shown are the other composite fixed point [325, 9, 118] (solid circle to right of basin boundary) and the saddle on the boundary (cross). D. Stochastic model time series from panel B, shown in composite state space (open triangles and circles), exhibits two basin boundary crossings corresponding to the phase shifts in panel B. E–H. Replicate 1 (panels E and G) and replicate 2 (panels F and H) of experimental control data from [7], same format as panels A–D. Phase shifts in time series (panels E and F) correspond to basin boundary crossings in composite state space (panels G and H, respectively).

In Figure 13.7A, the deterministic model time series with initial condition $[70, 35, 64]$ approaches the stable two-cycle. Composite (even) time steps are shown with solid triangles. Figure 13.7B shows a stochastic realization of the model, also with initial condition $[70, 35, 64]$. Phase shifts occur at $t = 7$ and $t = 15$. Directly underneath each of these two time series panels appear the corresponding composite state space plots (Figures 13.7C-D). The two fixed points corresponding to the two points on the original two-cycle attractor (triangle and circle) are separated by a basin boundary (surface), which hangs like a curtain between the two basins of attraction, visually indicated here by cross sections in the $P = 0, 60, 120$, and 180 planes. The saddle (cross) sits on the basin boundary. In Figure 13.7C, the deterministic trajectory of the composite model approaches fixed point $[18, 158, 106]$, remaining to the left of the basin boundary. In Figure 13.7D, the stochastic model time series exhibits two basin boundary crossings in composite state space, at $t = 7$ and $t = 15$. These crossings correspond to the phase shifts in the associated time series in Figure 13.7B.

Figures 13.7E-H show the experimental data from [7] in the same format. The two experimental replicate time series from Figure 13.5A are shown in panels E and F. The phase shifts in these two replicates correspond to the basin boundary crossings in composite state space (panels G and H, respectively).

From these cycling *Tribolium* data, we see that an understanding of phase switching in a constant habitat leads, surprisingly, to a discussion of multiple attractors, basins of attraction, basin boundaries, and saddles. In the next chapter, we will see how these concepts further lead to an understanding of resonance and attenuation in the advent of habitat periodicity.

For further exploration, see the examples for this section in the online LPA Simulator (https://lpasim.org/?c=13&s=2).

13.3 CONCLUSIONS

Cycles can be caused by inherent nonlinear mechanisms in a constant environment or can be periodically forced by the environment. In constant habitats, flour beetle populations often cycle [8, 10, 14] due to the Ricker-type nonlinear mechanisms of cannibalism [5], and they often stochastically switch cycle phase.

In this chapter, we discussed inherent cycles caused by nonlinear mechanisms in constant environments, how they are manifest in population time series, and how they differ quantitatively from purely random fluctuations. We also addressed the nonlinear phenomenon of phase switching and showed how it can be explained within the inherited structure of composite state space.

Why is it important to understand how cyclic mechanisms are structurally manifest in noisy time series, and why does the state space explanation of phase switching matter? The point is that stochastic perturbations of populations do not happen "in a vacuum." The underlying biological mechanisms establish the architecture of state space, including its attractors, basins of attraction, basin boundaries, saddles, and other invariant sets. These entities play a key role in how noise and pattern are manifested in experimental time series. *A complete understanding of the behavior of stochastic time series, even down to the prosaic phenomenon of phase switching, requires an understanding of biological mechanisms and the geometry of state space as it is shaped by those nonlinear forces.*

Bibliography

[1] Atkins GJ, Hayward JL, Henson SM (2021). How do gulls synchronize every-other-day egg laying? Wilson Journal of Ornithology 133:226-235. https://doi.org/10.1676/20-00019

[2] Chatfield C, Xing H (2019) The Analysis of Time Series: An Introduction with R. Seventh Edition. Boca Raton, Florida, Chapman & Hall/CRC.

[3] Costantino RF, Cushing JM, Dennis B, and Desharnais RA (1995) Experimentally induced transitions in the dynamic behavior of insect populations. Nature 375:227–230. https://doi.org/10.1038/375227a0

[4] Costantino RF, Desharnais RA, Cushing JM, Dennis B (1997) Chaotic dynamics in an insect population. Science 275:389–391. https://doi.org/10.1126/science.275.5298.389

[5] Costantino RF, Desharnais RA, Cushing JM, Dennis B, Henson SM, King AA (2005) Nonlinear stochastic population dynamics: The flour beetle *Tribolium* as an effective tool of discovery. In Desharnais RA (ed.), Advances in Ecological Research, pp. 101–141. New York, Academic Press.

[6] Dennis B, Desharnais RA, Cushing JM, Costantino RF (1995) Nonlinear demographic dynamics: Mathematical models, statistical methods, and biological experiments. Ecological Monographs 65:261–282. http://dx.doi.org/10.2307/2937060

[7] Desharnais RA, Costantino RF (1980) Genetic analysis of a population of *Tribolium*: VII. Stability: Response to genetic and demographic perturbations. Canadian Journal of Genetics and Cytology 22:577–589. https://doi.org/10.1139/g80-063

[8] Desharnais RA, Liu L (1987) Stable demographic limit cycles in laboratory populations of *Tribolium castaneum*. Journal of Animal Ecology 56:885–906. https://doi.org/10.2307/4955

[9] Gause G (1964) The Struggle for Existence. Baltimore, Williams and Wilkins.

[10] Henson SM, Cushing JM, Costantino RF, Dennis B, Desharnais RA (1998) Phase switching in population cycles. Proceedings of the Royal Society of London B 265:2229–2234. https://doi.org/10.1098/rspb.1998.0564

[11] Henson SM, Hayward JL, Cushing JM, Galusha JG (2010) Socially induced synchronization of every-other-day egg laying in a seabird colony. The Auk 127:571–580. https://doi.org/10.1525/auk.2010.09202

[12] Jillson DA (1980) Insect populations respond to fluctuating environments. Nature 288:699–700. https://doi.org/10.1038/288699a0

[13] King AA, Costantino RF, Cushing JM, Henson SM, Desharnais RA, Dennis B (2004) Anatomy of a chaotic attractor: Subtle model-predicted patterns revealed in population data. Proceedings of the National Academy of Sciences USA 101:408–413. https://doi.org/10.1073/pnas.2237266100

[14] Mertz DB (1969) Age-distribution and abundance in populations of flour beetles. I. Experimental studies. Ecological Monographs 39:1–31. https://doi.org/10.2307/1948563

[15] Nicholson AJ (1957) The self-adjustment of populations to change. Cold Spring Harbor Symposia on Quantitative Biology 22:153–173. https://doi.org/10.1101/sqb.1957.022.01.017

[16] Sandler AG, Megna LC, Hayward JL, Henson SM, Tkachuck CM, Tkachuck RD (2016) Every-other-day clutch-initiation synchrony in ring-billed gulls (*Larus delawarensis*). Wilson Journal of Ornithology 128:760–765. https://doi.org/10.1676/15-121.1

[17] Utida S (1957) Cyclic fluctuations of population density intrinsic to the host-parasite system. Ecology 38:442–449. https://doi.org/10.2307/1929888

CHAPTER 14

Resonance, Attenuance, and Multiple Attractors in Fluctuating Habitats

In an experiment conducted by Jillson [7], *Tribolium* populations were subjected to regularly-fluctuating habitats of various periodicities. Jillson was investigating how the "characteristic response time," a measure of how quickly a system returns to equilibrium following a disturbance, was related to the period of environmental fluctuation. May [10] had hypothesized that if the characteristic response time was long relative to the period of environmental fluctuation, then the population would average out environmental fluctuations, whereas if the response time was short relative to the habitat period, the population would track environmental variations. More importantly for our discussion here, *May had also hypothesized that periodic environmental fluctuations would decrease mean population size*, a prediction based on the logistic model.

Jillson noted that his constant habitat populations, cultured in $20g$ of flour, were in fact not at equilibrium but fluctuated due to cannibalistic interactions, and *he noted that periodically varying the habitat could increase the average population size, contrary to May's prediction.* In particular, he noted that the four-week habitat oscillation between 32 g for two weeks and 8 g for two weeks seemed ideally suited to boost population size because of the life stage durations and cannibalistic life-cycle interactions among *Tribolium*. Earlier, Wool and Sverdlov [13] had similarly reported that *Tribolium* productivity increased in temporally variable habitats. Jillson noted with clarity that "predictions of a reduction in mean population size were premised on the logistic model, which fails to incorporate the effects of cannibalism, an important factor in *Tribolium* population dynamics" [7].

Resonance is a well-known phenomenon in naturally oscillating systems in which the oscillator increases in amplitude or cycle average when subjected to external periodic forcing of similar frequency. In fact, Jillson's experiment had given a clear experimental demonstration of resonance in a population. Given the success of the LPA model, we wondered whether it could explain the resonance observed by Jillson and whether it could provide further testable hypotheses for varying environments.

In this chapter we discuss our theoretical and empirical studies of resonance using a periodically-forced version of the LPA model. Surprisingly, that model also predicts phenomena of attenuation and multiple attractors, which we were able to document in the laboratory.

DOI: 10.1201/9781003467120-14

14.1 A RICKER EXAMPLE

We begin by considering a generalized Ricker model with overlapping generations and periodic seasonal recruitment rate:

$$x_{t+1} = b\left(1 + \alpha(-1)^t\right) x_t \exp(-cx_t) + (1-\mu)\, x_t. \tag{14.1}$$

Here x_t is the population size at time t, the parameter $\mu \in (0,1)$ is the per capita probability of dying during one time interval, $\exp(-cx_t)$ is the fractional reduction of new recruits due to density dependent effects, and $b\left(1 + \alpha(-1)^t\right)$ is the seasonal recruitment rate that fluctuates around an average b with period two and amplitude $\alpha \in [0,1)$.

Model (14.1) predicts two-cycles for many values of its parameters. For purposes of illustration, consider parameter values $b = 40$, $c = 1$, and $\mu = 0.93$. For these parameters, the *autonomous*[1] version of model (14.1), that is, the model with $\alpha = 0$, predicts a stable two-cycle attractor. The sequence $x_0 = 1.035$, $x_1 = 14.78$, $\cdots$ and its phase-shifted sibling $x_0 = 14.78$, $x_1 = 1.035$, $\cdots$ are both locally-stable two-cycle solutions. Model (14.1) also has an unstable saddle equilibrium $x = c^{-1}\ln(b/\mu) \approx 3.761$, which can be obtained by solving the fixed-point equation

$$x = bx\exp(-cx) + (1-\mu)\, x.$$

As shown in Figure 14.1, as α is increased from $\alpha = 0$, one of these out-of-phase stable two-cycles increases in average (solid curves), and the other decreases in average (dashed curves). We designate these as the *resonant* and *attenuant* cycles, respectively. Furthermore, the (unstable) saddle equilibrium $x \approx 3.761$ is perturbed into a branch of unstable saddle two-cycles (dotted curves) as α is increased from zero. As α continues to increase, this branch of saddle cycles connects to and annihilates the branch of attenuant stable cycles (dashed curves) in a saddle-node bifurcation at about $\alpha = 0.57$ for these parameter values. The branch of resonant cycles (solid curves) persists for all $\alpha < 1$ [5].

For our discussion below, in which we discuss similar phenomena in the LPA system, it will be useful to visualize the cycles in three-dimensional LPA state space. To do this, we need to think in terms of the *composite* model, that is, the model composed with itself which looks at every-other time step.[2] Central to our discussion is the fact that, if an autonomous model is cycling on a two-cycle solution with values x_0, x_1, then its composite (think every-other-step) is at equilibrium, either with value x_0 or with value x_1, depending on whether we are looking at even or odd time steps. Furthermore, any equilibrium of the original model is also an equilibrium of its composite. Thus, the composite of model (14.1) with $\alpha = 0$ has three equilibria: two stable equilibria $x = 1.035$ and $x = 14.78$ corresponding to the two points on the two-cycle of the original model, and one unstable saddle equilibrium $x = 3.761$ which the composite inherits from the original model.

14.2 RESONANCE IN PERIODIC ENVIRONMENTS

In order to extend the LPA model to account for periodic habitat fluctuations, we recall from Chapter 4 that the cannibalism coefficients are inversely proportional to habitat size V, that is,

$$c_{ij} = \frac{k_{ij}}{V},$$

[1] An autonomous model is one that does not depend explicitly on time.

[2] The composite of $x_{t+1} = f(x_t)$ is the model $x_{t+1} = f(f(x_t))$.

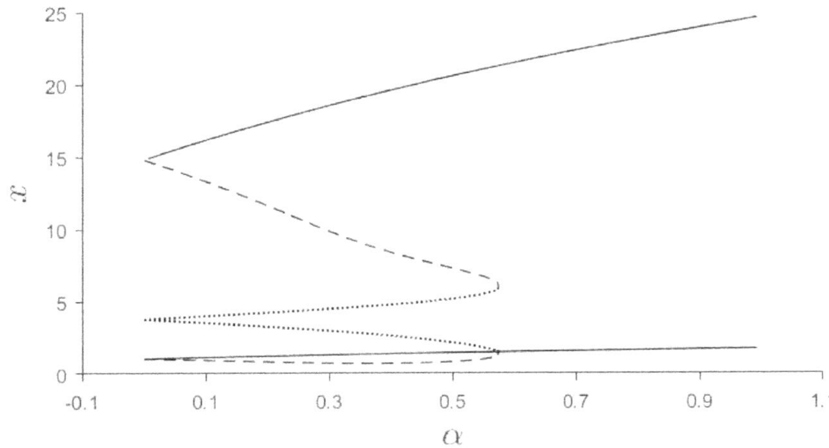

Figure 14.1 Two-cycle solutions of the periodically forced Ricker-type model (14.1) with $b = 40$, $c = 1$, and $\mu = 0.93$, shown as functions of amplitude α. When $\alpha = 0$, there are two stable two-cycles which are simply time shifts of each other, and an unstable fixed point. As α increases from zero, one of the two locally stable two-cycles increases in average (solid curves), while the other decreases in average (dashed curves). The unstable fixed point is perturbed into an unstable two-cycle (dotted curves). The branch of attenuant stable cycles (dashed curves) and the branch of unstable cycles annihilate each other in a saddle-node bifurcation at about $\alpha = 0.57$, while the resonant stable cycle persists for all $\alpha < 1$.

for constants $k_{ij} > 0$. To introduce periodicity in V, we take

$$V = V_{ave}(1 + \alpha(-1)^t),$$

where $\alpha \in [0, 1)$ denotes the amplitude of two-periodic oscillation about the average volume V_{ave}. If we take $V_{ave} = 1$ to be the standard volume occupied by 20 g of medium, and $\alpha = 0$, we see that $c_{ij} = k_{ij}$. Hence, the periodically-forced LPA model can be expressed as

$$\begin{aligned} L_{t+1} &= bA_t \exp\left(\frac{-c_{el}L_t - c_{ea}A_t}{1 + \alpha(-1)^t}\right) \\ P_{t+1} &= (1 - \mu_l)\, L_t \\ A_{t+1} &= P_t \exp\left(\frac{-c_{pa}A_t}{1 + \alpha(-1)^t}\right) + (1 - \mu_a)\, A_t. \end{aligned} \tag{14.2}$$

The periodically-forced model (14.2) collapses to the LPA model when $\alpha = 0$. For Jillson's experiment in which he varied the habitat between 32 g and 8 g of media, the amplitude was $\alpha = 0.6$, with volume V varying between $V = 0.4$ and $V = 1.6$.

The conditional least squares (CLS) parameter estimates for the autonomous LPA model obtained by Costantino et al. in [1] from Jillson's four constant-habitat cultures in [7] are

$$\begin{aligned} &b = 4.445, \qquad \mu_l = 0.4794, \qquad c_{pa} = 0.01053, \\ &c_{el} = 0.05841, \qquad c_{ea} = 0.005785, \qquad \mu_a = 0.1524. \end{aligned} \tag{14.3}$$

For these parameter values, the LPA model, that is model (14.2) with $\alpha = 0$, predicts a two-cycle attractor and (unstable) saddle equilibrium. Two phase-shifted stable two-cycle solutions "live" on this two-cycle attractor. In the advent of two-periodic forcing, that is, as α is increased from zero, these out-of-phase two-cycles are perturbed into:

1) a stable *resonant* two-cycle whose cycle average *increases* with forcing amplitude α (Figure 14.2A, solid curves), and

2) an out-of-phase stable *attenuant* two-cycle whose average *decreases* with forcing amplitude α (Figure 14.2A, dashed curves).

As the forcing amplitude α continues to increase, the attenuant stable two-cycle disappears in a saddle-node bifurcation at about $\alpha = 0.45$, whereas the resonant stable two-cycle persists as α continues to increase (Figure 14.2A). *The experimental value of $\alpha = 0.6$ in the Jillson experiment was well beyond the point at which the attenuant cycle is predicted to disappear. Thus, the periodically-forced LPA model (14.2) predicts resonance in the Jillson experiment.*

Figures 14.2B,C show deterministic model time series with $\alpha = 0$ and $\alpha = 0.6$, respectively. Note that the predicted two-cycle in panel C ($\alpha = 0.6$) is resonant; that is, it has larger average than the one in panel B ($\alpha = 0$). Directly below these figures are corresponding data time series from the Jillson experiment. The data in Figure 14.2D are from a constant habitat culture ($\alpha = 0$) and the data in Figure 14.2E are from a culture in which $\alpha = 0$ for weeks 0–16 ($t = 0$–8) and $\alpha = 0.6$ starting at week 18 ($t = 9$). The experimental data in the periodic habitat (panel E) exhibit a resonant cycle, which is consistent with the model prediction for $\alpha = 0.6$ (panel C).

The Jillson experiment, interpreted by the periodically-forced LPA model (14.2), demonstrates that *a population with a natural oscillation can increase in average population size in the advent of periodic habitat fluctuation, if the habitat oscillation has frequency similar to that of the natural oscillation.* More suprisingly, model (14.2) suggests that, *if the amplitude of habitat oscillation is small, both resonance and attenuance phenomena are possible!* In particular, *the model predicts the possibility of multiple, attracting, out-of-phase cycles, one resonant and one attentuant.* The prediction of these multiple attractors afforded an excellent testable hypothesis for the laboratory, which we now discuss in the next section.

For further exploration, see the examples for this section in the online LPA Simulator (https://lpasim.org/?c=14&s=2).

14.3 MULTIPLE ATTRACTORS IN PERIODIC ENVIRONMENTS

From our re-investigation of the historic Jillson experiment in the framework of the LPA model, as discussed above, we discovered that the periodically-forced LPA model (14.2) suggests non-intuitive predictions, not only of resonance and attenuance, but also of the phenomena associated with multiple attractors in composite state space, including the geometries of basins of attraction, saddles, and stable and unstable manifolds. We decided to use the periodically-forced LPA model (14.2) to construct testable hypotheses regarding the existence of these dynamic entities. To do so, we used the environmental-noise periodic LPA model

$$
\begin{aligned}
L_{t+1} &= bA_t \exp\left(\frac{-c_{el}L_t - c_{ea}A_t}{1+\alpha(-1)^t} + E_{1t}\right) \\
P_{t+1} &= (1-\mu_l)\, L_t \exp\left(E_{2t}\right) \\
A_{t+1} &= \left[P_t \exp\left(\frac{-c_{pa}A_t}{1+\alpha(-1)^t}\right) + (1-\mu_a)\, A_t\right] \exp\left(E_{3t}\right),
\end{aligned}
\tag{14.4}
$$

where the E_{it} are random variables having a multivariate normal distribution with mean zero and variance-covariance matrix Σ. We estimated the following maximum likelihood

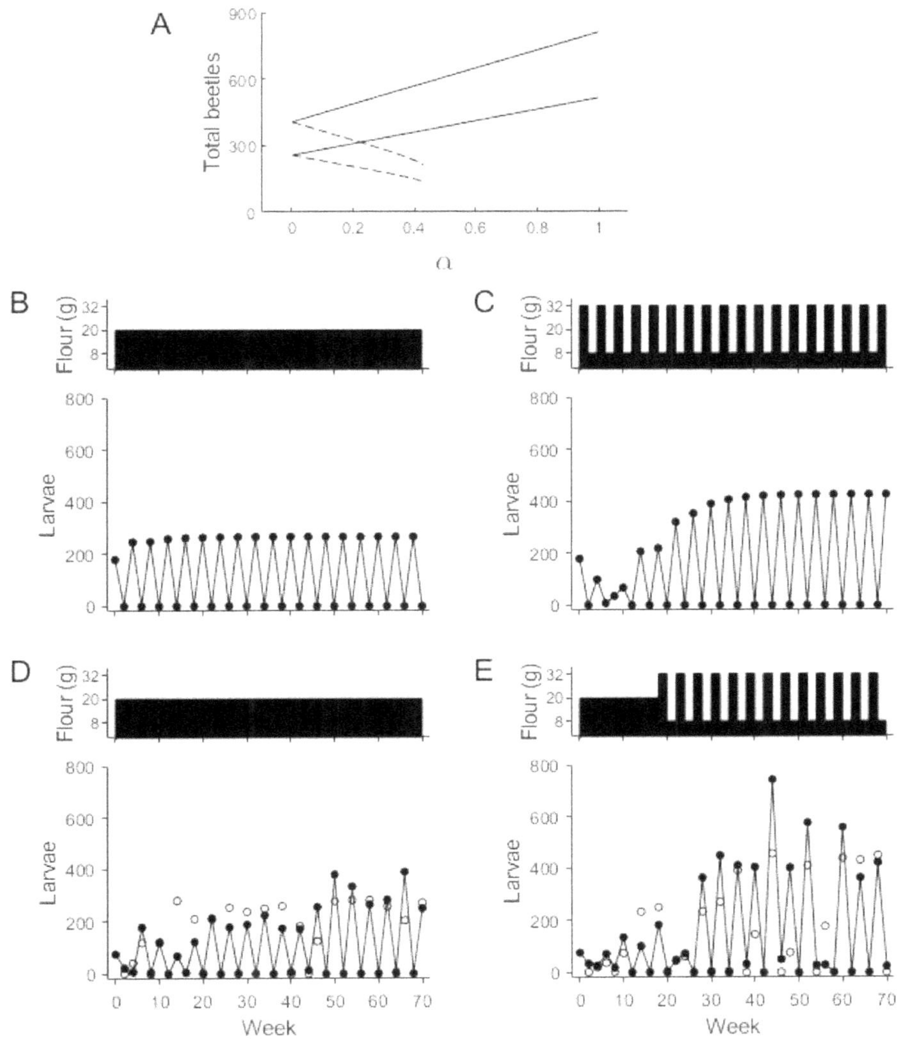

Figure 14.2 Model predictions and experimental data. A. Asymptotic predictions of the periodically-forced LPA model (14.2), using the Jillson parameters (14.3), with total numbers of animals plotted against relative amplitude of the flour volume oscillation α. The resonant two-cycle ($\alpha > 0$, solid curves) has a larger average than that of the two-cycle in the constant habitat ($\alpha = 0$). The attenuant two-cycle ($\alpha > 0$, dashed curves) has a smaller average. B. Deterministic time series of periodically-forced LPA model with relative amplitude $\alpha = 0$. C. Deterministic time series of periodically-forced LPA model with relative amplitude $\alpha = 0.6$. D. Experimental time series data from replicate 23 of [1] (solid circles), and one-step model predictions (open circles) in a constant habitat ($\alpha = 0$). E. Experimental time series data from replicate 25 of [1] (solid circles), and one-step model predictions (open circles) in a two-periodic habitat. Here $\alpha = 0$ for weeks 0–16 ($t = 0 - 8$) and $\alpha = 0.6$ starting at week 18 ($t = 9$).

(ML) parameter values from historical constant habitat data in [2]:

$$b = 6.598, \qquad \mu_l = 0.2055, \qquad c_{pa} = 0.004700, \tag{14.5}$$
$$c_{el} = 0.01209, \qquad c_{ea} = 0.01155, \qquad \mu_a = 0.007629, \tag{14.6}$$

$$\Sigma = \begin{pmatrix} 0.3411 & 0.0731 & -0.0017 \\ 0.0731 & 0.2488 & 0.0003 \\ -0.0017 & 0.0003 & 0.0002 \end{pmatrix}. \tag{14.7}$$

Using these parameters, we explored nearby parameter space for testable hypotheses. We chose to focus on

$$\begin{aligned} b &= 6.598, \qquad \mu_l = 0.2055, \qquad c_{pa} = 0.004700, \\ c_{el} &= 0.1, \qquad c_{ea} = 0.01, \qquad \mu_a = 0.1, \end{aligned} \tag{14.8}$$

where b, μ_l, and c_{pa} were the ML parameters (14.5), and c_{el}, c_{ea}, and μ_a were to be manipulated in the laboratory. The experimental protocol we used to manipulate c_{el}, c_{ea}, and μ_a is described in detail in [4].

For the parameters (14.8), we obtained the following deterministic model predictions. The autonomous LPA model, that is, model (14.2) with $\alpha = 0$, predicts a stable two-cycle (rounded to the nearest beetle)[3]

$$\begin{aligned} [L_0, P_0, A_0] &= [162, 0, 243] \\ [L_1, P_1, A_1] &= [0, 129, 219] \end{aligned}$$

along with its phase shift

$$\begin{aligned} [L_0, P_0, A_0] &= [0, 129, 219] \\ [L_1, P_1, A_1] &= [162, 0, 243], \end{aligned}$$

as well as an (unstable) saddle equilibrium

$$[23, 19, 110].$$

In the advent of periodic habitat forcing, that is, as amplitude α increases from zero, the first stable two-cycle increases in average (resonates), while the second decreases in average (attenuates). The unstable saddle equilibrium becomes an unstable saddle two-cycle. At a critical amplitude $\alpha_0 \approx 0.42$, the branch of stable attenuant cycles and the branch of unstable two-cycles annihilate each other in a saddle-node bifurcation, whereas the branch of stable resonant cycles persists for all $\alpha < 1$ (Figure 14.3).

The model therefore predicts three dynamic regimes as a function of the amplitude of habitat periodicity:

1. The zero-amplitude (constant habitat, autonomous) regime, with $\alpha = 0$, has a stable two-cycle attractor consisting of a stable two-cycle and its out-of-phase sibling;

2. The low-amplitude regime, with $0 < \alpha < \alpha_0 \simeq 0.42$, has multiple attractors consisting of a resonant two-cycle and an out-of-phase attenuant two-cycle;

3. The high-amplitude regime, with $\alpha > \alpha_0 \approx 0.42$, has a single attractor consisting of the resonant two-cycle. The attenuant two-cycle does not exist in this regime.

---

[3] We use square-bracket notation $[x, y, z]$ to denote the column vector $(x, y, z)^T$, which is the transpose of the row vector (x, y, z).

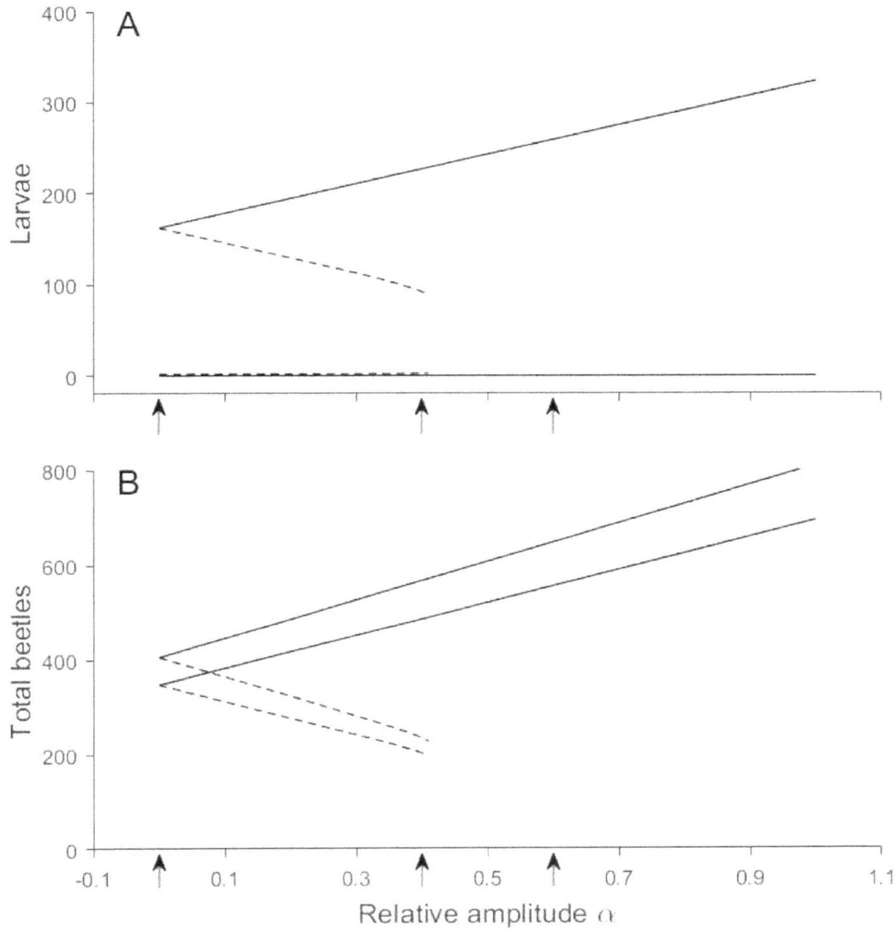

Figure 14.3 Two-cycle solutions of the periodically-forced LPA model (14.2), shown as functions of amplitude α. When $\alpha = 0$, there are two stable two-cycles which are simply time shifts of each other, and an unstable fixed point. As α increases from zero, one of the two locally stable two-cycles increases in average (solid curves), while the other decreases in average (dashed curves). The unstable fixed point is perturbed into an unstable two-cycle (not shown). At about $\alpha = 0.42$, the attenuant stable cycle and the unstable cycle annihilate each other in a saddle-node bifurcation, while the resonant stable cycle persists for all $\alpha < 1$. Arrows indicate placement of experimental cultures ($\alpha = 0$, 0.4, and 0.6). A. Larvae. B. Total population size.

We designed two experiments to study the predictions of these three dynamic regimes. In the first experiment, we chose three amplitudes (arrows, Figure 14.3), one in each of the three regimes, each with two initial conditions, in order to test the two-cycle attractor predictions of the periodically-forced LPA model (14.2). In the second experiment, we focused on the middle regime (the low-amplitude, multiple-attractor regime) with three different initial conditions, in order to test the geometry of the multiple attractors in composite state space.

We describe the experimental protocols in detail in [4].

TABLE 14.1 First experiment: predicted attractors of model (14.2) for each periodic habitat amplitude α and initial condition, with replicate numbers.

Initial condition	$\alpha = 0$ constant 20 g	$\alpha = 0.4$ alt 28 g, 12 g	$\alpha = 0.6$ alt 32 g, 8 g
$[150, 200, 150]$	autonomous two-cycle $[162, 0, 243]$ $[0, 129, 219]$ reps 1,10,14	resonant two-cycle $[0, 180, 306]$ $[227, 0, 340]$ reps 3,9,18	resonant two-cycle $[0, 206, 350]$ $[259, 0, 388]$ reps 4,11,16
$[150, 0, 150]$	same as above reps 6,12,17	attenuant two-cycle $[92, 0, 142]$ $[0, 73, 128]$ reps 5,8,13	same as above reps 2,7,15

14.3.1 Model Predictions and Experimental Design

14.3.1.1 First experiment

In the first experiment, we chose three amplitudes ($\alpha = 0$, $\alpha = 0.4$, and $\alpha = 0.6$) and two initial conditions ($[150, 200, 150]$ and $[150, 0, 150]$) in order to evaluate the asymptotic predictions in the three dynamic regimes. The experimental design and model-predicted two-cycle attractors are shown in Table 14.1. For $\alpha = 0$, both initial conditions are predicted to lead to trajectories that approach the autonomous stable two-cycle attractor. For $\alpha = 0.4$, the trajectory with initial condition having $P = 200$ approaches the resonant two-cycle attractor, whereas the trajectory with initial condition having $P = 0$ approaches the attenuant two-cycle attractor. For $\alpha = 0.6$, both initial conditions lead to trajectories that approach the resonant two-cycle attractor.

The model-predicted trajectories toward the attractors in Table 14.1 are shown in Figure 14.4 both as time series and in composite state space. Figure 14.4A shows the predictions for $\alpha = 0$. Both initial conditions lead to the same stable two-cycle; and in composite state space, every other step of these trajectories (squares) approach the fixed point $[162, 0, 243]$. For $\alpha = 0.4$ (Figure 14.4B), the initial condition with $P = 0$ leads to the attenuant two-cycle, with larval numbers in phase with the periodic habitat. The initial condition with $P = 200$ leads to the resonant two-cycle, with larval numbers out of phase with the habitat. In composite state space, the trajectory with initial $P = 0$ approaches the composite fixed point $[92, 0, 142]$, which is on the attenuant two-cycle, whereas the trajectory with $P = 200$ leads to the composite fixed point $[0, 180, 306]$, which is on the resonant two-cycle. For $\alpha = 0.6$ (Figure 14.4C), both time series approach the resonant two-cycle, and in composite state space, both trajectories approach the associated fixed point $[0, 206, 350]$. In this latter case, there is no attenuant cycle, because it was annihilated in the saddle-node bifurcation at approximately $\alpha = 0.45$.

Let's concentrate for a moment on the predictions for $\alpha = 0.4$. The transient dynamics predicted for both initial conditions, shown in composite state space in Figure 14.4B, are of particular interest. These two trajectories are shown again in Figure 14.5A from a slightly different angle, this time within the context of the nonlinear configuration of composite state space. The resonant and attenuant stable two-cycles shown in Table 14.1 for $\alpha = 0.4$ correspond to "upper" and "lower" stable fixed points $[0, 180, 306]$ and $[92, 0, 142]$ of the composite model (Figure 14.5A, solid circles). The unstable saddle two-cycle corresponds to an (unstable) saddle fixed point $[64, 3, 121]$ of the composite model (Figure 14.5A, cross). Numerical computations indicate the basins of attraction of the two stable fixed points are

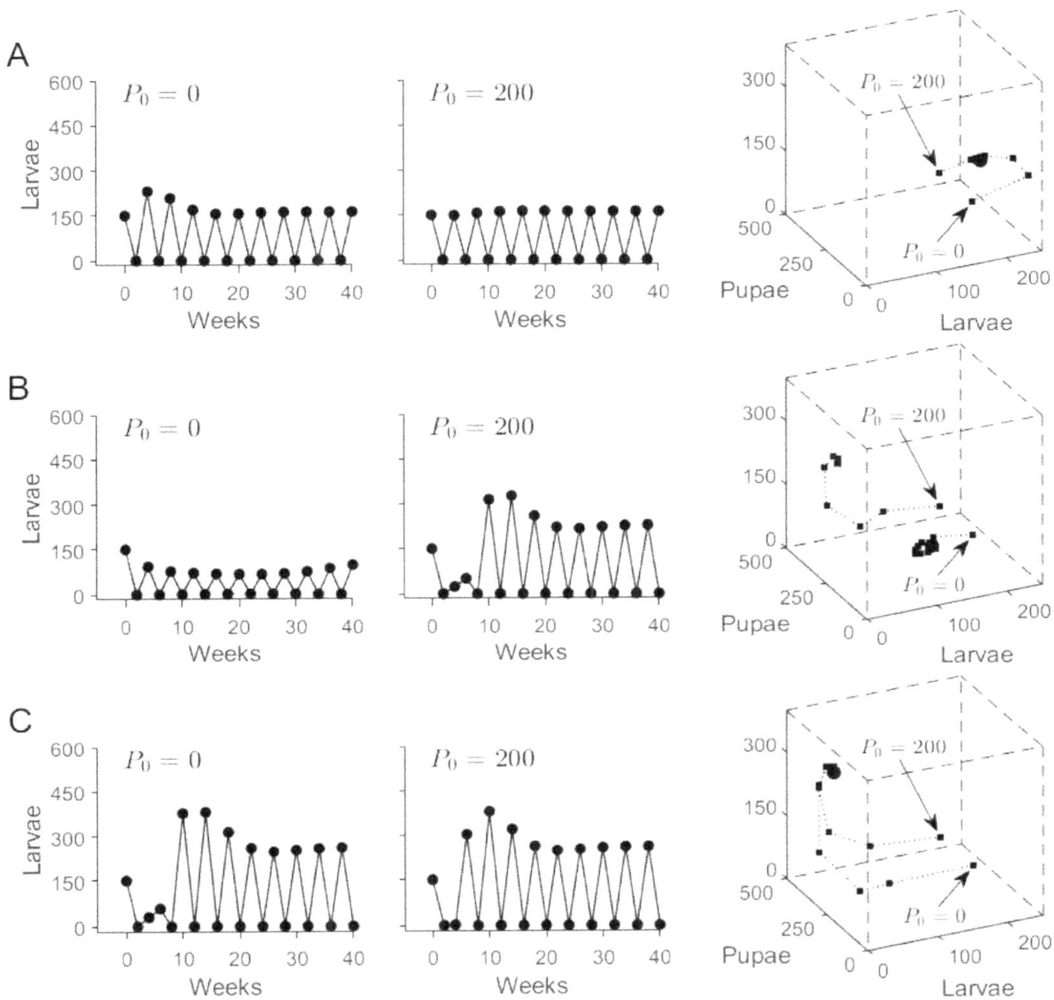

Figure 14.4 Model predictions for the first experiment, including transient dynamics. On the left are L-stage predicted time series; on the right are the trajectories in *composite state space* with fixed points shown as large solid circles. Solid squares correspond to the every-other-step times when the flour volume is high. A. Amplitude $\alpha = 0$ (constant 20 g habitat). Both initial conditions lead to a two-cycle with larvae oscillating in phase with the habitat. In composite state space, both trajectories approach the associated fixed point $[162, 0, 243]$. B. Amplitude $\alpha = 0.4$ (28 g-12 g habitat). The initial condition with $P = 0$ leads to the attenuant two-cycle, with larval numbers in phase with the habitat. The initial condition with $P = 200$ leads to the resonant two-cycle, with larval numbers out of phase with the habitat. In composite state space, the trajectory with initial $P = 0$ approaches the composite fixed point $[92, 0, 142]$, which is on the attenuant two-cycle, whereas the trajectory with $P = 200$ leads to the the composite fixed point $[0, 180, 306]$, which is on the resonant two-cycle. C. Amplitude $\alpha = 0.6$ (32 g-8 g). Both time series approach the resonant two-cycle, and in composite state space, both trajectories approach the associated fixed point $[0, 206, 350]$. In this latter case, no attenuant cycle exists.

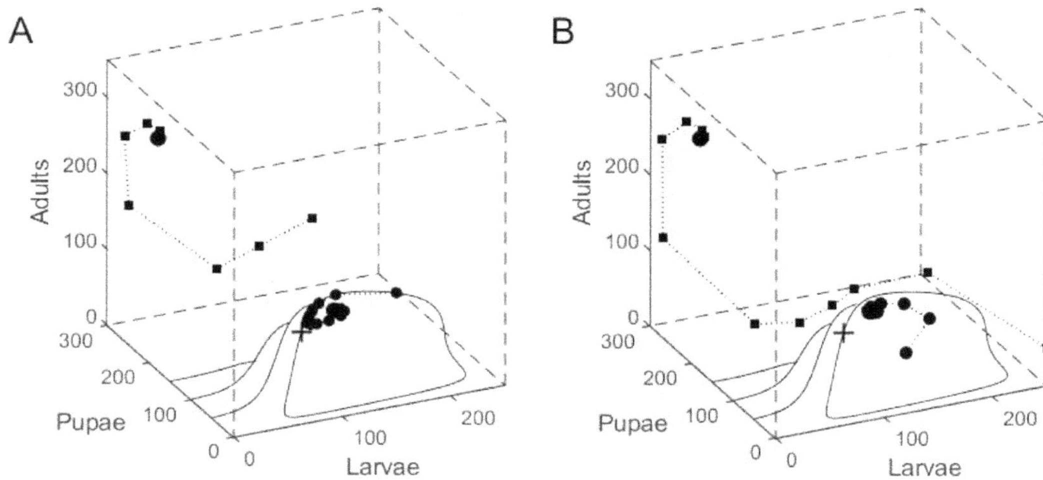

Figure 14.5 Deterministic model predictions. The configuration of composite state space for $\alpha = 0.4$ consists of two fixed points (large circles) and a saddle fixed point (cross) lying on the basin boundary (surface). A. Predicted deterministic trajectories for the first experiment. The trajectory starting at $[150, 200, 150]$ (squares) tends to the upper fixed point. The trajectory starting at $[150, 0, 150]$ (small circles) follows the basin boundary, flies by the saddle, then approaches the lower fixed point. B. Predicted deterministic trajectories for the second experiment. The trajectory starting at $[250, 0, 50]$ (squares) follows the boundary and flies by the saddle before approaching the upper fixed point. The trajectory starting at $[120, 0, 80]$ (small circles) spirals into the lower fixed point. The trajectory starting on the lower fixed point $[92, 0, 142]$ itself remains on the lower fixed point for all time.

"simple"; the boundary is a "smooth" surface in state space (Figure 14.5A, surface). The saddle (Figure 14.5A, cross) lies on the basin boundary, which is also the stable manifold of the saddle. The initial condition with $P = 200$ lies in the basin of the upper fixed point. The initial condition with $P = 0$ is in the basin of the lower fixed point but lies close to the basin boundary, that is, close to the stable manifold of the saddle. The ensuing trajectory follows the basin boundary and flies by the saddle before exiting toward the lower fixed point (Figure 14.5A).

In fact, the treatment with $\alpha = 0.4$ and initial condition $[150, 0, 150]$ deserves further special attention because of the way the trajectory follows the basin boundary and flies by the saddle. How might stochasticity alter this prediction? The eigenvalues and eigenspace basis vectors for the linearizations of the composite LPA model about its saddle fixed point $[63.60, 2.84, 120.79]$ are

$$\begin{aligned}
\lambda &= 2.66; \quad [1.00, -0.08, 0.15] \\
\lambda &= 0.70; \quad [0.23, -0.06, 1.00] \\
\lambda &= -1.87 \times 10^{-4}; \quad [-0.01, 1.00, -0.74] .
\end{aligned}$$

Thus, the saddle has a one-dimensional unstable manifold curve that is tangent to the line determined by the vector $[1.00, -0.08, 0.15]$, and a two-dimensional stable manifold surface (which is the basin boundary in Figure 14.5) that is tangent to the plane determined by the vectors $[0.23, -0.06, 1.00]$ and $[-0.01, 1.00, -0.74]$.

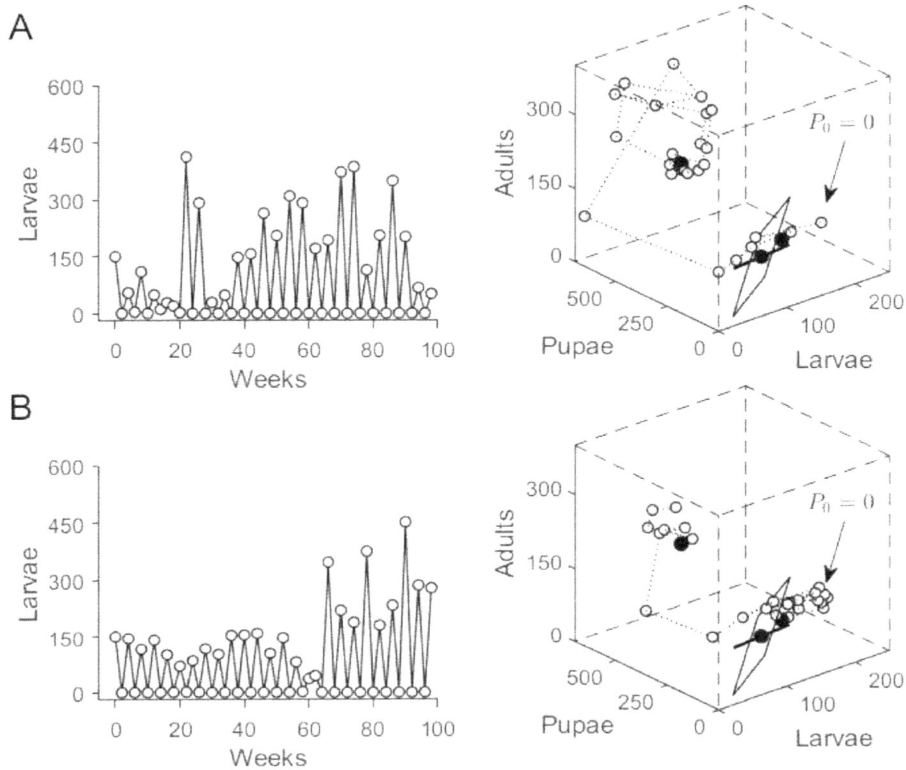

Figure 14.6 Stochastic model trajectories as time series and in composite state space in the 28 g-12 g ($\alpha = 0.4$) habitat, with initial condition $[150, 0, 150]$. A. Stochastic time series that changes phase at week 20 and exhibits the resonant cycle. In composite state space, the trajectory moves along the stable manifold, flies by the saddle, and exits along the unstable manifold toward the upper fixed point. B. Stochastic time series that exhibits the attenuant cycle for 60 weeks, changes phase at week 62, and then exhibits the resonant cycle. In composite state space, the trajectory lingers about the lower fixed point before being knocked into the basin of the upper fixed point. Parameters are from (14.8) and (14.7).

These stable and unstable manifolds are shown in Figure 14.6, along with two representative stochastic model trajectories for the treatment with $\alpha = 0.4$ and initial condition $[150, 0, 150]$, using the noise parameters (14.7). This figure should be compared to the deterministic predictions in Figure 14.4B and Figure 14.5A. In Figure 14.6A, the stochastic time series begins a depressed-amplitude cycling whose larval component is in phase with the attenuant cycle. The amplitude begins to decrease, reminiscent of time-series signature of a classic saddle flyby (Chapter 12). At week 20, the time series switches phase and the amplitude begins to increase. The stochastic time series then exhibits the resonant cycle. In composite state space (Figure 14.6A), the stochastic trajectory moves along the stable manifold, flies by the saddle, and exits along the unstable manifold toward the upper fixed point. In Figure 14.6B, the stochastic time series exhibits the attenuant cycle for 60 weeks, changes phase at week 62, and then exhibits the resonant cycle. In composite state space (Figure 14.6B), the trajectory lingers about the lower fixed point before being knocked into the basin of the upper fixed point.

TABLE 14.2 Second experiment: predicted attractors of model (14.2) and its composite transients for each initial condition in the alternating 28 g-12 g habitat ($\alpha = 0.4$), with replicate numbers.

Initial condition	Predicted attractor of model (14.2)	Deterministic transients of composite model
$[250, 0, 50]$ reps 24,25,28,31	Resonant two-cycle $[0, 180, 306]$ $[227, 0, 340]$	Move along basin boundary toward saddle, exit toward $[0, 180, 306]$
$[120, 0, 80]$ reps 21,26,29,30	Attenuant two-cycle $[92, 0, 142]$ $[0, 73, 128]$	Move along basin boundary toward saddle, exit toward $[92, 0, 142]$
$[92, 0, 142]$ reps 22,23,27,32	Attenuant two-cycle $[92, 0, 142]$ $[0, 73, 128]$	Deterministic trajectory remains on $[92, 0, 142]$

Thus, stochastic simulations illustrate that, for $\alpha = 0.4$:

1. The attenuant cycle can influence the stochastic dynamics for a long time;
2. Eventually such a stochastic attenuant cycle will switch phase and resonate as it is knocked into the (larger) basin of attraction of the resonant cycle;
3. This change can happen in a sudden, dramatic way, or more slowly via a saddle flyby, manifested in the time series as an episode of depressed amplitude.

14.3.1.2 Second experiment

In the second experiment, we focused on the low-amplitude ($\alpha = 0.4$), multiple-attractor regime. Our goal was to further test the model-predicted geometry of the multiple attractors in composite state space as shown in Figure 14.5.

In order to explore the geometry of state space for $\alpha = 0.4$, we chose initial conditions $[250, 0, 50]$, $[120, 0, 80]$, and $[92, 0, 142]$. Table 14.2 summarizes the experimental design and deterministic model predictions for the second experiment. Initial condition $[250, 0, 50]$ leads to the resonant two-cycle; initial condition $[120, 0, 80]$ leads to the attenuant two-cycle; and initial condition $[92, 0, 142]$ starts right on the attenuant two-cycle and thus is predicted to remain there for all time in the absence of noise.

Figure 14.5B shows the deterministic model-predicted trajectories toward the attractors in composite state space. The trajectory with initial condition $[250, 0, 50]$ moves along the basin boundary before exiting toward the upper fixed point. The trajectory with initial condition $[120, 0, 80]$ moves along the basin boundary before moving toward the lower fixed point.

14.3.2 Results

We now turn our attention to the experimental data and compare them to the model predictions. As we do so, it is important to recall that the model predictions are based on parameters estimated from an independent historical data set; that is, *the model was not re-parameterized with the new data.*

14.3.2.1 First experiment

Data from six representative replicates in the first experiment are shown in Figures 14.7–14.8. Figure 14.7 shows data for initial condition $[150, 0, 150]$, and Figure 14.8 shows data

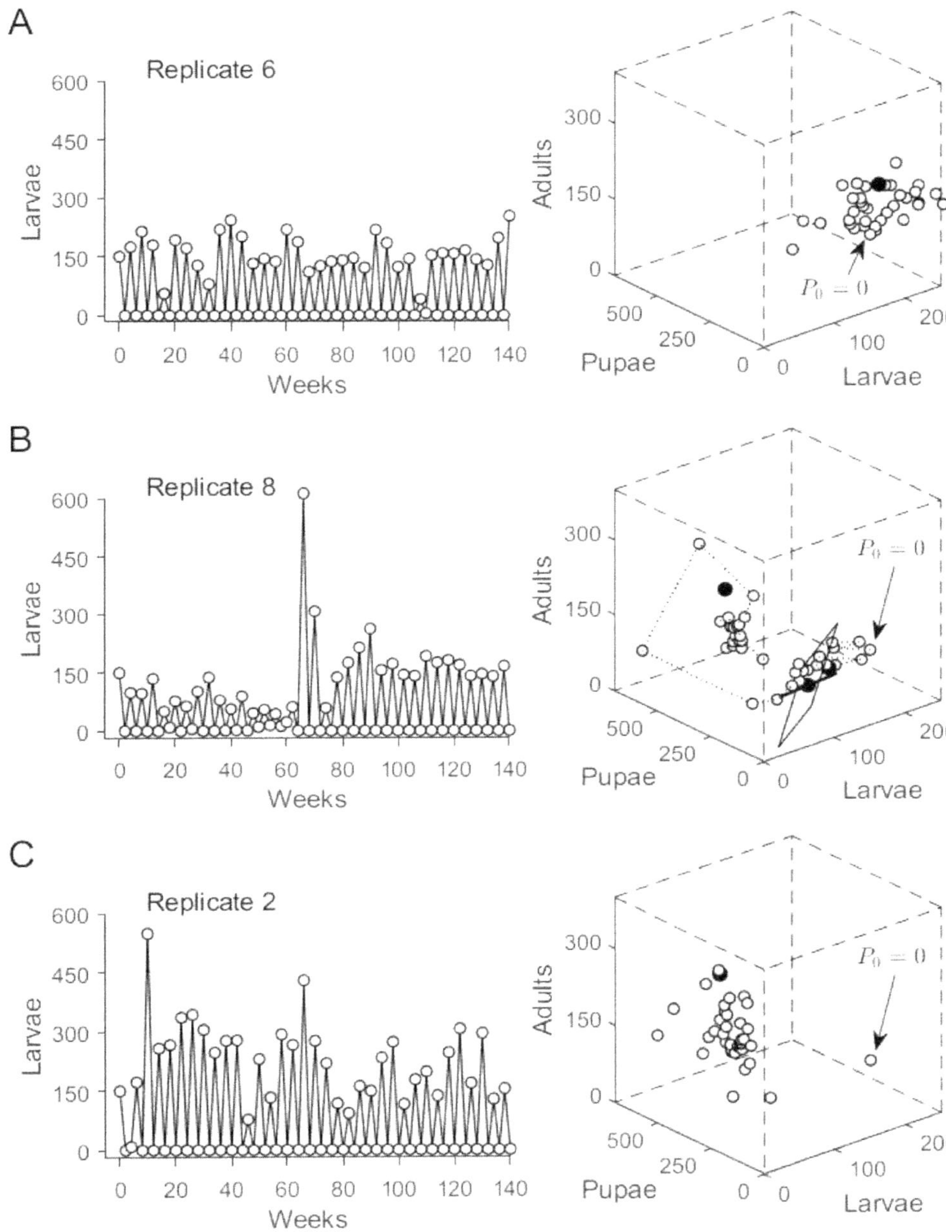

Figure 14.7 Data from three representative replicates from the first experiment with initial condition $[150, 0, 150]$. The data are shown as time series on the left, and in composite state space on the right. A. $\alpha = 0$ (constant $20g$), replicate 6. B. $\alpha = 0.4$ (alternating $28g$, $12g$), replicate 8. C. $\alpha = 0.6$ (alternating 32 g, 8 g), replicate 2.

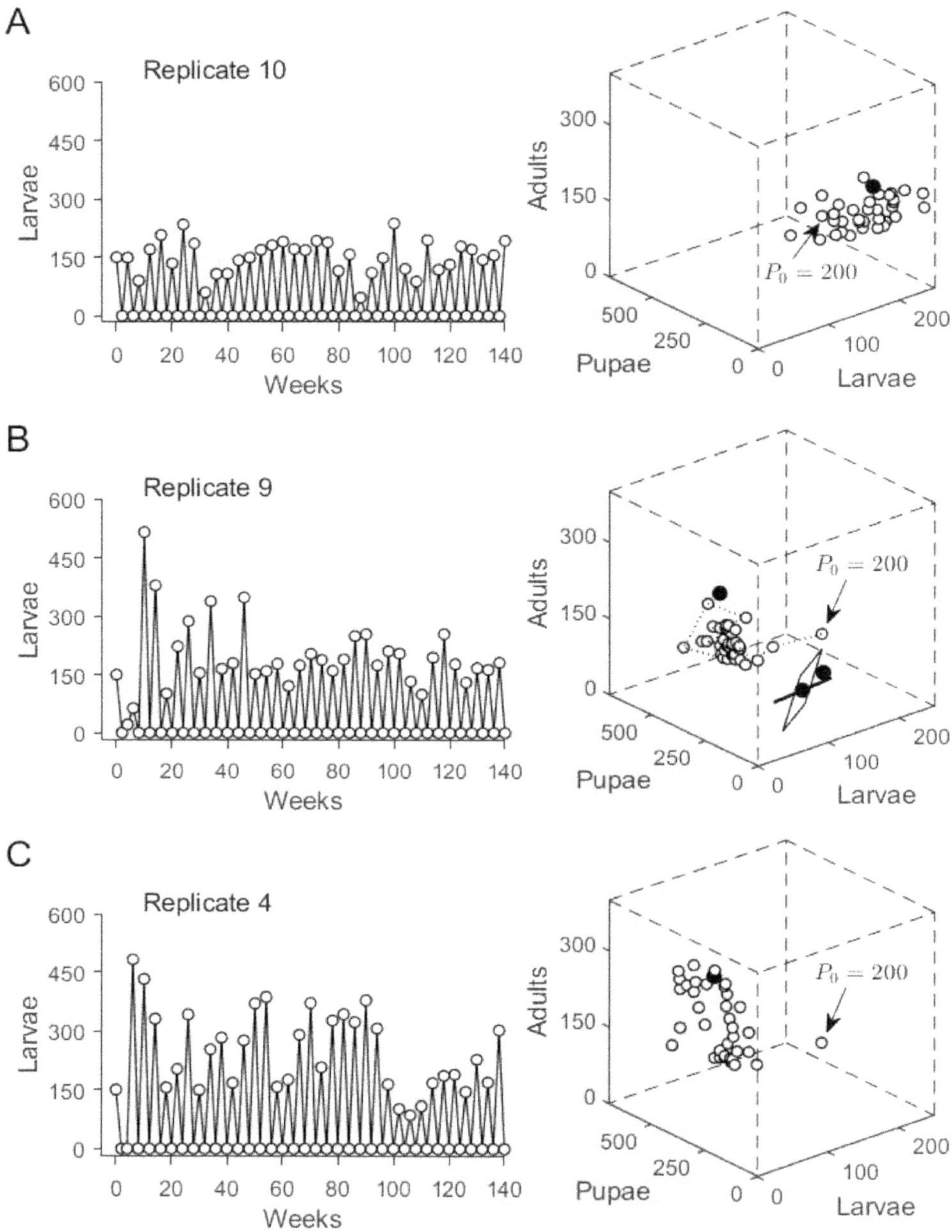

Figure 14.8 Data from three representative replicates from the first experiment with initial condition [150, 200, 150]. The data are shown as time series on the left, and in composite state space on the right. A. $\alpha = 0$ (constant 20 g), replicate 10. B. $\alpha = 0.4$ (alternating 28 g, 12 g), replicate 9. C. $\alpha = 0.6$ (alternating 32 g, 8 g), replicate 4.

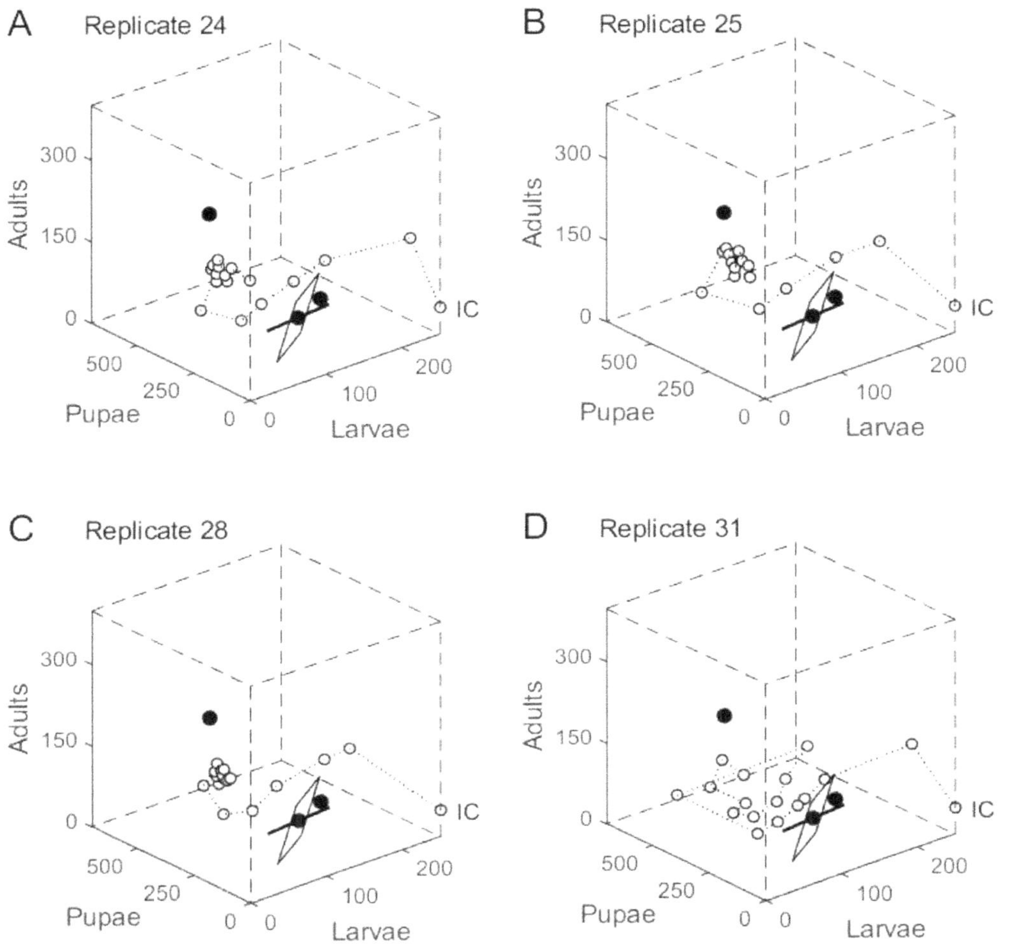

Figure 14.9 Data from all four replicates of the second experiment with initial condition [250, 0, 50] in composite state space. This initial condition is in the basin of attraction of the upper fixed point.

for initial condition [150, 200, 150]. These figures should be compared to the deterministic predictions in Figures 14.4 and 14.5A, and, for the $\alpha = 0.4$ and initial condition [150, 0, 150] treatment, the stochastic predictions in Figure 14.6.

In Figures 14.7A and 14.8A, the experimental data for both initial conditions ($P_{IC} = 0$ and $P_{IC} = 200$) cluster about the autonomous two-cycle as predicted in Figure 14.4A. Similarly, in Figures 14.7C and 14.8C, the experimental data for both initial conditions cluster about the resonant two-cycle as predicted in Figure 14.4C. Furthermore, the data in Figure 14.8B, which started with $P_{IC} = 200$, cluster about the resonant two-cycle, as predicted in Figure 14.4B.

The most interesting case is in Figure 14.7B (replicate 8). These data begin with initial $P_{IC} = 0$ in the $\alpha = 0.4$ habitat. The data hover about the attenuant attractor for some 60 weeks before flying by the saddle and exiting along the unstable manifold toward the resonant attractor. Compare these data to the stochastic predictions in Figure 14.6.

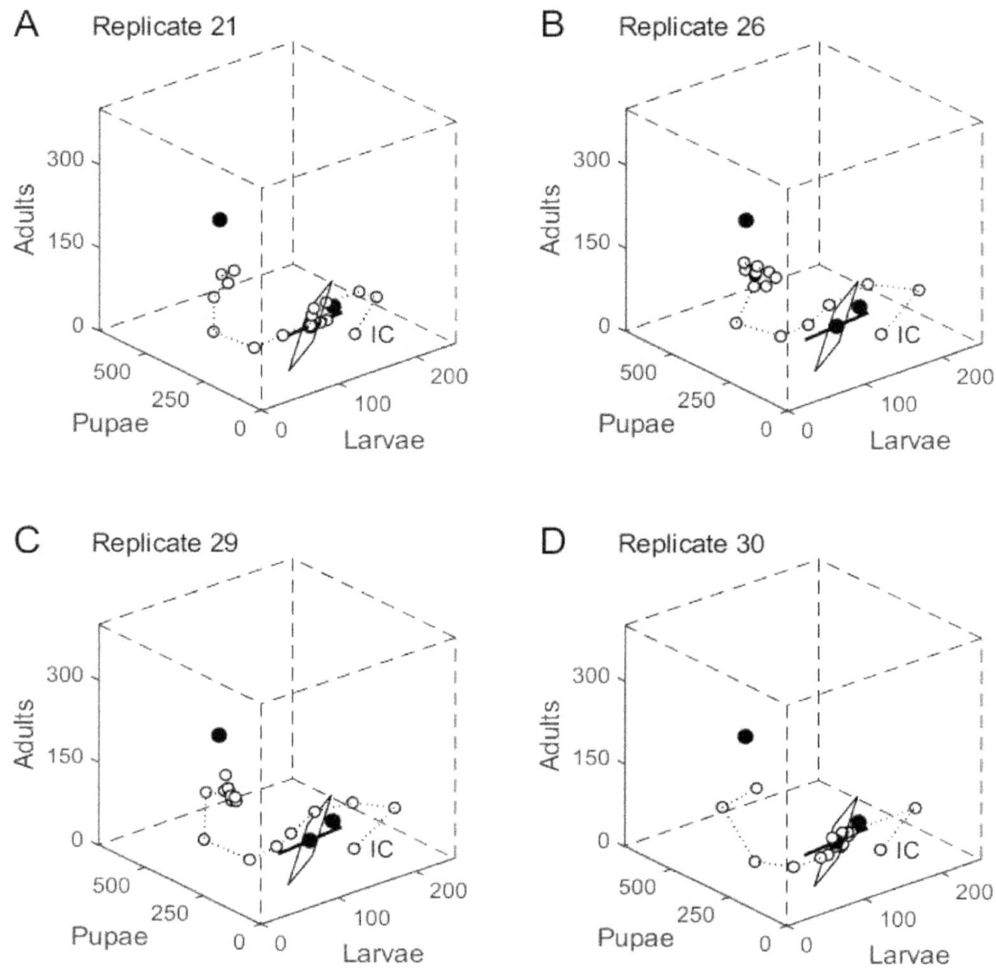

Figure 14.10 Data from all four replicates of the second experiment with initial condition [120, 0, 80] in composite state space. This initial condition is in the basin of attraction of the lower fixed point.

14.3.2.2 Second experiment

Data from all 12 replicates in the second experiment are shown in Figures 14.9–14.11, grouped according to the three initial condition treatments. These figures should be compared to the deterministic predictions in Figure 14.5B.

For $\alpha = 0.4$ and initial condition [250, 0, 50], all four replicates move toward the upper fixed point in composite state space (Figure 14.9), as predicted by the deterministic trajectories in Figure 14.5B. Replicate 31 (Figure 14.9D), in particular, follows the deterministic prediction as it moves along the basin boundary, flies by the saddle, and exits along the unstable manifold toward the upper fixed point.

For $\alpha = 0.4$ and initial condition [120, 0, 80], the deterministic prediction is that the trajectories will approach the lower fixed point. Two of the experimental replicates (26 and 29, Figure 14.10B,C) follow the basin boundary, are stochastically knocked out of the basin of the lower fixed point, exit along the unstable manifold, and approach the upper fixed point. The other two replicates (21 and 30, Figure 14.10A,D) hover about the lower fixed point before being knocked out of its basin and approaching the upper fixed point.

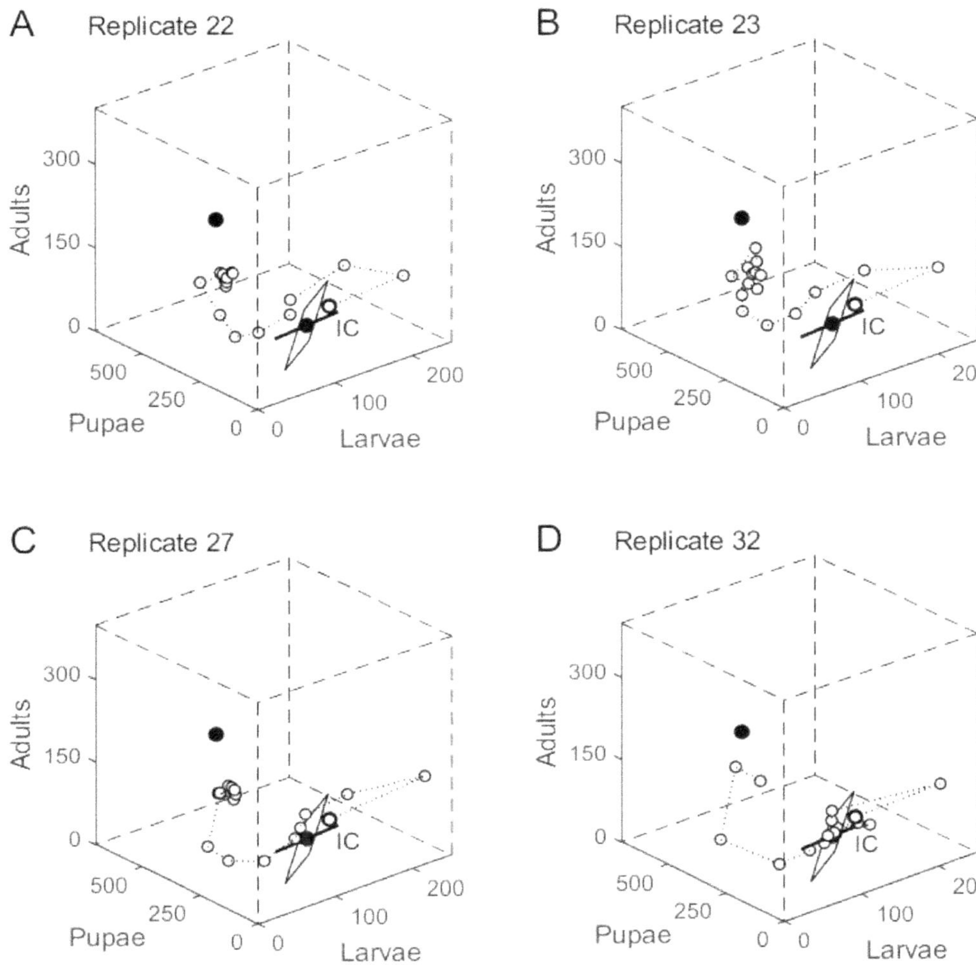

Figure 14.11 Data from all four replicates of the second experiment with initial condition [92, 0, 142] in composite state space. These replicates started directly on the lower fixed point.

For $\alpha = 0.4$ and initial condition [92, 0, 142], the deterministic prediction is for the trajectory to stay on the lower fixed point [92, 0, 142] for all time. Given stochasticity, of course this will not happen, and in fact all four replicates eventually approach the upper fixed point. Three of the four replicates moved away from the lower attractor quickly (within 20 weeks). Replicate 32 (Figure 14.11D) hovered about the lower fixed point for 48 weeks before being knocked out of its basin and following the now familiar path to the upper fixed point.

For further exploration, see the examples for this section in the online LPA Simulator (https://lpasim.org/?c=14&s=3).

14.4 CONCLUSIONS

The Jillson experiment confirmed experimentally, and the LPA model supported theoretically, that increased population biomass can result from resonance between a natural oscillation and periodic habitat oscillation of similar frequency. Jillson noted that the four-week habitat oscillation between 32 g for two weeks and 8 g for two weeks seemed ideally suited to boost population size because of the life stage durations and cannibalistic life-cycle interactions among *Tribolium* [7]. Jillson's findings did not agree with earlier predictions in the literature that fluctuations in carrying capacity would reduce the long-term average population size below that attained in a constant environment [9, 10, 11]. Predictions of a reduction in mean population size were premised on the logistic model, which does not incorporate life stage interactions such as cannibalism, an important factor in *Tribolium* population dynamics [3, 8].

Fluctuating habitats are of great importance in ecology. In this chapter, we saw how an exploration of resonance in the two-periodically-forced LPA model (14.2) led to non-intuitive predictions of other types of nonlinear phenomena in periodic habitats, including attenuant cycles, multiple attractors, basins of attraction, saddles, stable and unstable manifolds, and saddle flybys. *All of these phenomena were verified experimentally in the laboratory.*

Here, we have focused on two-periodic forcing of a natural two-cycle oscillator, but mathematically, all of these phenomena can occur for inherent cycles and periodic forcing of higher periods. In general, an autonomous discrete dynamical system with an oscillatory solution develops multiple oscillatory solutions when perturbed by periodic forcing. The period of the resulting cycles is the least common multiple of the inherent and forcing periods, and the resulting cycles are out of phase and may differ in average. Generically, the average of at least one of the cycles will resonate with the forcing, while the average of at least one other will attenuate [5, 6]. That is to say, the advent of a periodic habitat may enhance or depress average total population size, depending on the initial population vector. In an even more general setting, these results hold true for other types of attractors, as well; and for small amplitude periodic forcing, the attractors for the forced system inherit their structure from the attractor of the original autonomous system and the forcing frequency [12]. The architecture of these multiple attractors in composite state space gives rise to the possibility that *a population system can stochastically switch between resonant and attenuant modes in the presence of fluctuating habitats.*

Bibliography

[1] Costantino RF, Cushing JM, Dennis B, Desharnais RA, Henson SM (1998) Resonant population cycles in temporally fluctuating habitats. Bulletin of Mathematical Biology 60:247–273. https://doi.org/10.1006/bulm.1997.0017

[2] Costantino RF, Desharnais RA, Cushing JM, Dennis B (1997) Chaotic dynamics in an insect population. Science 275:389–391. https://doi.org/10.1126/science.275.5298.389

[3] Desharnais RA, Costantino RF (1982) The approach to equilibrium and the steady-state probability distribution of adult numbers in *Tribolium brevicornis*. American Naturalist 119:102–111. https://doi.org/10.1086/283893

[4] Henson SM, Costantino RF, Cushing JM, Dennis B, Desharnais RA (1999) Multiple attractors, saddles, and population dynamics in periodic habitats. Bulletin of Mathematical Biology 61:1121–1149. https://doi.org/10.1006/bulm.1999.0136

[5] Henson SM (2000) Multiple attractors and resonance in periodically forced population models. Physica D: Nonlinear Phenomena 140:33–49. https://doi.org/10.1016/S0167-2789(99)00231-6

[6] Henson SM, Costantino RF, Desharnais RA, Cushing JM, Dennis B (2002) Basins of attraction: Population dynamics with two stable 4-cycles. Oikos 98:17–24. https://doi.org/10.1034/j.1600-0706.2002.980102.x

[7] Jillson DA (1980) Insect populations respond to fluctuating environments. Nature 288:699–700. https://doi.org/10.1038/288699a0

[8] Lloyd M (1968) Self regulation of adult numbers by cannibalism in two laboratory strains of flour beetles (*Tribolium castaneum*). Ecology 49:245–259. https://doi.org/10.2307/1934453

[9] May RM (1973) Stability and Complexity in Model Ecosystems. Princeton, Princeton University Press.

[10] May RM (1976) Theoretical Ecology: Principles and Applications. Philadephia, Saunders.

[11] Nisbet RM, Gurney WSC (1976) Population dynamics in a periodically varying environment. Journal of Theoretical Biology 56:459–475. https://doi.org/10.1016/S0022-5193(76)80086-0

[12] Selgrade JF, Roberds JH. On the structure of attractors for discrete, periodically forced systems with applications to population models. Physica D: Nonlinear Phenomena 158:69–82. https://doi.org/10.1016/S0167-2789(01)00324-4

[13] Wool D, Sverdlov E (1976) Sib-mating populations in an unpredictable environment: Effects on components of fitness. Evolution 30:119–129. https://doi.org/10.1111/j.1558-5646.1976.tb00889.x

V

Further Exploration

CHAPTER 15

Competition: Park Revisited

Species competition and the conditions for coexistence are major topics in ecology. A key premise, the "competitive exclusion principle," states that two ecologically similar species cannot coexist. Thomas Park and his colleagues conducted experiments using two closely related species, *Tribolium castaneum* and *Tribolium confusum*, over a 20-year period, looking at a variety of experimental conditions. Except in rare circumstances, they found that one species excluded the other. In this chapter, we explore *Tribolium* competition using a two-species version of the LPA model and demonstrate that coexistence, as well as competitive exclusion, can occur when interspecific competition is increased, a prediction that contradicts the classical tenets of competition theory.

15.1 COMPETITION THEORY IN ECOLOGY

Competition theory in ecology attempts to explain how two or more species sharing the same limited resources can coexist. A mathematical treatment of competition theory began with the pioneering work of Alfred J. Lotka [10] and Vito Volterra [25].

15.1.1 Lotka-Volterra Equations

The Lotka-Volterra equations describe the population dynamics of two interacting species. In the case of two competitors, the model consists of two ordinary autonomous differential equations:

$$\begin{aligned} dN_1(t)/dt &= N_1(t)\left(r_1 - \alpha_{11}N_1(t) - \alpha_{12}N_2(t)\right) \\ dN_2(t)/dt &= N_2(t)\left(r_2 - \alpha_{22}N_2(t) - \alpha_{21}N_1(t)\right). \end{aligned} \tag{15.1}$$

The parameters r_1 and r_2 are the *intrinsic rates of increase* for species 1 and species 2, respectively. The *competition coefficients*, α_{ij} $(i, j = 1.2)$, represent the *per capita* change in the growth rate of species i due to changes in the density of species j. The coefficients α_{11} and α_{22} represent the density-dependent effects of intraspecific competition, and the coefficients α_{12} and α_{21} represent the interspecific effects of competition over shared resources. When each species is alone, its dynamics follow the continuous-time *logistic model* of population growth [22, 24]. Despite its simplicity, the Lotka-Volterra model (15.1) has played an important role in ecological theory [11, 12].

15.1.2 Species Isoclines and Model Predictions

The dynamics of the classic competition model (15.1) can be analyzed graphically by examining the *isoclines* for the two species, which are found by setting the growth rates equal

DOI: 10.1201/9781003467120-15

to zero:

$$r_1 - \alpha_{11}N_1 - \alpha_{12}N_2 = 0 \qquad (15.2)$$
$$r_2 - \alpha_{22}N_2 - \alpha_{21}N_1 = 0.$$

Solving these equations for the state variables, we find the coexistence equilibrium:

$$N_1^* = \frac{\alpha_{22}r_1 - \alpha_{12}r_2}{\alpha_{11}\alpha_{22} - \alpha_{12}\alpha_{21}} \qquad (15.3)$$
$$N_2^* = \frac{\alpha_{11}r_2 - \alpha_{21}r_1}{\alpha_{11}\alpha_{22} - \alpha_{12}\alpha_{21}}.$$

This equilibrium is biologically meaningful only if $N_1^* > 0$ and $N_2^* > 0$. Additional equilibria occur when each species is alone: $(N_1^*, N_2^*) = (r_1/\alpha_{11}, 0)$ and $(N_1^*, N_2^*) = (0, r_2/\alpha_{22})$.

The isoclines (15.2) can be plotted in state space and used to determine the trajectory of the system. When the system is below an isocline, the growth rate for that species is positive; above the isocline it is negative. Since the isoclines (15.2) are linear, it is helpful to consider the intercepts of the axes. Figure 15.1 illustrates the four possible outcomes for the competitive system (15.1)[1]:

- Case 1. If $r_1/\alpha_{12} > r_2/\alpha_{22}$ and $r_2/\alpha_{21} > r_1/\alpha_{11}$, then the isoclines cross and the coexistence equilibrium (15.3) exists. All positive initial conditions are attracted to this globally stable point (Figure 15.1A).
- Case 2. If $r_1/\alpha_{12} > r_2/\alpha_{22}$ and $r_2/\alpha_{21} < r_1/\alpha_{11}$, then the isocline for species 1 is above the isocline for species 2. All positive initial conditions are attracted to the single species equilibrium $(N_1^*, N_2^*) = (r_1/\alpha_{11}, 0)$ and species 1 eliminates species 2 (Figure 15.1B).
- Case 3. If $r_1/\alpha_{12} < r_2/\alpha_{22}$ and $r_2/\alpha_{21} > r_1/\alpha_{11}$, then the isocline for species 2 is above the isocline for species 1. All positive initial conditions are attracted to the single species equilibrium $(N_1^*, N_2^*) = (0, r_2/\alpha_{22})$ and species 2 eliminates species 1 (Figure 15.1C).
- Case 4. If $r_1/\alpha_{12} < r_2/\alpha_{22}$ and $r_2/\alpha_{21} < r_1/\alpha_{11}$, then the isoclines cross and the coexistence equilibrium (15.3) exists. However, this coexistence equilibrium is unstable. The two single species equilibria are locally stable. The unstable coexistence equilibrium has a stable manifold which is the line that contains the equilibrium point and the origin (Figure 15.1D). For initial conditions below this manifold, species 1 will eliminate species 2. For initial conditions above this manifold, species 2 will eliminate species 1. Initial conditions near the stable manifold will do a flyby of the unstable equilibrium. Initial conditions exactly on the manifold will approach the unstable equilibrium.

Case 1 is of particular interest to ecologists. The conditions in Case 1 imply the following inequality:

$$\alpha_{21}\alpha_{12} < \alpha_{11}\alpha_{22}. \qquad (15.4)$$

Thus, for two competing species to coexist, it is necessary (but not sufficient) that the overall intensity of interspecific competition must be less than the intensity of intraspecific competition.[2]

[1]In the degenerate case in which the two isoclines are coincident, every point on the isoclines is a coexistence equilibrium. This case is ecologically unrealistic and will not be considered.

[2]Mathematically, the conditions for the coexistence equilibrium to be both positive and stable are exactly those given in Case 1, which can be expressed as $\alpha_{12}\alpha_{21} < (r_1/r_2)\,\alpha_{21}\alpha_{22} < \alpha_{11}\alpha_{22}$. Note that a large difference in intrinsic growth rates makes coexistence more difficult to achieve.

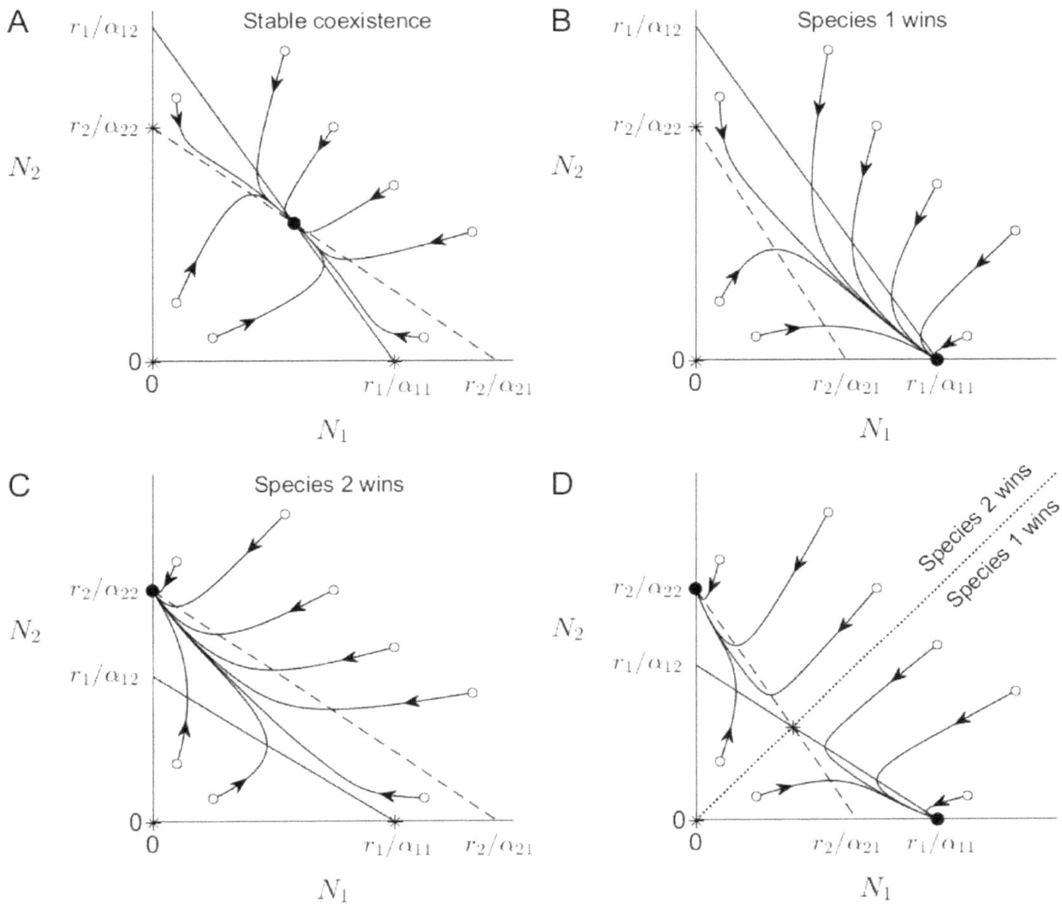

Figure 15.1 Dynamics of the continuous-time Lotka-Volterra model (15.1) for two competing species. In each state space plot, the solid line is the isocline for species 1 and the dashed line is the isocline for species 2. The solid circles are the stable equilibria and the asterisks are the unstable equilibria. The open circles are the initial conditions for sample trajectories. In panel D, the dotted line is the stable manifold of the unstable coexistence equilibrium which separates the basins of attraction for the two stable single-species equilibria.

15.2 THOMAS PARK'S SPECIES COMPETITION STUDIES USING *TRIBOLIUM*

Thomas Park and his colleagues conducted a 20-year research program studying competition between two closely-related species of flour beetles: *Tribolium castaneum* and *Tribolium confusum*. Their laboratory experiments looked at a number of factors: different genetic strains of each species [19], different environmental conditions [16], different combinations of initial numbers of beetles [9, 17], and different degrees of genetic variability among the founding individuals [13]. With a few exceptions, one species eliminated the other; coexistence was rarely achieved. In many of the treatments, the surviving species varied among the replicates.

15.2.1 Competitive Exclusion Principle

The *competitive exclusion principle* has been an influential concept in ecology. Often credited to Gause [5], the principle contends that "two species with similar ecology cannot live together in the same place," or, to put it more alliteratively, "complete competitors cannot coexist" [6]. The competition experiments of Park and his colleagues are often cited as support for this principle. After all, the two species must survive together on a single resource, the flour medium. However, Park was not convinced [17]. He wrote,

> The general conclusion is that these two species, rather similar to each other in their ecology, do not coexist when exploiting a shared habitat in the laboratory ... This, now rather hoary, conclusion runs the risk of becoming dogma. Indeed, it would be quite an exciting counterirritant if exceptions, derived from careful study and meaningful design, could be found.

After extensive experiments using several genetic strains of both species, Park and his colleagues wrote [19],

> It is true that one species always excluded the other when these particular strains were competing together. But, if we accept the logical consequences of the way two species may be interrelated in terms of the model, it would follow that other strains of the two *Tribolium* might in time be found, which would exhibit the phenomenon of coexistence of both species in a competitive system. Theoretically speaking, the so called Gause's Principle does not necessarily hold in all possible cases.

Clearly, they felt that coexistence of *T. castaneum* and *T. confusum* might be possible, given the right experimental conditions. The Lotka-Volterra model (15.1) has been used to support the competitive exclusion principle. If two species have similar ecological requirements, then the interspecific competitive effects, α_{21} and α_{12}, will be similar to or greater than the intraspecific competitive effects, α_{11} and α_{22}, and the inequality (15.4) necessary for stable coexistence will be difficult to satisfy. One species will eliminate the other, depending on the values of the intrinsic rates of increase, r_1 and r_2, and, possibly, the initial conditions (Fig 15.1).

15.2.2 Indeterminate Competition

A remarkable observation, made repeatedly by Park and his colleagues, is that the winner of the competitive contest was not always the same species. In most cases, each experimental treatment had 20–30 replicate populations. Although coexistence was not observed, in some treatment groups, in a subset of replicates *T. castaneum* would be the winner and *T. confusum* would prevail in the remaining cultures. This phenomenon was referred to as *competitive indeterminacy*.

One of Park's experiments [16] provides an example of competitive indeterminacy. In this experiment, mixed species cultures were maintained at three temperatures and two relative humidities in a two-way experimental design. Each population was initiated with two males and two females of each species and there were 30 replicate cultures for each treatment group. One species always eliminated the other. Table 15.1 shows the competitive outcomes in terms of the percentage of cultures where *T. castaneum* was the victor. Hot-wet climates favored *T. castaneum* whereas cold-dry climates favored *T. confusum*. The competitive outcome was indeterminate in the intermediate environmental conditions.

TABLE 15.1 Percentage of outcomes with *T. castaneum* the winner.

	Temperature		
Relative humidity	Hot (34°C)	Temperate (29°C)	Cold (24°C)
Wet (70% RH)	100%	86%	29%
Dry (30% RH)	10%	13%	100%

Competitive indeterminacy was the subject of debate and further experimentation. Two hypotheses emerged. The first, proposed by Lerner and Dempster [7] and others, held that competitive indeterminacy was the result of a "founder effect." Random genetic differences among the beetles which were sampled to start the mixed species cultures resulted in differences in the attributes of the populations, affecting the winner of competitive contest. In terms of the Lotka-Volterra model (15.1), random genetic differences produced variation in the parameters of the two species, so that in some cultures, species 1 was competitively superior (Figure 15.1B), and vice-versa in other cultures (Figure 15.1C). The second hypothesis, supported by Park and his colleagues [8, 14], proposed that competitive indeterminacy was the product of demographic stochasticity. They noted that, if an unstable coexistence equilibrium existed, deterministic theory predicted that the species with an advantage in its initial numbers would outcompete the other species. In terms of the Lotka-Voterra theory, they realized that, if the experiments were initiated near the separatrix between the two stable exclusion equilibria, then in the presence of stochasticity the system could wander between the domains of attraction, and the outcome would thus have an element of chance. In a clever experimental study, Mertz, Cawthon, and Park [13] looked at both factors—founder effect (genetic variability among the adults chosen to initiate the population) and demographic stochasticity (the ratio of the initial numbers of the two species)—simultaneously using a two-way design. They wrote,

> The data reported in the rows of Table 2 clearly show that species survival in the face of competition is much more affected by population densities achieved during competition than by founder effects. We reassert: regardless of the founder treatment, the populations appear to exhibit an unstable stationary state of the type predicted by stochastic competition theory. ... In sum, we believe on the basis of *empirical* evidence, that we have helped to restore the concept of demographic stochasticity to its rightful status as an integral part of population biology.

15.2.3 A Case of Coexistence?

In three of his early experimental studies, Park found that one of the two species, *T. castaneum* or *T. confusum*, became extinct when cultured together [15, 16, 17]. Ecologists concluded that his experimental results supported the idea that two species cannot coexist on a single limiting resource. However, Park himself was not convinced.

In a later study, Park and his colleagues looked at competition between four genetic strains of *T. castaneum* (cI, cII, cIII, cIV) and four strains of *T. confusum* (bI, bI, bII, bIV) [19]. There were 10 replicates for each of the 16 pairwise combinations. Coexistence was not realized in any of the 160 mixed species populations, although, in several replicates, the time to competitive exclusion was long (Figure 15.2A). They referred to these populations as "recalcirant." In his final experimental study [9], Park and his colleagues conducted an experiment where the initial number of each species was varied. There were 25 replicates.

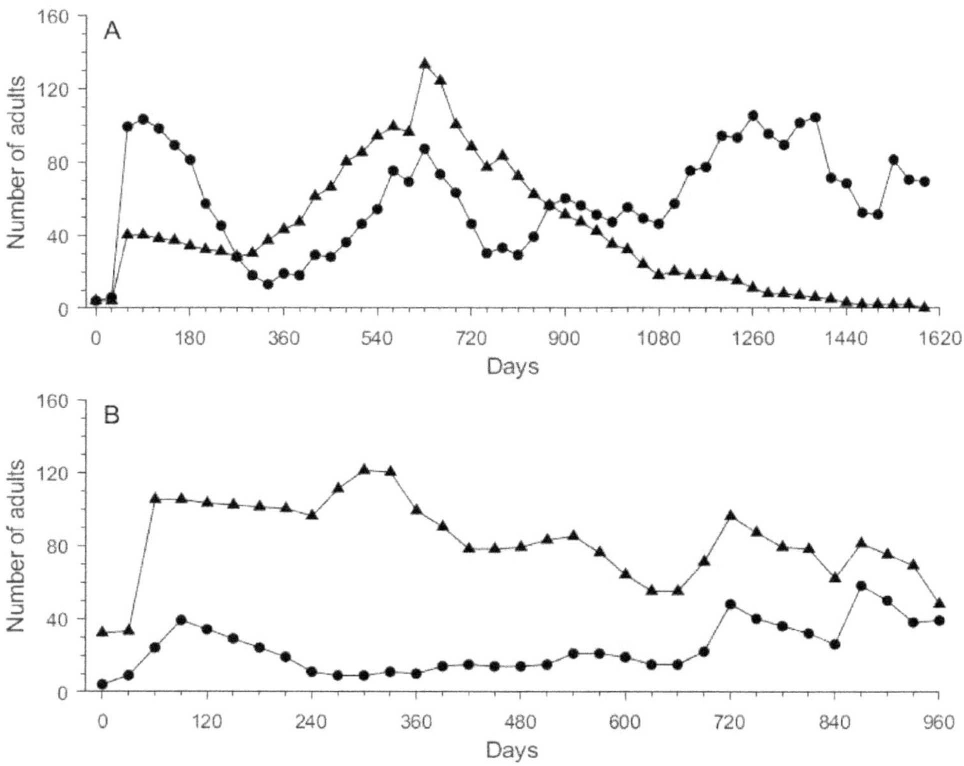

Figure 15.2 Adult time series for two mixed species cultures. A. Culture from Park et al. [19] started with four adults of *T. castaneum* strain cIV (circles) and 4 adults of *T. confusum* strain bII (triangles). In this "recalcitrant" culture, *T. confusum* went extinct at day 1590. B. Culture from Leslie *et al.* [9] started with four adults of *T. castaneum* strain cIV-a (circles) and 32 adults of *T. confusum* strain bI (triangles). The two species coexisted for 960 days, until termination of the experiment, at which time both species were still doing well.

In 24 of the replicates one species or the other prevailed. However, in one of the replicates, the two species coexisted for 960 days (over 30 generations). By the time this "coexistence culture" was terminated, both species were doing well (Figure 15.2B).

15.3 AN LPA SPECIES COMPETITION MODEL

From classic competition theory, the argument that two *Tribolium* species cannot coexist is based on the fact that both populations utilize a single resource: the flour medium [17]. However, as with single species populations, the dynamics of mixed species flour beetle populations are determined by cannibalism [21, 23].

In this section, we present a two-species version of the LPA model. We look at two examples: one that predicts a coexistence two-cycle [4] and a second that predicts coexistence on an invariant loop [3]. In both cases, coexistence comes about as the intensity of the interspecific interactions increase—a result that is at odds with classic competition theory.

15.3.1 Two-Species LPA Model

In mixed-species *Tribolium* cultures, cannibalism of eggs by larvae and adults and cannibalism of pupae by adults happens within and between species. This necessitates the addition of interspecific cannibalism terms. Let $c_{xy,ij}$ represent the rate at which life stage y of species j consumes life stage x of species i. Additionally, let b_i be the egg production rate for species i and let $\mu_{x,i}$ represent the probability of mortality for life stage x of species i during one time step. The two-species LPA model is

$$\begin{aligned}
L_{1,t+1} &= b_1 A_{1,t} \exp\left(-\frac{c_{el,11}}{V} L_{1,t} - \frac{c_{el,12}}{V} L_{2,t} - \frac{c_{ea,11}}{V} A_{1,t} - \frac{c_{ea,12}}{V} A_{2,t}\right) \\
P_{1,t+1} &= L_{1,t}\left(1-\mu_{l,1}\right) \\
A_{1,t+1} &= P_{1,t} \exp\left(-\frac{c_{pa,11}}{V} A_{1,t} - \frac{c_{pa,12}}{V} A_{2,t}\right) + A_{1,t}\left(1-\mu_{a,1}\right) \\
L_{2,t+1} &= b_2 A_{2,t} \exp\left(-\frac{c_{el,22}}{V} L_{2,t} - \frac{c_{el,21}}{V} L_{1,t} - \frac{c_{ea,22}}{V} A_{2,t} - \frac{c_{ea,21}}{V} A_{1,t}\right) \\
P_{2,t+1} &= L_{2,t}\left(1-\mu_{l,2}\right) \\
A_{2,t+1} &= P_{2,t} \exp\left(-\frac{c_{pa,22}}{V} A_{2,t} - \frac{c_{pa,21}}{V} A_{1,t}\right) + A_{2,t}\left(1-\mu_{a,2}\right),
\end{aligned} \tag{15.5}$$

where $L_{i,t}$, $P_{i,t}$, and $A_{i,t}$ are the numbers of larvae, pupae, and adults, respectively, for species i, and V is the habitat size relative to 20 g of flour medium. We will refer this extension of the LPA model to species competition as the "cLPA model."

Unlike the Lotka-Volterra model (15.1) for two competing species, the cLPA model (15.5) is a mechanistic model that considers how the two species interact when confined to a common habitat. The model has 18 parameters (not including habitat size) and can exhibit a diverse range of dynamical behaviors.

15.3.2 Two Exclusion Equilibria and a Coexistence Cycle

Edmunds et al. [4] examined the dynamics of the cLPA model (15.5) for the parameter values listed in Table 15.2, where species 1 represents *T. castaneum* and species 2 represents *T. confusum*. Unfortunately, it was not possible to obtain cLPA parameter estimates directly from Park's data for several reasons: (1) except for a few examples, the time series were not published;[3] (2) in the few cases where the time series are available, the results are for adults only; and (3) the cultures were counted every 30 days instead of the 14-day time interval of the cLPA model. Instead, Edmunds et al. [4] used rounded parameter values estimated from the controls in the route-to-chaos experiment discussed in Chapter 7 [2]. Since Park and his colleagues conducted all of their experiments using 8 g of flour medium, the parameter values in Table 15.2 assume that $V = 8/20 = 0.4$.

The effects of interspecific cannibalism in the cLPA model (15.5) were examined by setting the interspecific rates of larva-on-egg cannibalism equal ($c_{el,12} = c_{el,21}$) and examining the predicted dynamics when these rates are changed. Figure 15.3 shows a bifurcation diagram for total numbers of each species as these parameters are increased from a value of 0.016 through 0.032 with all remaining parameters as specified in Table 15.2. Five regions are identified (Figure 15.3). A stable coexistence equilibrium exists in region I when the interspecific egg cannibalism rates are low. At a value of $c_{el,12} = c_{el,21} = 0.0201$, this gives way to a region II of competitive exclusion where species 1 outcompetes species 2. This is followed by a small region III, from 0.0243 to 0.0253, where a stable coexistence equilibrium and a stable competitive exclusion equilibrium exist together. In region IV (0.0253

[3]Park's original physical data records were destroyed after his retirement (David B. Mertz, personal communication).

TABLE 15.2 Parameter values from Edmunds et al. [4].

Description	Species 1 *T. castaneum*	Species 2 *T. confusum*
Vital rates	$b_1 = 10$ $\mu_{l,1} = 0.20$ $\mu_{a,1} = 0.02$	$b_2 = 10$ $\mu_{l,2} = 0.20$ $\mu_{a,2} = 0.02$
Intraspecific cannibalism	$c_{el,11} = 0.010$ $c_{ea,11} = 0.010$ $c_{pa,11} = 0.012$	$c_{el,22} = 0.010$ $c_{ea,22} = 0.010$ $c_{pa,22} = 0.010$
Interspecific cannibalism	$c_{el,21} = 0.010$ $c_{ea,21} = 0.010$ $c_{pa,21} = 0.008$	$c_{el,12} = 0.010$ $c_{ea,12} = 0.010$ $c_{pa,12} = 0.005$

to 0.0288) coexistence is lost and two stable species exclusion equilibria exist. Finally, for $c_{el,12} = c_{el,21} > 0.0288$ (region V), there are three attractors: a stable coexistence two-cycle and two stable species exclusion equilibria. Region V persists for arbitrarily large values of $c_{el,12} = c_{el,21}$.

The stable coexistence two-cycle has a large basin of attraction. Figure 15.4 shows a state space plot with dashed curves that delineate the domains of attraction for $c_{el,12} = c_{el,21} = 0.032$, in region V. Sample system trajectories also are shown. Interestingly, in this case, *increased rates of interspecific effects promote the possibility of coexistence*, contrary to the predictions of the classic theory of competition as exemplified by the Lotka-Volterra model (15.1) and Figure 15.1D.

15.3.3 Coexistence on a Double-Invariant Loop

Another example of the dynamics of the cLPA model (15.5) was provided by Desharnais et al. [3]. They used information from several sources to obtain parameter values: (1) the statistical analyses of time series data on *T. castaneum* [1], (2) assay experiments on fecundity, fertility, rate of development, and adult longevity of the two species [18] and, (3) assay experiments on the cannibalistic interactions of adults eating eggs, adults eating pupae, and larvae eating eggs of the two species [20]. The parameter values, adjusted for $V = 0.4$, are given in Table 15.3.

The effects of interspecific cannibalism in the cLPA model (15.5) was examined by varying a single parameter, $c_{el,21}$, the cannibalism rate of larvae of species 1 on the eggs of species 2, while keeping the remaining parameters at the values given in Table 15.3. Figure 15.5 shows a bifurcation diagram for total numbers of each species. Throughout the range of $c_{el,21} \geq 0$, there is a stable single-species equilibrium where species 2 excludes species 1. This is the only attractor for $0 \leq c_{el,21} < 0.0245$. A second attractor, a coexistence invariant loop, appears as $c_{el,21}$ increases beyond 0.0245 and persists until $c_{el,21} = 0.0368$, at which point it collapses into a coexistence two-cycle. The two-cycle persists for $0.0368 < c_{el,21} \leq 0.0682$. For $c_{el,21} > 0.0682$, there are two exclusion equilibria; a coexistence attractor no longer exists.

The coexistence invariant loop has some interesting properties. As shown in Figure 15.6, there are oscillations on two time scales. There are short-time-scale oscillations of period-two, with small amplitudes in the adults and large amplitudes in the immature stages. These short-time-scale oscillations are superimposed onto long time scale oscillations with a period of approximately 145 time units. When viewed in state space, one has the impression that

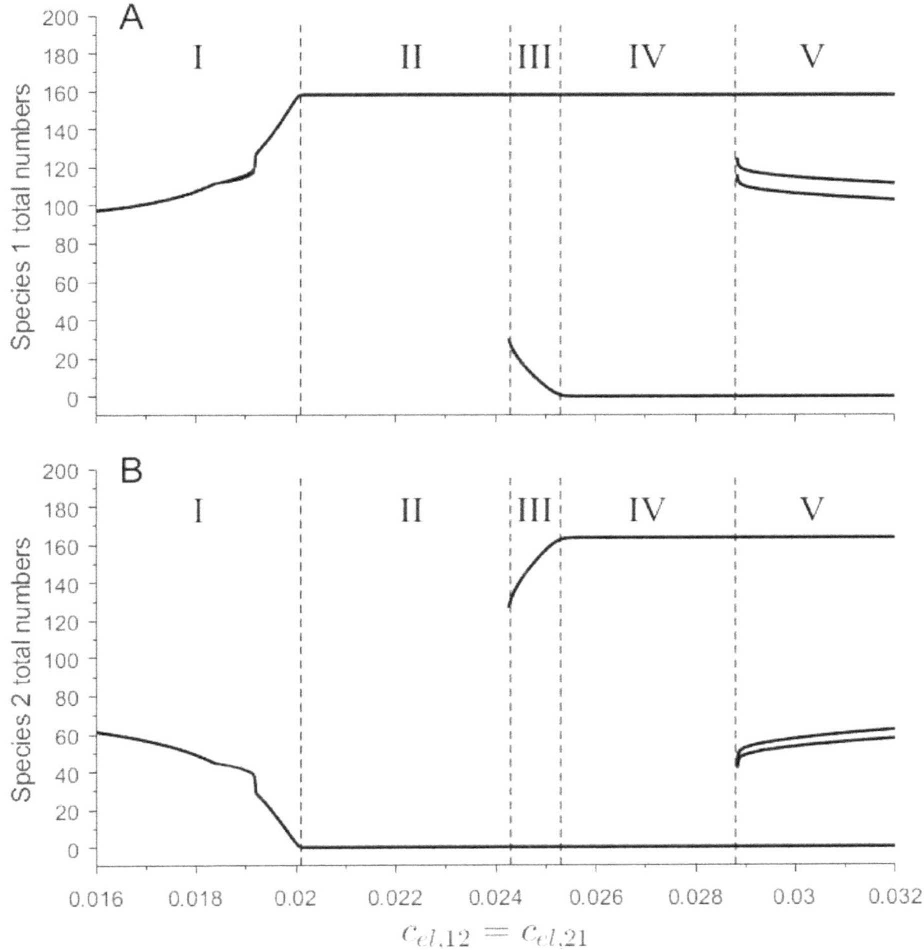

Figure 15.3 Bifurcation plot for total numbers of the cLPA model (15.5) using $c_{el,12} = c_{el,21}$ as the bifurcation parameter and with all remaining parameters as specified in Table 15.2. The vertical dashed lines demarcate qualitative changes in the system dynamics. These regions are identified with roman numerals. See the main text for details. A. Species 1. B. Species 2.

the system is bouncing between a double invariant loop (Figure 15.7). The total numbers of species 2 reach small values for a significant portion of the long-time-scale cycles (Figure 15.6). This feature is sometimes present in the time series data of Park and his colleagues (Figure 15.2).

15.3.4 A Stochastic LPA Competition Model

Many of the experimental results of Thomas Park and his colleagues were characterized by competitive indeterminacy. Often, the winner of the competitive contest in mixed cultures varied among replicates from the same treatment group. The most accepted explanation

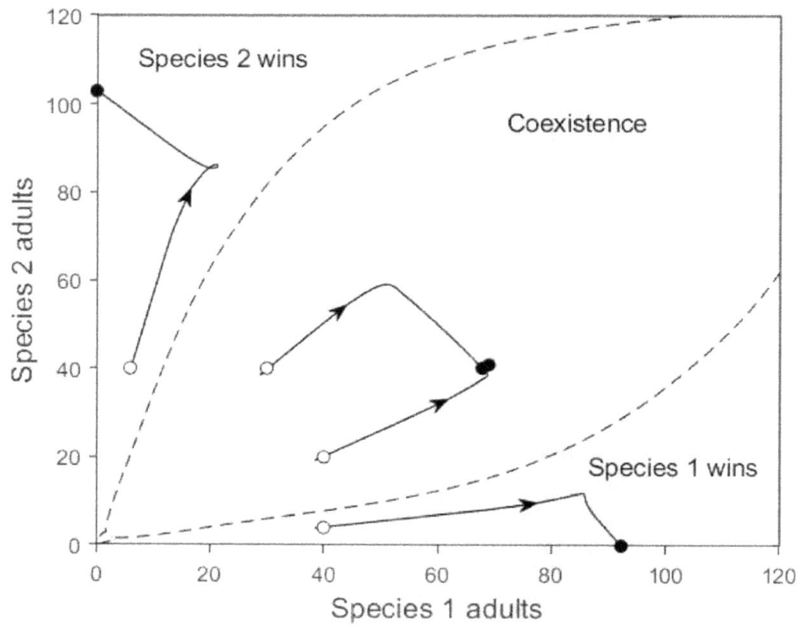

Figure 15.4 State space for adult numbers in the cLPA model (15.5) using $c_{el,12} = c_{el,21} = 0.032$ and with all remaining parameters as specified in Table 15.2. The solid circles represent the attractors. The open circles are the initial conditions for sample trajectories. The dashed curves demarcate the basins of attraction for the stable single-species exclusion equilibria and the stable coexistence two-cycle.

for this result was demographic stochasticity [13]. It seems appropriate, then, to examine a stochastic version of the cLPA model (15.5).

Expanding on the approach used in Chapter 10, we can use the Poisson and binomial probability distributions to model variation in reproduction and survival. Let $p_{i,t}$ and $q_{i,t}$ represent the probabilities that an egg and pupa of species i, respectively, survive

TABLE 15.3 Parameter values from Desharnais et al. [3].

Description	Species 1 *T. castaneum*	Species 2 *T. confusum*
Vital rates	$b_1 = 10$	$b_2 = 10$
	$\mu_{l,1} = 0.20$	$\mu_{l,2} = 0.20$
	$\mu_{a,1} = 0.14$	$\mu_{a,2} = 0.14$
Intraspecific cannibalism	$c_{el,11} = 0.0048$	$c_{el,22} = 0.0012$
	$c_{ea,11} = 0.0024$	$c_{ea,22} = 0.0060$
	$c_{pa,11} = 0.0256$	$c_{pa,22} = 0.0040$
Interspecific cannibalism	$c_{el,21} = 0.0248$	$c_{el,12} = 0.0040$
	$c_{ea,21} = 0.0028$	$c_{ea,12} = 0.0044$
	$c_{pa,21} = 0.0120$	$c_{pa,12} = 0.0068$

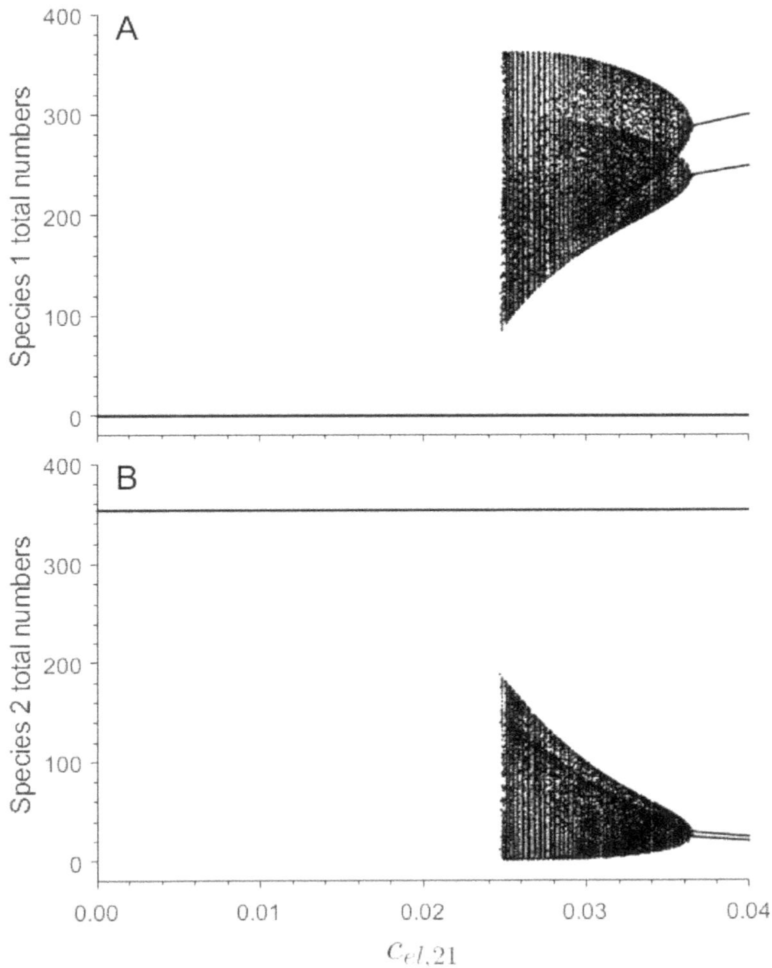

Figure 15.5 Bifurcation plot for total numbers of the cLPA model (15.5) using $c_{el,21}$ as the bifurcation parameter. All remaining parameters have the values specified in Table 15.3. A. Species 1. B. Species 2.

intraspecific cannibalism from species i and interspecific predation from species j:

$$p_{i,t} = \exp\left(-\frac{c_{el,ii}}{V}L_{i,t} - \frac{c_{el,ij}}{V}L_{j,t} - \frac{c_{ea,ii}}{V}A_{i,t} - \frac{c_{ea,ij}}{V}A_{j,t}\right)$$
$$q_{i,t} = \exp\left(-\frac{c_{pa,ii}}{V}A_{i,t} - \frac{c_{pa,ij}}{V}A_{j,t}\right),$$

for $i, j = 1, 2$ $(i \neq j)$. The Poisson-binomial version of the cLPA model is

$$\begin{aligned} L_{i,t+1} &\sim \text{Pois}\left(b_i A_{i,t} p_{i,t}\right) \\ P_{i,t+1} &\sim \text{Bin}\left(L_{i,t}, 1 - \mu_{l,i}\right) \\ R_{i,t+1} &\sim \text{Bin}\left(P_{i,t}, q_{i,t}\right) \\ S_{i,t+1} &\sim \text{Bin}\left(A_{i,t}, 1 - \mu_{a,i}\right) \\ A_{i,t+1} &= R_{i,t+1} + S_{i,t+1}, \end{aligned} \tag{15.6}$$

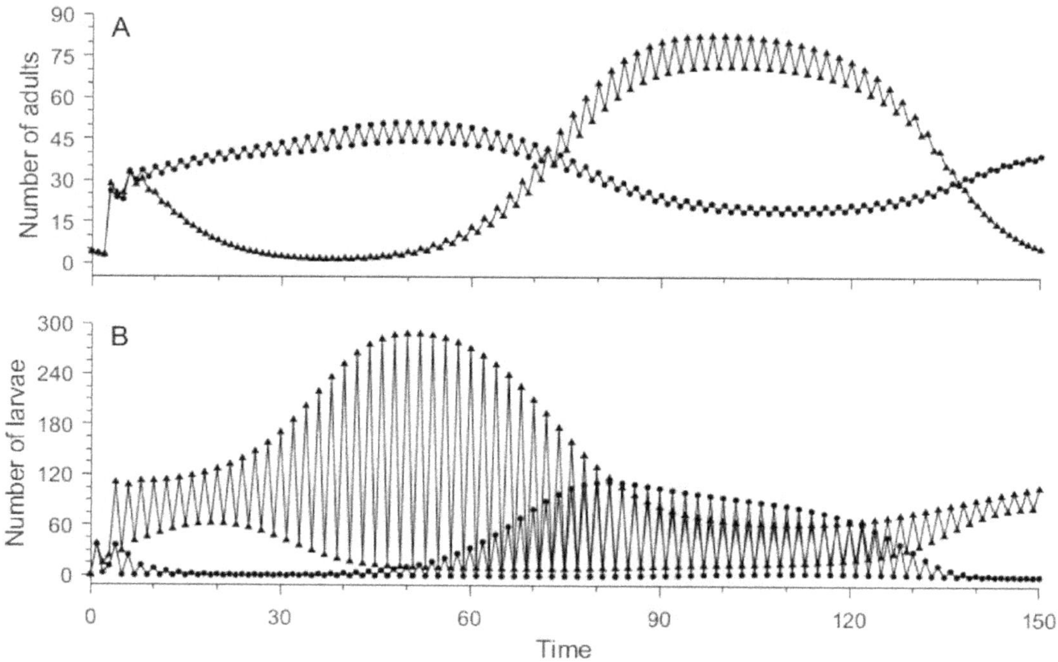

Figure 15.6 Time-series plots for a simulation of the cLPA model (15.5) using parameter values as specified in Table 15.3. The circles are for species 1 and the triangles are for species 2. A. Adult numbers. B. Larval numbers. The initial condition is four adults of each species.

for $i = 1, 2$. Here, Pois (λ) represents a Poisson distribution with mean value λ, Bin (n, p) represents a binomial distribution for n Bernoulli trails, each with probability p, and "$\sim$" means "is distributed as." The equation for adults involves the sum of two independent random variables, one for the recruitment of new adults, $R_{i,t+1}$, and another for the survival of adults from the previous time interval, $S_{i,t+1}$. The stochastic cLPA model (15.6) has the same parameters as the deterministic cLPA model (15.5).

Desharnais et al. [3] examined numerous realizations of the stochastic cLPA model (15.6) for different values of the bifurcation parameter, $c_{el,21}$, used in Figure 15.5 with the remaining parameter values as in Table 15.3. Each simulation used four adults of each species as the initial condition. They ran 10,000 simulations for each value of $c_{el,21}$ ranging from 0 to 0.04 in increments of 0.001. They recorded the frequency of occurrences of five time series patterns, based on the time a species goes extinct: the early or late loss of species one, the early or late loss of species two, and coexistence. Early elimination was defined as the loss of a species in the time interval $0 < t \leq 60$, late elimination as a loss in the interval $60 < t \leq 150$, and coexistence as the presence of both species at $t > 150$. This was used to construct the "stochastic outcome diagram" shown in Figure 15.8. It is a stochastic counterpart to the deterministic bifurcation diagram shown in Figure 15.5. As $c_{el,21}$ increases, there is a reduction in the likelihood of the loss of species one and a corresponding increase in the likelihood of the loss of species two. Indeterminate competition, including coexistence, is predicted. For example, in the range $0.02 \leq c_{el,21} \leq 0.03$, Figure 15.8 shows positive probabilities of occurrence for the two exclusion outcomes and coexistence.

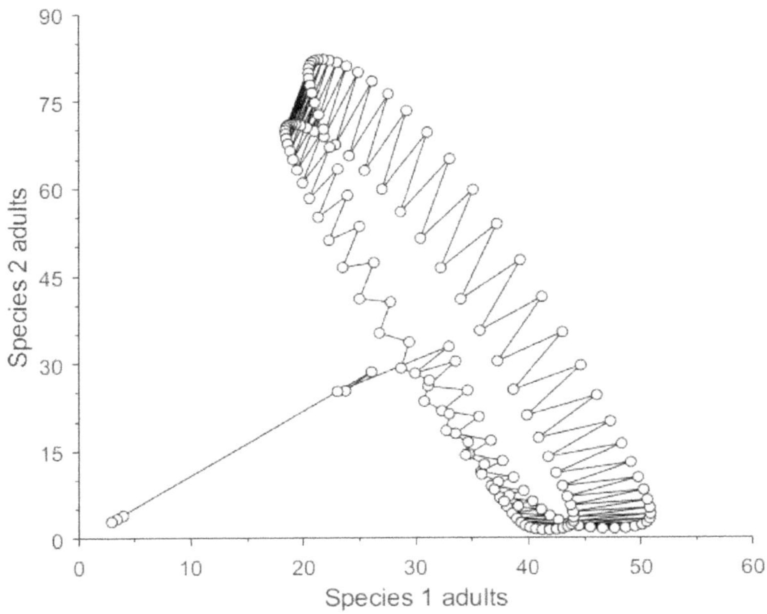

Figure 15.7 State space plot of adult numbers for a simulation of the cLPA model (15.5) using the parameter values specified in Table 15.3. The initial condition is four adults of each species.

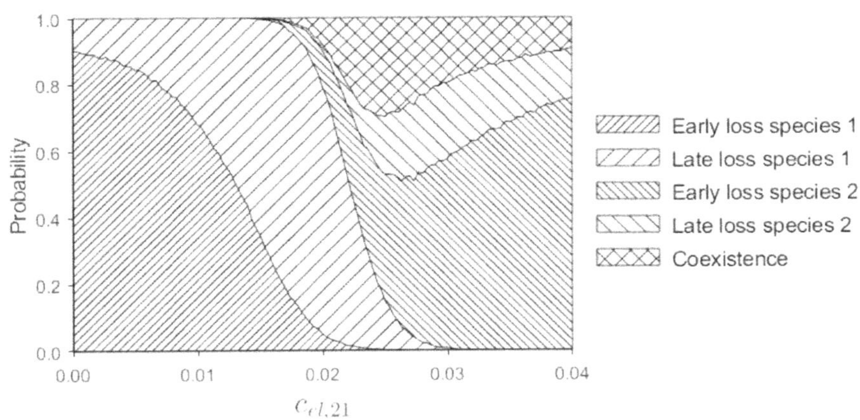

Figure 15.8 Outcome probabilities for the cLPA model of demographic stochasticity (15.6) as $c_{el,21}$ is increased. Other parameter values are as specified in Table 15.3. The initial condition is four adults of each species. Early and late loss is defined as a species going extinct in the time intervals $0 < t \leq 60$ and $60 < t \leq 150$, respectively. Coexistence is defined as the presence of both species for $t > 150$.

15.4 CONCLUSIONS

An interesting prediction of the cLPA (15.5) model is that, in some cases, increased intensities of species interactions can enhance coexistence. For example, it was shown that in the stochastic version of the cLPA model (15.6), an increase in the rate of cannibalism by larvae of species 1 on the eggs of species 2 can enhance the likelihood of coexistence of the two species (Figure 15.8). The simulation time series for the deterministic model reveals that the period-two component of the larval oscillation for the two species are in phase (Figure 15.6B). The same is true for the coexistence two-cycle predicted in region V of Figure 15.3. It may be the case that having small amounts of larval cannibalism at the troughs of these cycles benefits both species, whereas higher cannibalism rates at the peaks is not detrimental since few eggs of either species escape cannibalism.[4]

The research reported in this chapter, based on the cLPA model (15.5), demonstrates how low-dimensional, mechanistic nonlinear models coupled with stochasticity can lead to more parsimonious explanations of ecological data than generic models like the Lotka-Volterra equations (15.1).

Bibliography

[1] Dennis B, Desharnais RA, Cushing JM, Costantino RF (1995) Nonlinear demographic dynamics: Mathematical models, statistical methods, and biological experiments. Ecological Monographs 65:26–282. https://doi.org/10.2307/2937060

[2] Dennis B, Desharnais RA, Cushing JM, Henson SM, Costantino RF (2001) Estimating chaos and complex dynamics in an insect population. Ecological Monographs 71:277–303. https://doi.org/10.1890/0012-9615(2001)071[0277:ECACDI]2.0.CO;2

[3] Desharnais RA, Edmunds J, Costantino RF, Henson SM (2005) Species competition: Uncertainty on a double invariant loop. Journal of Difference Equations and Applications 11:311–325. https://doi.org/10.1080/10236190412331335391

[4] Edmunds J, Cushing JM, Costantino RF, Henson SM, Dennis B, Desharnais RA (2003) Park's *Tribolium* competition experiments: A non-equilibrium species coexistence hypothesis. Journal of Animal Ecology 72:703–712. https://doi.org/10.1046/j.1365-2656.2003.00743.x

[5] Gause GF (1934) The Struggle for Existence. Baltimore, Williams and Wilkins.

[6] Hardin G (1960) The competitive exclusion principle: An idea that took a century to be born has implications in ecology, economics, and genetics. Science 131:1292–1297. https://doi.org/10.1126/science.131.3409.1292

[7] Lerner IM, Dempster ER (1962) Indeterminism in interspecific competition. Proceedings of the National Academy of Sciences 48:821–26. https://doi.org/10.1073/pnas.48.5.821

[8] Leslie PH (1962) A stochastic model for two competing species of *Tribolium* and its application to some experimental data. Biometrika 49:1–25. https://www.jstor.org/stable/2333464

[9] Leslie PH, Park T, Mertz DB. (1968) The effect of varying the initial numbers on the outcome of competition between two *Tribolium* species. Journal of Animal Ecology 37:9–23. https://www.jstor.org/stable/2708

[10] Lotka AJ (1926) Elements of Physical Biology. Baltimore, Williams and Wilkins.

[4]In support of this explanation, we imposed a perturbation on the simulated coexistence two-cycle by phase shifting the numbers of one species. This caused a transient decrease in the numbers of both species followed by a quick return to the phase-synchronized cycle.

[11] MacArthur RH (1972) Geographical Ecology: Patterns in the Distribution of Species. New York, Harper and Row.

[12] May RM (1973) Stability and Complexity in Model Ecosystems. Princeton, Princeton University Press.

[13] Mertz DB, Cawthon DA, Park T (1976) An experimental analysis of competitive indeterminacy in *Tribolium*. Proceedings of the National Academy of Sciences 73:1368–1372. https://doi.org/10.1073/pnas.73.4.1368

[14] Neyman J, Park T, Scott EL (1956) Struggle for existence. The *Tribolium* model: Biological and statistical aspects. In Proceedings of the Third Berkeley Symposium on Mathematical Statistics and Probability. Volume 4. pp 41–79. Berkeley, University of California Press.

[15] Park T (1948) Experimental studies of interspecific competition. I. Competition between populations of the flour beetles *Tribolium confusum* Duval and *Tribolium castaneum* Herbst. Ecological Monographs 18:265–308. https://www.jstor.org/stable/1948641

[16] Park T (1954) Experimental studies of interspecies competition. II. Temperature, humidity, and competition in two species of *Tribolium*. Physiological Zoology 27:177–238. https://www.jstor.org/stable/30152164

[17] Park T (1957) Experimental studies of interspecies competition. III. Relation of initial species proportion to competitive outcome in populations of *Tribolium*. Physiological Zoology 30:22–40. https://www.jstor.org/stable/30166306

[18] Park T, Mertz DB, Petrusewicz K (1961) Genetic strains of *Tribolium*: Their primary characteristics. Physiological Zoology 34:62–80. https://doi.org/10.1086/physzool.34.1.30152682

[19] Park T, Leslie PH, Mertz DB (1964) Genetic strains and competition in populations of *Tribolium*. Physiological Zoology 37:97–162. https://doi.org/10.1086/physzool.37.2.30152328.

[20] Park T, Mertz DB, Grodzinski W, Prus T (1965) Cannibalistic predation in populations of flour beetles. Physiological Zoology 38:289–321. https://doi.org/10.1086/physzool.38.3.30152840

[21] Park T, Nathanson M, Ziegler JR, Mertz DB (1970) Cannibalism of pupae by mixed-species populations of adult *Tribolium*. Physiological Zoology 43:166–184. https://doi.org/10.1086/physzool.43.3.30155527

[22] Pearl R, Reed LJ (1920) On the rate of growth of the population of the United States since 1790 and its mathematical representation. Proceedings of the National Academy of Sciences 6:275–288. https://doi.org/10.1073/pnas.6.6.275

[23] Sokoloff A, Lerner IM (1967) Laboratory ecology and mutual predation of *Tribolium* species. American Naturalist 101:261–276. https://doi.org/10.1086/282489

[24] Verhulst PF (1838) Notice sur la loi que la population suit dans son accroissement. Correspondence Mathematique et Physique 10:113–121.

[25] Volterra V (1926) Variazioni e fluttuazioni del numero d'individui in specie animali conviventi. Memorie dell'Accademia dei Lincei 2:31–113.

CHAPTER 16

Spectral Analyses of Population Dynamics

Spectral methods can be used to analyze ecological data in the frequency domain, providing estimates of the dominant periodicities of a population time series. These tools can predict and assess the impact of stochasticity on population dynamics. In this chapter, we apply these spectral methods to the LPA model to obtain testable predictions and evaluate these hypotheses using experimental data.

16.1 FREQUENCY DOMAIN ANALYSES OF POPULATION FLUCTUATIONS

Most analyses of ecological time series examine data in the time domain. The previous chapters of this book look at *Tribolium* life stage numbers as they change over time. In some cases, it is also possible to look at data in the frequency domain. This is appropriate for data that is stationary, that is, the time series represents random fluctuations around a stable attractor. In cases like these, one can apply spectral methods to look at the distribution of periodic cycles in the data.

16.1.1 The Power Spectrum

Let x_t, $t = 0, \pm 1, \pm 2, \ldots$ represent a stochastic time series with a finite mean and variance. This time series is a *stationary process* if the joint probability distribution of any sample from the time series is invariant with respect to a shift in time.[1] Let $x_0, x_1, \ldots, x_{n-1}$ represent a sample of equally-spaced observations from a stationary time series. The *discrete Fourier transform*[2] of this sample is a series of complex numbers given by

$$\Im_x(k) = \sum_{j=0}^{n-1} x_j e^{-i2\pi jk/n}, \qquad k = 0, 2, \ldots, n-1,$$

where $i = \sqrt{-1}$. By Euler's formula $e^{i\theta} = \cos(\theta) + i\sin(\theta)$, so we can write the exponentials as trigonometric functions. The *power spectrum* of the observed time series at frequency f is given by the product of the Fourier transform with its complex conjugate:

$$S_x(f) = \Im_x(nf)\bar{\Im}_x(nf)/n, \qquad f = n^{-1}, 2n^{-1}, \ldots, mn^{-1},$$

[1] Formally, the stochastic process is *stationary* if the joint probability distribution of $\{x_{t_1}, x_{t_2}, \ldots, x_{t_n}\}$ and $\{x_{t_1+\tau}, x_{t_2+\tau}, \ldots, x_{t_n+\tau}\}$ is the same for all integer values of $t_1, t_2, \ldots, t_n$, n, and $\tau = \pm 1, \pm 2, \ldots$.

[2] A Fourier transform is a mathematical technique to convert a series from the time domain to a representation in the frequency domain.

DOI: 10.1201/9781003467120-16

where $\bar{\Im}_x$ is the complex conjugate of $\Im_x$, and $m = n/2$ if n is even, or $m = (n-1)/2$ if n is odd. Because the product of a complex number with its conjugate is a real number,[3] the power spectrum $S_x(f)$ is a real-valued nonnegative function.[4] For values of f where the periodicity of the trigonometric functions correlate with a periodicity in the times series x_t, the value of $S_x(f)$ will be large and a peak will occur in the power spectrum. For a discrete evenly-spaced time series, a cycle cannot be detected if the period is less than two. Since the period is the inverse of the frequency, the upper limit of the frequency is 0.5.

In practice, plots of power spectra often appear jagged, making cycle peaks harder to detect. Practitioners commonly apply a smoothing algorithm to the spectra. We used a simple modification of a method developed by Daniell [4] which uses a moving average across the spectral values.[5] The span of the averaging window, which must be an odd integer, can be varied. Often, two or more passes of the moving average procedure are used with the same or different window sizes. Longer time series usually require wider choices for the span(s) of the window(s).

To assist in the comparison of power spectra for different times series, we will use a normalized spectrum which has an approximate area of one. Although the terms "power spectrum" and "spectral density" are often used interchangeably, we will use *spectral density* to refer to the normalized spectrum. The spectral density of x_t is given by

$$\tilde{S}_x(f) = \frac{S_x(f)}{\Delta f \sum_f S_x(f)}, \tag{16.1}$$

where Δf is the spacing between sampled frequencies. In this way, $\tilde{S}_x(f)$ is somewhat analogous to a probability density function. It describes how the total spectral power is distributed across the range of frequencies.

16.1.2 Spectra for the Ricker Model

We will use a stochastic version of the Ricker model to illustrate the concept of spectral analysis for a univariate population time series. Adding environmental noise ε_t on a logarithmic scale, we have

$$N_{t+1} = bN_t e^{-cN_t} e^{\varepsilon_t}, \tag{16.2}$$

where σ^2 is the variance of the random variable ε_t. We will assume $\varepsilon_0, \varepsilon_1, \ldots, \varepsilon_{n-1}$ are independent identically distributed random variables with expectation $\mathrm{E}(\varepsilon_t) = 0$ for $t \geq 0$, a variance $\mathrm{Var}(\varepsilon_t) = \sigma^2$ for $t \geq 0$, and no serial autocorrelation. The deterministic model $(\sigma = 0)$ has an positive equilibrium $N^* = (\ln b)/c$ provided $b > 1$. For $\sigma = 0$, small perturbations from the equilibrium can be examined using the linearized system $w_{t+1} = \lambda w_t$, where $w_t = N^* - N_t$ and $\lambda = 1 - \ln b$ (Chapter 3). If $|\lambda| < 1$, the equilibrium is stable and small perturbations decay to zero. For $0 < \lambda < 1$, that is, for $1 < b < e$, the population approaches its equilibrium monotonically, a process called *undercompensatory dynamics*. If $-1 < \lambda < 0$, that is, if $e < b < e^2$, the population exhibits damped oscillations in its approach to equilibrium, a process called *overcompensatory dynamics*. These two domains are separated by the case where $\lambda = 0$, that is, $b = e$, a process called *compensatory dynamics*.

[3]The conjugate of a complex number $a+bi$ is $a-bi$. The product of a complex number with its conjugate is the real number a^2+b^2.

[4]An alternative formulation of the power spectrum uses the Fourier transform of the autocorrelations of the time series.

[5]In the unmodified version of the algorithm, the first and last windows are centered at $x_{(s+1)/2}$ and $x_{n-(s+1)/2}$, where s is the span of the averaging window. To avoid a shortening of the smoothed data series, the modified version uses lopsided shortened windows near the endpoints of the time series.

Figure 16.1 shows three examples of the model (16.2) for values of $b = 1.2$, $b = e \approx 2.7183$, and $b = 5$ with $c = 0.001$. The top three plots (Figs 16.1A,B,C) show trajectories of the deterministic model ($\sigma = 0$) with $N_0 = 10$. These three cases demonstrate undercompensatory, compensatory, and overcompensatory dynamics. For the stochastic model, we used $\sigma = 0.03$ (weak noise). Realizations of the stochastic model appear in the middle row of plots (Figs 16.1D,E,F). The bottom row of plots shows estimated spectral densities, $\tilde{S}_w(f)$, for long realizations of the stochastic model (Figs 16.1G,H,I). Each realization was a stationary time series of $n = 4096$ points after 100 transients were removed. Two passes of Daniell method were used to smooth the spectra, with window spans of 101 and 103. For the undercompensatory dynamics (Figure 16.1G), the model oscillations are shifted toward low frequencies (long time scales). For the compensatory example (Figure 16.1H), the model oscillations are distributed evenly across all frequencies. For the overcompensatory dynamics (Figure 16.1I), the model oscillations are shifted toward high frequencies (short time scales). By analogy with frequencies of light, these three cases are described as a "red-shifted spectrum," "white noise spectrum," and "blue-shifted spectrum," respectively.

For a discrete-time stochastic model with weak noise, one can derive the expected power spectrum for stationary dynamics, provided the deterministic version of the model has a stable equilibrium. This involves applying *linear filter theory* to a linearization of the model equations. We postpone the details until Section 16.3.1. For the linearized stochastic Ricker model, the theoretical power spectrum is

$$S_w(f) = \frac{\sigma^2 (N^*)^2}{1 + \lambda^2 - 2\lambda \cos(2\pi f)}. \tag{16.3}$$

The expected spectral densities for the three cases appear as dashed curves in Figs 16.1G,H,I. In general, there is a good correspondence between the spectral densities estimated from realizations of the nonlinear model and the theoretical curves for the model linearization.

Spectral analysis can also be applied to cases where the deterministic version of the stochastic model has a non-equilibrium attractor. Figure 16.2 shows three examples from the Ricker model (16.2) for values of $b = 10$, $b = 14$, and $b = 18.5$. The top three plots (Figs 16.2A,B,C) show trajectories of the deterministic model ($\sigma = 0$) with $N_0 = 10$. These three cases demonstrate a two-cycle, a four-cycle, and chaotic dynamics. For the stochastic model, we used $\sigma = 0.03$ (weak noise). Realizations of the stochastic model appear in the middle row of plots (Figs 16.2D,E,F). The bottom row of plots shows estimated spectral densities, $\tilde{S}_w(f)$, with the same methods as were used for the spectra in Figure 16.1. For the stochastic model with a deterministic two-cycle, a large spectral peak occurs at $f = 0.5$ (Figure 16.2G). Since the period is the inverse of the frequency, this corresponds to a cycles of period two. For the stochastic model with a deterministic four-cycle, spectral peaks occur at $f = 0.25$ and $f = 0.5$ (Figure 16.2H). This corresponds to a mixture of cycles with periodicities two and four. Although the deterministic attractor is a four-cycle, there is a strong covariance between the data and sinusoidal oscillations of period two. For the stochastic model with a deterministic chaotic attractor, the model oscillations are shifted toward high frequencies, but fairly evenly distributed across a wide range of intermediate frequencies (Figure 16.2I).

These results suggest that spectra can provide information about the dynamics of a population without an explicit model. What is needed is a stationary time series of population numbers or densities of sufficient length. In the case of populations with equilibrium dynamics, the spectra can indicate how the population responds to perturbations caused by stochasticity. For example, one could, in principle, distinguish between undercompensatory, compensatory, and overcompensatory dynamics. For populations with more complex dynamics, one might be able to detect the signatures of periodic cycles or chaos.

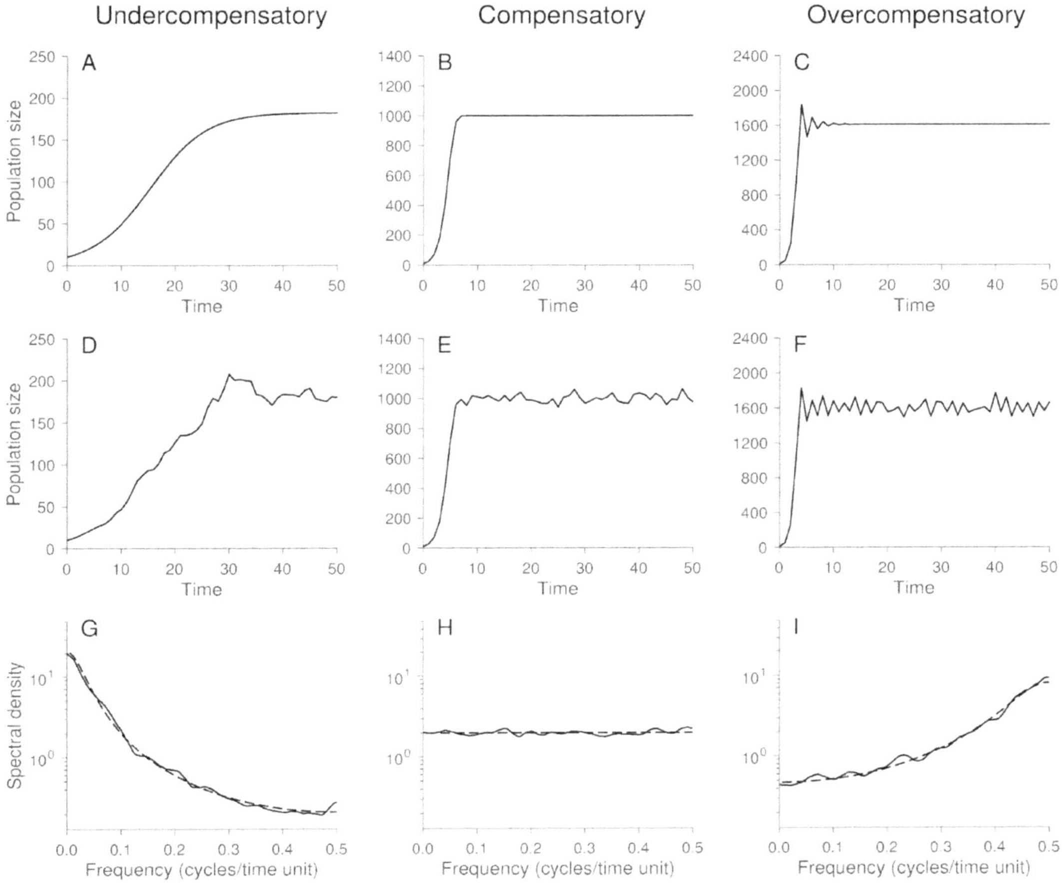

Figure 16.1 Simulations and spectral analyses for three cases of the Ricker model (16.2), each with a stable equilibrium. Panels A, B, and C show simulations of the deterministic model ($\sigma = 0$) with growth rate values of $b = 1.2$, $b = e \approx 2.7183$, and $b = 5$. Parameter $c = 0.001$ and the initial condition was $N_0 = 10$. Panels D, E, and F show realizations of the stochastic model ($\sigma = 0.03$) with the same parameter values and initial condition. Panels G, H, and I are plots of the spectral densities of the stochastic Ricker model (16.2) with parameter values as in panels D, E, and F. The solid curves are the spectral densities estimated from a long stationary time series of length $n = 4096$ after 100 initial transients were removed. Two passes of the Daniell method were used to smooth the spectra, with window spans of 101 and 103. The dashed curves are the theoretical spectral densities (16.3) for the parameter values used in the simulations.

16.2 SPECTRA FOR MULTIVARIATE TIME SERIES

Let $\mathbf{X}_t = [x_{1t}, x_{2t}, \ldots, x_{\omega t}]$, $t = 0, \pm 1, \pm 2, \ldots$, be a time series of vectors,[6] each with ω components. For ecological applications, these could be, for example, vectors of species, or they could be vectors of age-class or life-stage densities. As in the univariate case, we will assume

[6] We use square-bracket notation $[x_{1t}, x_{2t}, \ldots, x_{\omega t}]$ to mean the column vector $(x_{1t}, x_{2t}, \ldots, x_{\omega t})^T$, which is the transpose of the row vector $(x_{1t}, x_{2t}, \ldots, x_{\omega t})$.

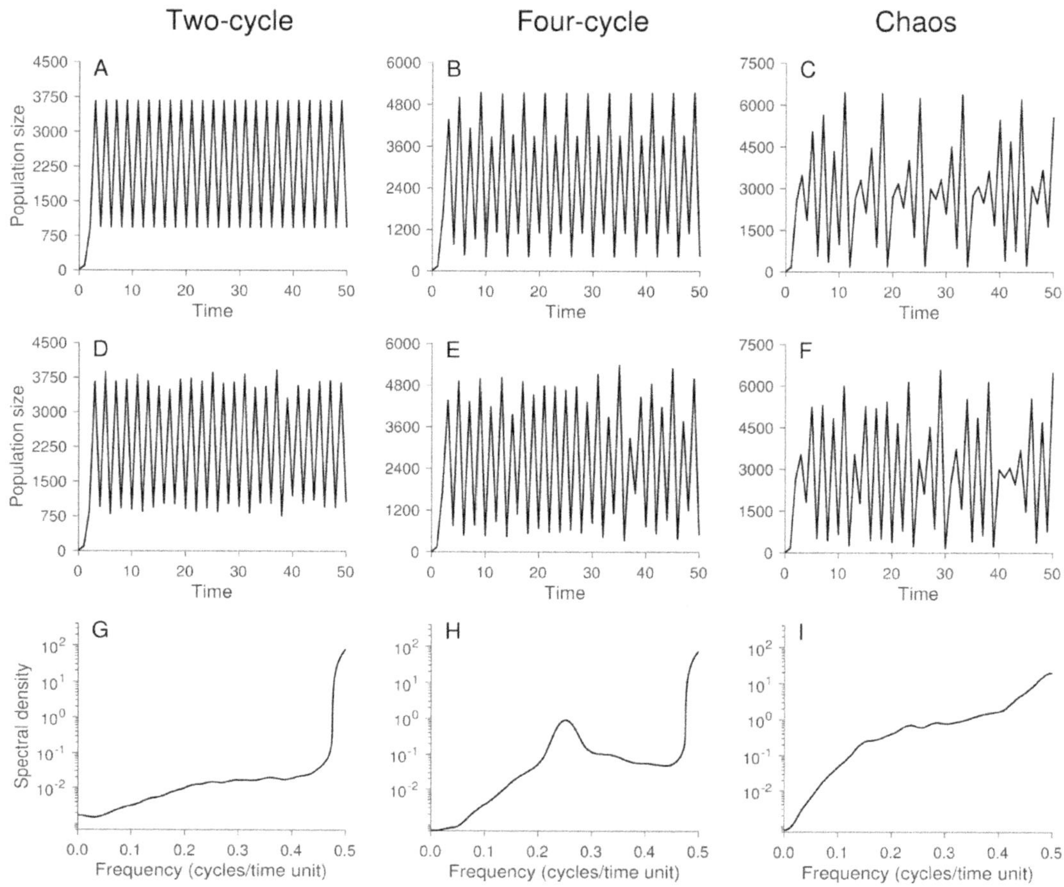

Figure 16.2 Simulations and spectral analyses for three cases of the Ricker model (16.2), each with a non-equilibrium deterministic attractor. Panels A, B, and C show simulations of the deterministic model ($\sigma = 0$) with $c = 0.001$ and an initial condition of $N_0 = 10$. In panel A, $b = 10$ and the attractor is a two-cycle. In panel B, $b = 14$ and the attractor is a four-cycle. In panel C, $b = 17.5$ and the dynamics exhibit a chaotic attractor. Panels D, E, and F show realizations of the stochastic model ($\sigma = 0.03$) with the same parameter values and initial condition. Panels G, H, and I are plots of the spectral densities of the stochastic Ricker model (16.2) with parameter values as in panels D, E, and F. Two passes of the Daniell method were used to smooth the spectra, with window spans of 101 and 103.

the process $\mathbf{X}_t$ is stationary. Let $\mathbf{X}_0, \mathbf{X}_2, \ldots, \mathbf{X}_{n-1}$ represent a sample of equally-spaced observations from a multivariate stationary time series. The discrete Fourier transform of this sample results in ω series of n complex numbers, one series for each component p of $\mathbf{X}_t$:

$$\Im_{\mathbf{X},p}(k) = \sum_{j=0}^{n-1} x_{pj} e^{-i2\pi jk/n}, \quad k = 0, 1, \ldots, n-1; \quad p = 1, 2, \ldots, \omega.$$

As we did above, let $m = n/2$ if n is even, or $m = (n-1)/2$ if n is odd. The $\omega \times \omega$ *spectral matrix*, $\mathbf{S}_{\mathbf{X}}(f) = (S_{\mathbf{X},pq}(f))$, of the observed multivariate time series at frequency f has elements given by

$$S_{\mathbf{X},pq}(f) = \Im_{\mathbf{X},p}(nf)\,\bar{\Im}_{\mathbf{X},q}(nf)/n\,, \qquad f = n^{-1}, 2n^{-1}, \ldots, mn^{-1}.$$

The diagonal elements ($p = q$) of the spectral matrix $\mathbf{S}_{\mathbf{X}}(f)$ are real nonnegative values that form the power spectrum of each component of $\mathbf{X}_t$. Each off-diagonal element $S_{\mathbf{X},pq}(f)$ of the spectral matrix ($p \neq q$), is called the *cross-spectrum* of component p with component q at frequency f. In general, these are complex functions. The real part of the cross-spectrum is called the *cospectrum*. It measures the in-phase frequency distribution of the covariance of x_{pt} with x_{qt}. The complex part of the cross-spectrum is called the *quadrature spectrum*. It measures the covariance of x_{pt} with x_{qt} lagged by a phase shift of 1/4 of a cycle at frequency f. Corresponding off-diagonal elements of the spectral matrix are complex conjugates: $S_{\mathbf{X},pq}(f) = \bar{S}_{\mathbf{X},qp}(f)$.[7] In this chapter, our focus will be on the spectra. We will consider cospectra when we discuss ecological synchrony in the next chapter.

We present an example of spectral methods applied to *Tribolium* data. Since each census has three life stages (larvae, pupae, adults), the data are multivariate. Our focus is on the spectra for the larval stage, where the oscillations are most pronounced. We used the control data for an experiment designed to test the effects of noise color on population dynamics [13]. We consider this topic in more detail in Section 16.3.

The deterministic LPA model with an explicit variable for habitat size was used to design the experiment:

$$\begin{aligned} L_{t+1} &= bA_t \exp\left(-\frac{c_{el}L_t}{V_t} - \frac{c_{ea}A_t}{V_t}\right) \\ P_{t+1} &= L_t\left(1-\mu_l\right) \\ A_{t+1} &= P_t \exp\left(-\frac{c_{pa}A_t}{V_t}\right) + A_t\left(1-\mu_a\right), \end{aligned} \tag{16.4}$$

where V_t is the habitat size in units of 20 g of flour medium and the remaining parameters have their usual meaning. The variable V_t can be used to introduce experimentally-manipulated environmental stochasticity. For the control data that we present below, habitat size was held constant at $V_t = 2.2$ for $t \geq 0$, which corresponds to 44 g of medium.

There were two control treatments, each cultured with laboratory protocols designed to yield different deterministic dynamics. In the first control treatment, rates of adult recruitment and adult mortality were manipulated to be $c_{pa} = 0.0029$ and $\mu_a = 0.3458$. When combined with the remaining parameter values estimated from the route-to-chaos experiment (Chapter 7; Table 7.2), this treatment is predicted by the deterministic LPA model (16.4) to yield a two-cycle attractor. For the second control treatment, the experimental protocol $c_{pa} = 0.53$ and $\mu_a = 0.91$ yields a predicted three-cycle attractor. For the both treatments, the initial life stage values were $[L_0, P_0, A_0] = [220, 4, 139]$. The experiment was continued for 80 weeks ($n = 41$). There were three replicates in each treatment group.

Deterministic model predictions and the experimental results for larval numbers are presented in Figure 16.3. The top two plots (Figs 16.3A,B) show the deterministic trajectories predicted by the LPA model (16.4) with $V_t = 2.2$ for $t \geq 0$. Asymptotically, large oscillations exist in larval numbers with periods of two and three time steps (four and six weeks), respectively. Figs 16.3C,D show experimental data for larval numbers from one replicate in each treatment group. Although stochasticity is present, the observed stationary dynamics are a good match to the model predictions. Figs 16.3E,F show the larval spectral densities estimated using Daniell smoothing windows with spans of three and five time steps. Initial transients for weeks 0–22 were removed for the spectral estimations (leaving $n = 30$). Thin

[7] In mathematical terms, $\mathbf{S}_{\mathbf{X}}(f)$ is a *Hermitian*, or *self-adjoint*, matrix.

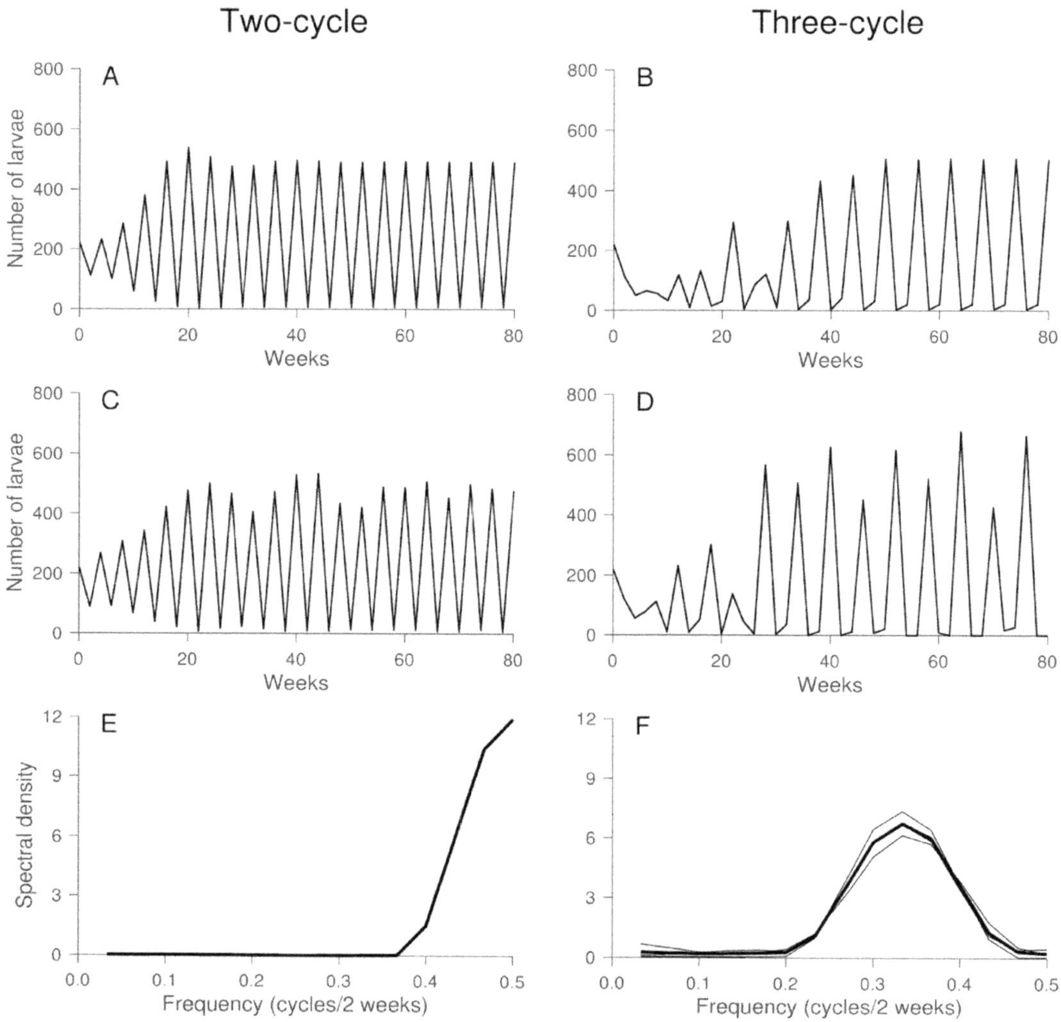

Figure 16.3 Deterministic model predictions, census data, and spectral analyses of larval numbers for two control treatment groups from an experiment designed to look at the effects of noise color on *Tribolium* population dynamics [13]. In these two groups, initial numbers were $[L_0, P_0, A_0] = [220, 4, 139]$ and habitat size was held constant at $V = 2.2$. Panels A, C, and E are based on an experimental protocol where two parameters were manipulated to have values $c_{pa} = 0.0029$ and $\mu_a = 0.3458$, which leads to a predicted two-cycle deterministic attractor when the remaining parameters in model (16.4) have values as specified in Table 7.2 (panel A). In panels B, D, and F, the two parameters were manipulated to have values $c_{pa} = 0.53$ and $\mu_a = 0.91$, which leads to a three-cycle attractor (panel B). Panels C and D show larval numbers for one representative replicate from each treatment group. Panels E and F are spectral density plots of larval numbers for weeks 22–80. Thin lines are for the three individual replicates and the thick line is the mean. Two passes of the Daniell method were used to smooth the spectra, with window spans of three and five.

lines show the results for each replicate; the thick curve is the average. Large spectral peaks occur near $f = 0.5$ and $f = 0.33$ for the two treatment groups. These correspond to periods $(1/f)$ of two and three time steps, respectively.

Spectral methods work best when the time series are long. In the route-to-chaos experiment discussed in Chapter 7 [3, 5], three replicate populations from the chaos treatment group ($c_{pa} = 0.35$ and $\mu_a = 0.96$) were maintained for 424 weeks ($n = 213$). A spectral analysis was performed on these data. The results appear in Figure 16.4. The top plot (Figure 16.4A) shows the predicted dynamics for larval numbers from the deterministic model (16.4) with $V_t = 1$ for $t \geq 0$, remaining parameters as given in Table 7.2, and the same initial numbers as were used in the experiment: $[L_0, P_0, A_0] = [250, 5, 100]$. The middle plot (Figure 16.4B) shows the larval numbers for one replicate from the chaos treatment group. The bottom plot (Figure 16.4C) is the estimated spectral density using Daniell smoothing windows with spans of three and five time steps. The largest peak is for $f = 0.33$ which corresponds to a three-cycle. This reflects the up-down-down motion which is a prominent feature of the data. Two smaller peaks appear at $f = 0.167$ and $f = 0.5$ which correspond to cycles of period six and two, respectively. The six-cycle is a prediction of the lattice version of the stochastic LPA model, discussed in Chapter 8 [9]. The lag metric comparison (LMC), discussed in Chapter 9, also provides evidence for both the three-cycle and the six-cycle dynamics [10]. The reason for two-cycle peak is less obvious. There are several occurrences of an up-down-up-down sequence in both the model predictions and the experimental data. For example, the model predictions (Figure 16.4A) show this pattern at weeks 6–12 (142.1, 1.7, 124.7, 1.4), weeks 50–56 (148.2, 1.6, 117.9, 2.0), and weeks 132–138 (149.6, 1.6, 116.1, 2.1), as well as several other places. The replicate population plotted in Figure 16.4B also shows this pattern several times, for example, at weeks 6–12 (161, 3, 173, 2), weeks 30–36 (113, 2, 103, 0), and weeks 46–52 (118, 1, 128, 1). The episodic appearance of these brief two-cycle patterns might be the reason for the small peak at $f = 0.5$.

One advantage of the frequency-based approach is that no model is needed for the identification of hypothetical cycles in the data. This is in contrast to the LMC approach which requires a model-predicted cycle sequence for comparison. On the other hand, spectral methods have weak power to detect cycles when the time series are short and/or the data are very noisy.

16.3 LINEAR FILTER THEORY

In the frequency domain, filtering is used to modify a signal in order to remove undesirable components or enhance desirable features. It is widely used in applications such electronics and telecommunication. Mathematical tools are well-developed for the case of linear filters, where the outputs are a linear function of the inputs. A *transfer function* models the outputs as a function of the inputs. In the case of linear filters, explicit formulations can be derived to relate the spectral properties of the outputs as a function of the input spectra.

How does linear filter theory relate to the dynamics of ecological populations? Figure 16.5 illustrates the concept. One can consider variation over time in environmental attributes, such as temperature or rainfall, as the extrinsic "input signals" that impact the population being considered. The intrinsic dynamics of the population then "filters" this input signal to produce "output signals" of population numbers or densities over time. If one assumes the environmental and population time series are stationary processes and that the population would have a stable equilibrium in the absence of noise, then one can linearize the population dynamics around its equilibrium to predict the effects of environmental variability on the population [7, 8]. What is needed is a spectral matrix for the environmental variables and a linearized population model that includes the random variables for environmental noise.

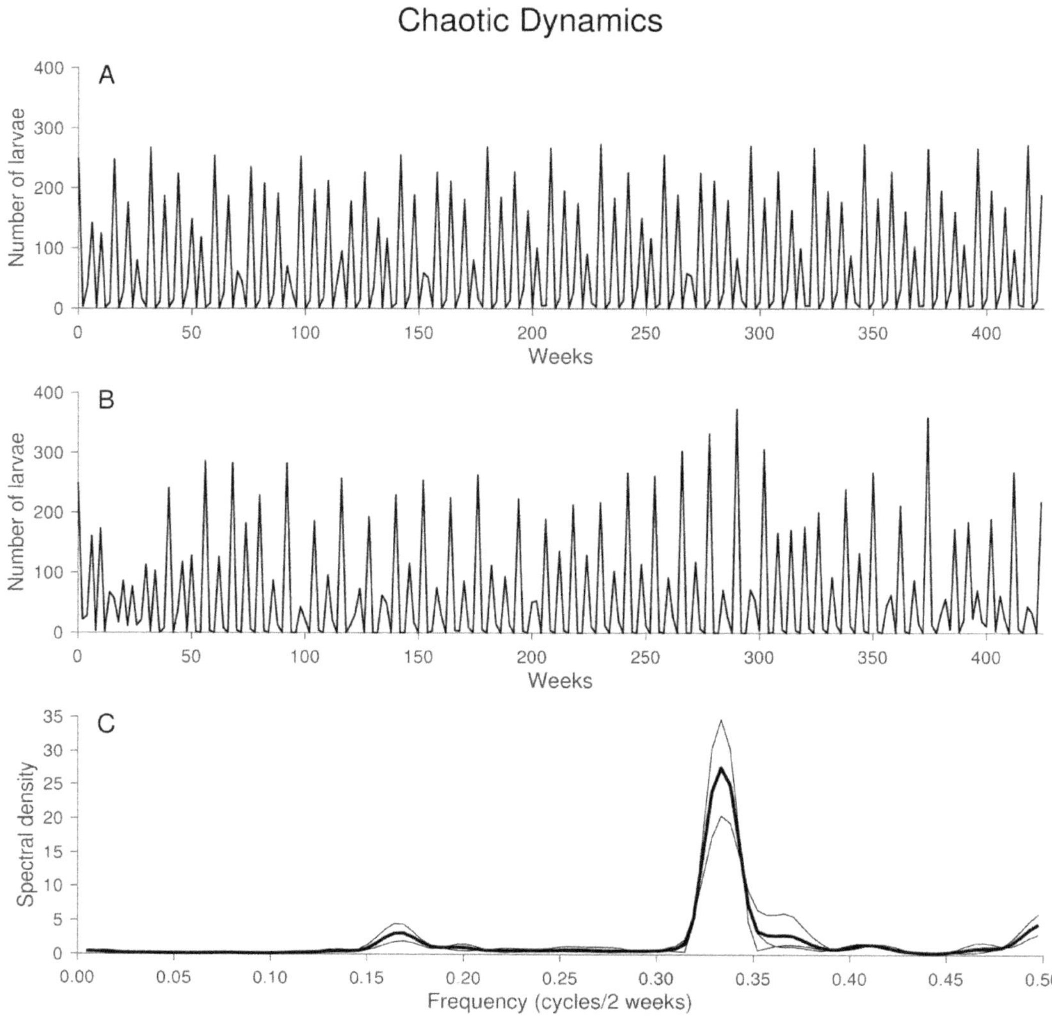

Figure 16.4 Deterministic model predictions, census data, and spectral analyses of larval numbers for the chaos treatment group ($c_{pa} = 0.35$, $\mu_a = 0.96$) from the route-to-chaos experiment (Chapter 7) [3, 5]. The three replicate populations were maintained for 424 weeks ($n = 213$). Initial numbers were $[L_0, P_0, A_0] = [250, 5, 100]$ and habitat size was held constant at $V = 1$. A. Deterministic model trajectory when the remaining parameters in model (16.4) have values as specified in Table 7.2. B. Larval numbers for one representative replicate. C. Spectral density plot of larval numbers. Thin lines are for the three individual replicates and the thick line is the mean. Two passes of the Daniell method were used to smooth the spectra, with window spans of three and five.

16.3.1 Filter Theory for Linear Equations

Let $\mathbf{x}_t = [x_{1t}, x_{2t}, \ldots, x_{\omega t}]$ be a vector of population variables (e.g., species densities, age-class numbers, life-stage numbers, etc.) whose dynamics are described by the model

$$\mathbf{x}_{t+1} = \mathbf{H}(\mathbf{x}_t, \boldsymbol{\varepsilon}_t),$$

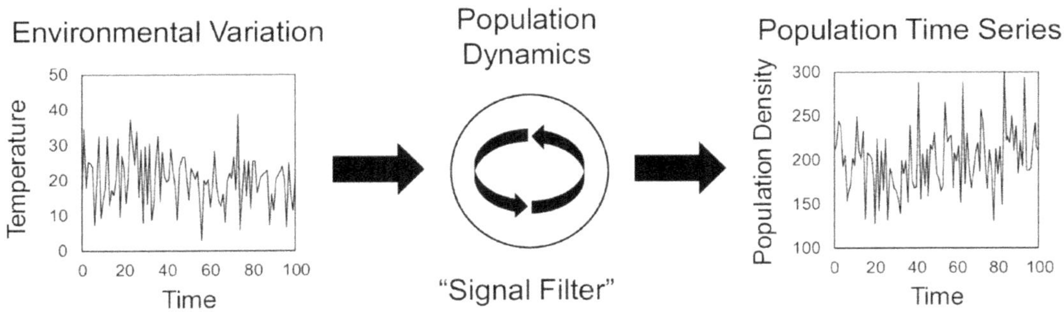

Figure 16.5 A conceptual representation of spectral filter theory applied to ecological population dynamics. Environmental variation can be treated as a "input signal" which is modified by the dynamics of the population to produce an "output signal." The latter is a time series of population numbers or densities. If the deterministic ecological model has a stable equilibrium, the model can be linearized around the equilibrium and linear filter theory can be used to predict the spectral properties of the population as a function of the model parameters and the spectra of the environmental noise.

where $\mathbf{H} = [h_1(\mathbf{x}_t, \boldsymbol{\varepsilon}_t), \ldots, h_\omega(\mathbf{x}_t, \boldsymbol{\varepsilon}_t)]$ is a vector-valued function, and $\boldsymbol{\varepsilon}_t$ is a vector of random environmental variables with expectation $\mathrm{E}(\boldsymbol{\varepsilon}_t) = \mathbf{0}$ for all $t \geq 0$. Assume $\mathbf{x}_t$ fluctuates around a stable equilibrium $\mathbf{x}^*$ due to noise. If the noise variables are of small magnitude, the dynamics of $\mathbf{x}(t)$ in the neighborhood of $\mathbf{x}^*$ can be described by the linearized stochastic model

$$\mathbf{w}_{t+1} = \mathbf{M}\mathbf{w}_t + \boldsymbol{\Lambda}\boldsymbol{\varepsilon}_t, \tag{16.5}$$

where $\mathbf{w}_t = \mathbf{x}_t - \mathbf{x}^*$. The $\omega \times \omega$ matrices $\mathbf{M} = \{m_{pq}\}$ and $\boldsymbol{\Lambda} = \{\gamma_{pq}\}$ have elements

$$\begin{aligned} m_{pq} &= \left.\frac{\partial h_p}{\partial x_{qt}}\right|_{\mathbf{x}_t=\mathbf{x}^*,\ \boldsymbol{\varepsilon}=\mathbf{0}} \\ \gamma_{pq} &= \left.\frac{\partial h_p}{\partial \varepsilon_{qt}}\right|_{\mathbf{x}_t=\mathbf{x}^*,\ \boldsymbol{\varepsilon}=\mathbf{0}}. \end{aligned} \tag{16.6}$$

If the roots of $\det(\mathbf{I} - \mathbf{M}z) = 0$ lie outside the unit circle in the complex plane (i.e., $|z| > 1$), it can be shown that (16.5) has a stationary solution [1]. Since, by assumption, the equilibrium $\mathbf{x}^*$ is stable, the roots of $\det(\mathbf{M} - \lambda\mathbf{I}) = 0$ lie within the unit circle ($|\lambda| < 1$; see Chapter 3), which implies $|z| > 1$.[8]

The linear transformation $\boldsymbol{\Lambda}^{-1}\mathbf{w}_t - \boldsymbol{\Lambda}^{-1}\mathbf{M}\mathbf{w}_{t-1}$ from (16.5) is a stable realizable filter[9] with a transfer function given by

$$\mathbf{T}(f) = \boldsymbol{\Lambda}^{-1}(\mathbf{I} - \mathbf{M}z), \tag{16.7}$$

where $z = e^{2\pi f i}$, $i = \sqrt{-1}$, and f is the frequency per unit time [1, 12]. The spectral matrices of $\boldsymbol{\varepsilon}_t$ and $\mathbf{w}_t$ have the relationship

$$\mathbf{S}_{\boldsymbol{\varepsilon}}(f) = \mathbf{T}(f)\,\mathbf{S}_{\mathbf{w}}(f)\,\bar{\mathbf{T}}(f)^{\mathrm{T}}, \tag{16.8}$$

where $\mathbf{S}_{\boldsymbol{\varepsilon}}(f)$ and $\mathbf{S}_{\mathbf{w}}(f)$ are the spectral matrices of the environmental noise and population fluctuations, respectively, and $\bar{\mathbf{T}}(f)^{\mathrm{T}}$ is the transpose of the complex conjugate of the

[8] An assumption here is that 0 is not an eigenvalue of M.

[9] A filter is realizable if it can be expressed in terms of only present and past values of $\mathbf{w}_t$.

transfer matrix [1, 12]. One can rearrange expression (16.8) to obtain a prediction for the spectral matrix of the population fluctuations:

$$\mathbf{S}_{\mathbf{w}}(f) = \mathbf{T}(f)^{-1}\mathbf{S}_{\boldsymbol{\varepsilon}}(f)\left(\bar{\mathbf{T}}(f)^{\mathrm{T}}\right)^{-1}. \tag{16.9}$$

The Ricker model (16.2) provides a simple univariate example. By assumption, the random noise variable ε_t describes a white noise process which has a flat power spectrum $S_\varepsilon(f) = \sigma^2$ for $0 \leq f \leq 0.5$. The coefficients for the linearization of the population and the noise are

$$M = \left.\frac{d\left(bN_t e^{-cN_t} e^{\varepsilon_t}\right)}{dN_t}\right|_{N_t=N^*,\ \varepsilon_t=0} = \lambda,$$
$$\Lambda = \left.\frac{d\left(bN_t e^{-cN_t} e^{\varepsilon_t}\right)}{d\varepsilon_t}\right|_{N_t=N^*,\ \varepsilon_t=0} = N^*,$$

where $\lambda = 1 - \ln b$. The transfer function is

$$T(f) = \Lambda^{-1} - \Lambda^{-1}Me^{2\pi i f} = \frac{1-\lambda e^{2\pi i f}}{N^*}.$$

From (16.9), the spectral density for the population fluctuations is

$$\begin{aligned} S_w(f) &= T(f)^{-1}S_\varepsilon(f)\bar{T}(f)^{-1} \\ &= \frac{N^*}{1-\lambda e^{2\pi i f}}\sigma^2\frac{N^*}{1-\lambda e^{-2\pi i f}} \\ &= \frac{(N^*)^2\sigma^2}{1+\lambda^2-\lambda e^{2\pi i f}-\lambda e^{-2\pi i f}}. \end{aligned} \tag{16.10}$$

Using Euler's identity, we obtain

$$e^{2\pi i f} + e^{-2\pi i f} = \cos(2\pi f) + i\sin(2\pi f) + \cos(-2\pi f) + i\sin(-2\pi f).$$

Since $\cos(2\pi f) = \cos(-2\pi f)$ and $\sin(-2\pi f) = -\sin(2\pi f)$, we have the result for $S_w(f)$ given in equation (16.3).

16.3.2 Predictions and Results for *Tribolium* Data

As discussed in Chapter 5, the LPA model for demographic noise added on a square root scale is

$$\begin{aligned} L_{t+1} &= \left(\sqrt{bA_t\exp\left(-\frac{c_{el}L_t}{V_t}-\frac{c_{ea}A_t}{V_t}\right)}+\varepsilon_{1t}\right)^2 \\ P_{t+1} &= \left(\sqrt{L_t(1-\mu_l)}+\varepsilon_{2t}\right)^2 \\ A_{t+1} &= \left(\sqrt{P_t\exp\left(-\frac{c_{pa}A_t}{V_t}\right)+A_t(1-\mu_a)}+\varepsilon_{3t}\right)^2, \end{aligned} \tag{16.11}$$

where the quantities to be squared are taken to be zero if negative, and $V_t = 1$ for $t \geq 0$. Although demographic stochasticity is considered an intrinsic property of a population, given the fact that it is being modeled by introducing a random vector $\boldsymbol{\varepsilon}_t = [\varepsilon_{1t}, \varepsilon_{2t}, \varepsilon_{3t}]$, we can use linear filter theory to obtain predictions for the population spectral matrix $\mathbf{S}_{\mathbf{w}}(f)$ as a function of the model parameters. By assumption, the random vectors $\boldsymbol{\varepsilon}_0, \boldsymbol{\varepsilon}_1, \ldots, \boldsymbol{\varepsilon}_{n-1}$ are independent and identically distributed with $\mathrm{E}(\boldsymbol{\varepsilon}_t) = \mathbf{0}$ and a constant variance-covariance matrix $\boldsymbol{\Sigma}$. This represents a stationary white noise process with a constant spectral matrix $\mathbf{S}_{\boldsymbol{\varepsilon}}(f) = \boldsymbol{\Sigma}$ for $0 \le f \le 0.5$. The matrix $\mathbf{M}$ of linearization coefficients is the same as the one for the deterministic LPA model (16.4) with $V_t = 1$:

$$\mathbf{M} = \begin{pmatrix} -c_{el}L^* & 0 & (L^*/A^*)(1 - c_{ea}A^*) \\ 1 - \mu_l & 0 & 0 \\ 0 & A^*(1/P^* + \mu_a - 1) & 1 - \mu_a c_{pa} A^* - \mu_a \end{pmatrix}, \tag{16.12}$$

where L^*, P^*, and A^* are the unique equilibrium values for the deterministic model. It is easy to show that $\boldsymbol{\Lambda}$ is a diagonal matrix:

$$\boldsymbol{\Lambda} = \begin{pmatrix} 2\sqrt{L^*} & 0 & 0 \\ 0 & 2\sqrt{P^*} & 0 \\ 0 & 0 & 2\sqrt{A^*} \end{pmatrix}.$$

The deterministic equilibrium values L^*, P^*, and A^* can be obtained numerically for a constant habitat size V by finding the positive value of α which solves the following nonlinear equation:

$$\beta_1 \exp\left(-\alpha\left(\beta_2 e^{\alpha c_{pa}/V} + c_{ea} + c_{pa}\right)/V\right) = 1, \tag{16.13}$$

where $\beta_1 = b(1 - \mu_l)/\mu_a$ and $\beta_2 = c_{el}\mu_a/(1 - \mu_l)$. Because the exponential term in equation (16.13) is monotonically decreasing as α increases from 0, equation (16.13) has a unique root if and only if $\beta_1 > 1$.[10] If we let α^* represent the root of equation (16.13), then the equilibrium life stages are $A^* = \alpha^*$, $P^* = \alpha^* \mu_a e^{\alpha^* c_{pa}/V}$, and $L^* = \alpha^* \mu_a e^{\alpha^* c_{pa}/V}/(1 - \mu_l)$. The equilibrium is asymptotically stable if the eigenvalues of $\mathbf{M}$ in matrix (16.12) lie within the unit circle of the complex plane.

As an example, we use the control cultures from the route-to-chaos experiment described in Chapter 7. Table 7.2 shows the conditional least squares parameter estimates for these populations. The predicted equilibrium is $L^* = 20.24$, $P^* = 16.20$, and $A^* = 375.23$. The matrices $\mathbf{M}$ and $\boldsymbol{\Lambda}$ are

$$\mathbf{M} = \begin{pmatrix} -0.3505 & 0 & -0.2113 \\ 0.8000 & 0 & 0 \\ 0 & 0.1767 & 0.9791 \end{pmatrix}$$

and

$$\boldsymbol{\Lambda} = \begin{pmatrix} 9.000 & 0 & 0 \\ 0 & 8.050 & 0 \\ 0 & 0 & 38.74 \end{pmatrix}.$$

The eigenvalues of $\mathbf{M}$ are $\lambda_1 = -0.4040$, $\lambda_2 = 0.0774$, and $\lambda_3 = 0.9552$, which implies that the equilibrium is stable. The spectral matrix $\mathbf{S}_{\boldsymbol{\varepsilon}}(f)$ for the white noise random process $\boldsymbol{\varepsilon}_0, \boldsymbol{\varepsilon}_1, \boldsymbol{\varepsilon}_2, \ldots$ is the estimated variance-covariance matrix for the control treatment group (Table 7.2):

$$\mathbf{S}_{\boldsymbol{\varepsilon}}(f) = \begin{pmatrix} 1.621 & -0.1336 & -0.01339 \\ -0.1336 & 0.7375 & -0.0009612 \\ -0.01339 & -0.0009612 & 0.01212 \end{pmatrix}, \quad 0 \le f \le 0.5. \tag{16.14}$$

[10] A similar result was obtained using an argument based on the net reproductive rate in Chapter 5.

The transfer function (16.7) is

$$\mathbf{T}(f) = \begin{pmatrix} 0.1111 & 0 & 0 \\ 0 & 0.1242 & 0 \\ 0 & 0 & 0.02581 \end{pmatrix} - \begin{pmatrix} -0.03894 & 0 & -0.02348 \\ 0.09938 & 0 & 0 \\ 0 & 0.004561 & 0.02527 \end{pmatrix} e^{2\pi i f}, \quad 0 \leq f \leq 0.5. \tag{16.15}$$

The matrices $\mathbf{T}(f)$ and $\mathbf{S}_{\boldsymbol{\varepsilon}}(f)$ can be used in equation (16.9) to compute the predicted spectral matrix of the control populations as a function of frequency.

Figure 16.6 shows the results for the control group from the route-to-chaos experiment (Chapter 7). The results for larvae and adults appear on the left and right sides, respectively. The top two panels (Figure 16.6A,B) are plots of the deterministic model (16.4) predictions for the parameter values in Table 7.2 and a constant habitat size $V = 1$. The middle plots show one representative replicate from the control treatment (Figure 16.6C,D). In both the model predictions and the census data, there is a transient period before the populations settle into an equilibrial behavior. For this reason, only the data from weeks 12–80 were used to estimate the spectral matrix function, $\mathbf{S_w}(f)$, for the three life stages. The spectral density functions for larvae, $\tilde{S}_{\mathbf{w},11}(f)$, and adults, $\tilde{S}_{\mathbf{w},33}(f)$, appear in Figures 16.6E,F. The thin unbroken curves are the spectra for the three replicates using Daniell smoothing windows with spans of three and five time steps, and the thick unbroken curve is the mean of the three curves. The dashed curve is the expected spectral density computed using the equations (16.14) and (16.15).

For larvae (Figure 16.6E), both the observed and expected spectra are U-shaped. For adults (Figure 16.6F), both the observed and expected spectra are skewed toward low frequencies. This difference demonstrates the possibility that the time-scales of the responses to stochastic perturbations of different components of a population may differ. Also, the two extreme peaks for larvae shows that the response to stochasticity may manifest itself on more than one time-scale.

For both larvae and adults, there are deviations between the mean of the observed spectral densities and the theoretical values. This is because the observed time series are short, especially after the removal of transients ($n = 35$). It is particularly difficult to estimate the spectral power of low frequency cycles in short time series because the cycle periods are long. As a check of the accuracy of linear filter methods, we simulated the stochastic model for $n = 2^{16}$ iterations after the removal of transients. The dotted curves in Figures 16.6E,F show the spectral densities for these model simulations using Daniell smoothing windows with spans of 1001 and 1003 time steps. There is a close agreement between the spectra from the model simulations and the theoretical predictions. This suggests that linear filter theory performs well at the realistic levels of noise used in the nonlinear model.

The relevance of linear filter theory for ecology is that it provides an analytical tool for connecting mechanistic demographic models to the stationary stochastic dynamics of population numbers. As an example, we can use the LPA model with parameters as given Figure 16.6. Setting the adult mortality rate to a value of $\mu_a = 0.010$ would yield a predicted adult spectrum where most of the power was concentrated at low frequencies, similar to the dashed curve in Figure 16.6F. At $\mu_a = 0.057$, a U-shaped spectrum is predicted, with the power distributed nearly equally at low and high frequencies. Increasing μ_a to a value of 0.074 gives a spectrum with a single strong peak at $f = 0.5$, which corresponds to a two-cycle. (In all three examples, the deterministic LPA model predicts a stable equilibrium.) Similar predictions could be made for the larval and pupal power spectra and for other model parameters. In general, linear filter theory can be used to predict the impact of

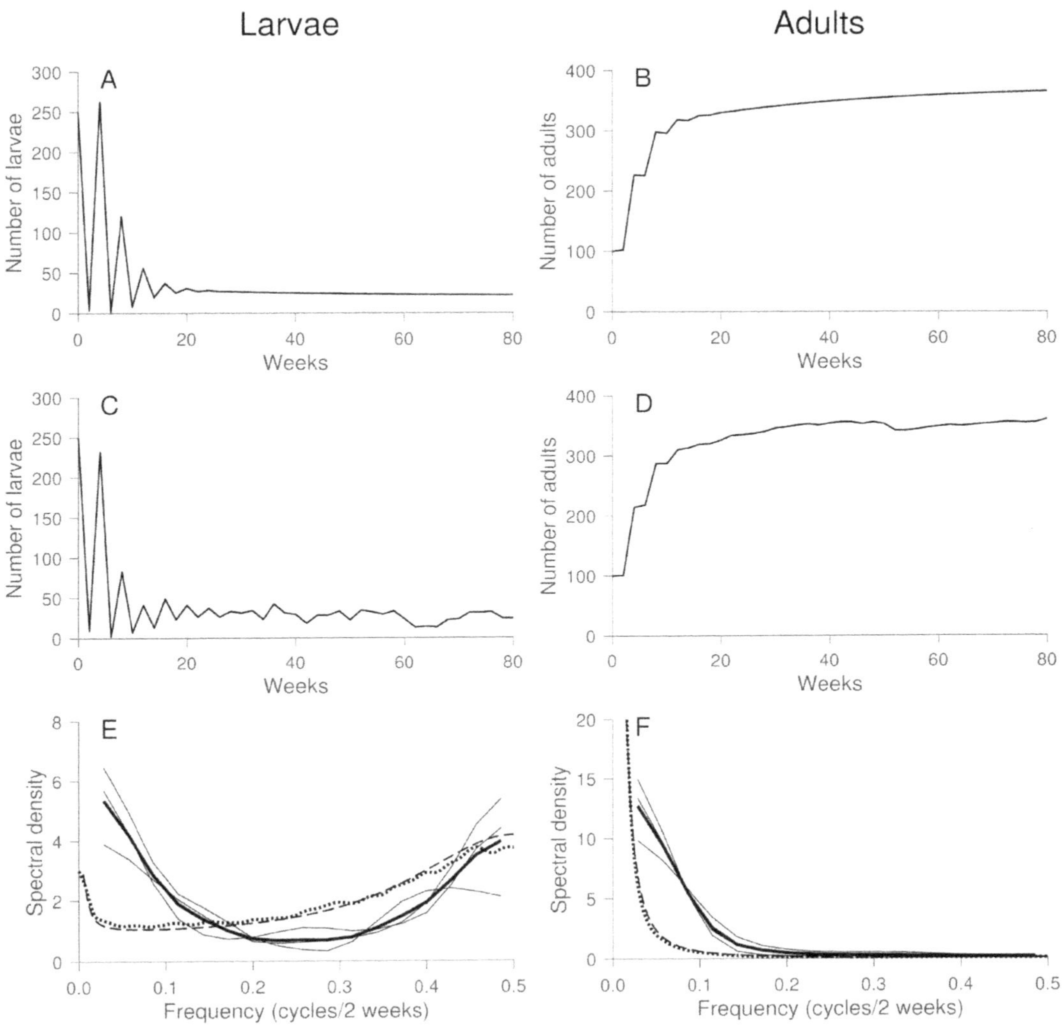

Figure 16.6 Deterministic model predictions, census data, and spectral analyses of larval and adult numbers for the control group from the route-to-chaos experiment (Chapter 7) [3, 5]. The three replicate populations were maintained for 80 weeks. Initial numbers were $[L_0, P_0, A_0] = [250, 5, 100]$ and habitat size was held constant at $V = 1$. Panels A and B show the deterministic model trajectories for larva and adults when the parameters in (16.4) have values as specified in Table 7.2. Panels C and D show larval and adult numbers for one representative replicate (replicate 24). Panels E and F are spectral density plots of larval and adult numbers for weeks 14–80. Thin lines are for the three individual replicates and the thick line is the mean. Two passes of Daniell method were used to smooth the spectra, with window spans of three and five. The dashed curves are the predicted spectral densities obtained by applying linear filter theory to the stochastic model (16.11) using the variance-covariance matrix estimated from the data (Table 7.2). The dotted curves are spectral densities for larvae and adults obtained by simulating the stochastic model (16.11) for $n = 65,536$ time steps, after removing 100 initial transients. Two passes of the Daniell method were used to smooth the spectra for the simulations, with window spans of 1001 and 1003.

changes in the values of life history parameters on the characteristics of the population power spectra, and, in particular, the time scales at which population numbers fluctuate.

16.4 COLORED ENVIRONMENTAL NOISE

Most ecological models that include environmental stochasticity use white noise random variables. However, there is evidence that some time series of environmental variables are reddened (environmental fluctuations are shifted toward low frequencies, that is, long time scales) [15] and that the reddening has increased in recent years due to climate change [6]. Colored noise can interact with the population dynamics in ways that differ from white noise [7, 8, 14]. Thus, it may be more realistic to include colored noise in stochastic ecological models, especially where predictions are needed for conservation efforts and wildlife management.

Linear filter theory allows for the inclusion of any arbitrary spectral matrix for the noise process in the system being modeled. In this section, we look at the impact of colored noise on the Ricker model and present some predictions and results from a *Tribolium* experiment that looked at the effects of colored noise on population time series.

16.4.1 Ricker Model Examples

Linear filter theory was applied to the Ricker model (16.2) in Section 16.2.1. In the case of white noise, the random variable $\varepsilon(t)$ has a constant power spectrum, $S_\varepsilon(f) = \sigma^2$, where σ^2 is the variance of the noise. We will now consider a situation where the environmental noise variable has a power spectrum for a first-order autoregressive process:

$$S_\varepsilon(f) = \frac{\sigma^2}{1 + a^2 - 2a\cos(2\pi f)}, \tag{16.16}$$

where a is the autoregression coefficient. For $a > 0$, the power spectrum is skewed toward low frequencies (red noise), and for $a < 0$ the power spectrum is skewed toward high frequencies (blue noise). White noise is represented by $a = 0$. Replacing the σ^2 in equation (16.10) with the expression (16.16) gives the following power spectrum for the population fluctuations around its stable equilibrium:

$$S_w(f) = \frac{(N^*)^2\sigma^2}{(1 + a^2 - 2a\cos(2\pi f))\left(1 + \lambda^2 - 2\lambda\cos(2\pi f)\right)}, \tag{16.17}$$

where $N^* = (\ln b)/c$ and $\lambda = 1 - \ln b$.

Figure 16.7 shows the spectral densities for all possible combinations of reproductive rates $b = 1.2, 2.1783, 5.0$ paired with noise color parameters $a = 0, 0.9, -0.9$. The remaining parameters were $c = 0.001$, $\sigma = 0.03$, and $N_0 = 10$. In each case, the Ricker model (16.2) was iterated 100 times to remove transients and then 4096 iterations were used to produce a long stationary time series. The solid curves show the spectral densities estimated from the stochastic model time series and the dashed curves show the linear filter predictions obtained using equation (16.17). In general, there is a good agreement in the frequency domain between the nonlinear stochastic model realizations and the linearized theoretical results. For slow population growth rates ($b = 1.2$), white noise and red noise give similar results (Figs 16.7A,D) which are skewed toward low frequencies (long time scales), whereas blue noise gives spectral peaks at both high and low frequencies (Figure 16.7G). For $b = e \approx 2.7183$ ($\lambda = 0$), the population spectrum reflects the color of the environmental noise (Figure 16.7B,E,H). For fast population growth rates ($b = 5$), white noise and blue noise give similar results (Figs 16.7C,I) which are skewed toward high frequencies (short time scales), whereas red noise gives spectral peaks at both high and low frequencies (Figure 16.7F).

Clearly, there is an interaction between the stability characteristics of the population and the color of the environmental noise. That is to say, Figure 16.7 demonstrates the interplay between the influence of underlying biological mechanism and the frequency of environmental perturbation.

When the deterministic tendency toward equilibrium is weak (Figure 16.7A,D,G), any environmental perturbations will accumulate, contributing to long-term stochastic fluctu-

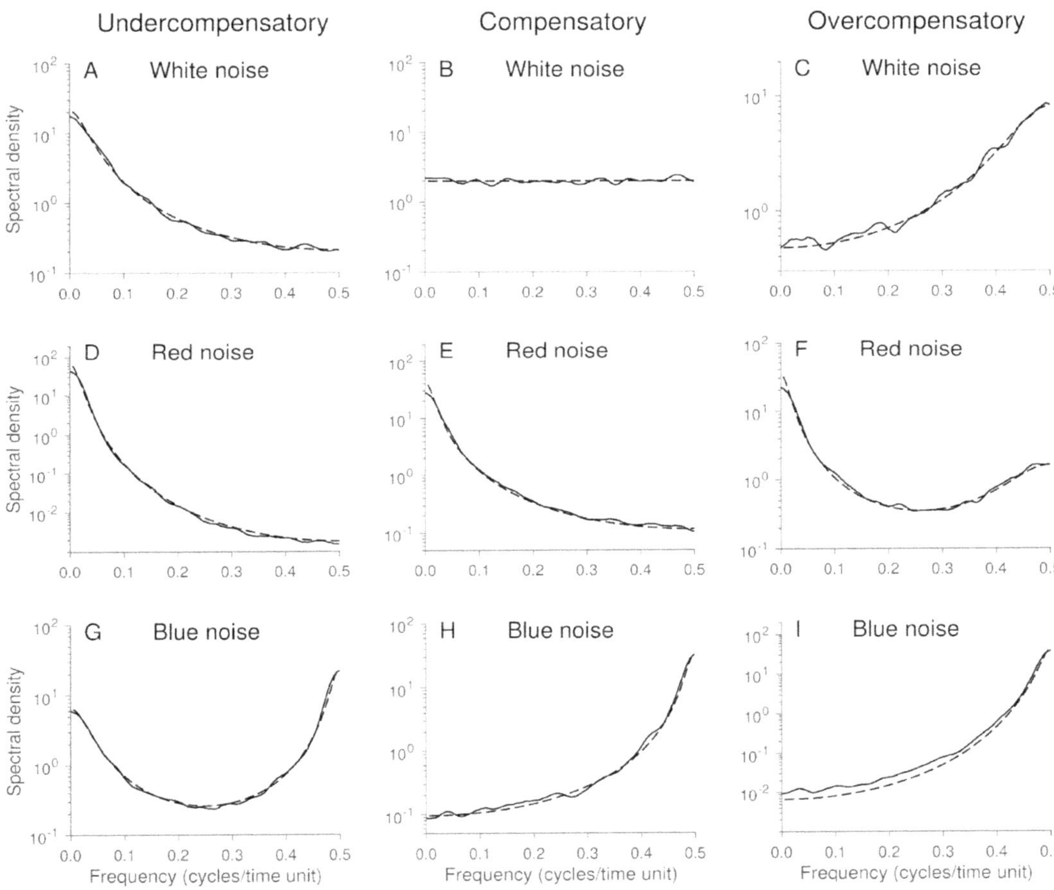

Figure 16.7 Spectral analyses for the Ricker model (16.2) with colored environmental noise. Three growth rate values were chosen: $b = 1.2$, $b = e \approx 2.7183$, and $b = 5$. Each of these were paired with three parameter values used in equation (16.16) to set the "spectral color" of the environmental noise: white ($a = 0$), red ($a = 0.9$), and blue ($a = -0.9$). Parameter $c = 0.001$ and the initial condition was $N_0 = 10$. Panels A, B, and C show the spectral densities for white noise paired with the three population growth rates $b = 1.2$, e, and 5, respectively. Panels D–F and G–I show the equivalent spectral densities for red and blue noise, respectively. The solid curves are the spectral densities estimated from a long stationary time series of length $n = 4096$ after 100 initial transients were removed. Two passes of the Daniell method were used to smooth the spectra, with window spans of 101 and 103. The dashed curves are the theoretical spectral densities (16.17) from linear filter theory for the parameter values used in the simulations.

ations in population numbers. This corresponds to the low-frequency peaks in panels A, D, and G. In addition, however, if the environmental perturbations are frequent (panel G), they will force high-frequency fluctuations in the population, as well, which is why we see both low- and high-frequency peaks in the spectral densities in panel G.

By contrast, when the deterministic tendency toward equilibrium is strong (Figure 16.7C,F,I), the effects of environmental perturbations are quickly damped out, and so resulting stochastic fluctuations in the population will tend to look like the deterministic transient tendencies, which in this case are like (damped) two-cycles. Thus, the population will show high-frequency stochastic fluctuations (panels C and I, and to some degree panel F). In addition, however, environmental perturbations that occur over longer periods (panel F) will impart more persistent and less frequent stochastic bumps in population numbers, which is why there are two peaks in the spectral density in panel F.

16.4.2 A *Tribolium* Experiment

We conducted a *Tribolium* experiment to evaluate the effects of noise color on the spectral properties of population dynamics, using habitat size V_t as the random environmental variable. There were four treatment groups, each with four replicates: the controls with a constant habitat size of $V = 2.7$ (54 g of medium), and three experimental groups for which the habitat size was varied randomly around a mean of $V^* = 2.7$ with a standard deviation of 1.2. In the experimental groups, we used *spectral mimicry* [2] to vary the sequence of habitat sizes to have a white, red, or blue power spectrum. Specifically, four random sequences of 40 habitat sizes were generated from a normal distribution with a mean of 2.7 and a standard deviation of 1.2, where the individual elements of each sequence were independently and identically distributed (white noise). Habitat sizes less than 0.4 (8 g medium) were set at 0.4 and values above 5.0 (100 g medium) were set at 5.0.[11] For red noise, four sequences of length 40 were generated using a first-order autoregressive model (AR1) $y_{t+1} = ay_t + \eta_t$, where y_0 and η_t are independently identically distributed standard normal variables, and $a = 0.9$. This model has a power spectrum given by equation (16.16) with $\sigma = 1$. For each of the four sequences, the ranks of $y_0, y_1, \ldots, y_{n-1}$ for $n = 41$ were obtained. To produce the red noise habitat sequences, the four white noise sequences were rearranged so that each had the same rank ordering as the four AR1 sequences. For the four sequences of blue noise habitat sizes, this process was repeated using $a = -0.9$. Thus, the four sets of habitat sizes for the red and blue noise treatments had the same as elements as for the white noise treatment, but with different orderings. Asymptotically, as $n \to \infty$, this method of spectral mimicry produces power spectra that have the same shape as the spectrum for the stationary time series being mimicked [2].

For each of the replicates in the experiment described above, we manipulated two LPA model parameters to have values of $c_{pa} = 0.0842$ and $\mu_a = 0.3$ using the same experimental protocols as in the route-to-chaos experiment (Chapter 7). Using previously estimated values for the remaining parameters (Table 7.2), the predicted attractor is a stable equilibrium with life-stage numbers $L^* = 166.4$, $P^* = 133.1$, and $A^* = 62.73$. To minimize transients, each population was initiated with life stage values near the predicted equilibrium: $L_0 = 166$, $P_0 = 133$, and $A_0 = 63$.

To apply linear filter theory to our experimental data, we ignored demographic stochasticity and assumed that the fluctuations in the life-stage numbers were dominated by the random fluctuations in habitat size, $V_t = V^* + \varepsilon_t$. To evaluate the validity of this assumption, we looked at the observed variances in life stage numbers for the control populations compared to the replicates where habitat size was varied. For larvae, the controls had a

[11] 3.1% of the habitat sizes required adjustment.

variance of 766.7 across replicates, whereas the populations with fluctuations in habitat size had a variance of 36,019.7, giving a ratio of 2.1%. The ratios for pupae and adults were 2.3% and 1.0%, respectively. Therefore, we assumed the only source of random noise in the stochastic model (16.4) was environmental stochasticity.

Equation (16.9) was used to generate predictions for the spectral properties of the population fluctuations from the linearized model. The elements were computed using equations (16.6):

$$\mathbf{M} = \begin{pmatrix} -c_{el}L^*/V^* & 0 & (L^*/A^*)\left(1 - c_{ea}A^*/V^*\right) \\ 1-\mu_l & 0 & 0 \\ 0 & A^*\left(1/P^* + \mu_a - 1\right) & 1 - \mu_a c_{pa}A^*/V^* - \mu_a \end{pmatrix}. \tag{16.18}$$

The eigenvalues of $\mathbf{M}$ are $\lambda_1 = -0.7033$, $\lambda_2 = -0.6841$, and $\lambda_3 = 0.4338$, each with a modulus less than one, implying that the equilibrium is stable and the linearized stochastic process has a stationary solution. Since the three life stages experience the same habitat sequence and the equation for pupa is unaffected by habitat size, $\varepsilon_{1t} = \varepsilon_{3t} = \varepsilon_t$, and $\varepsilon_{2t} = 0$ for $t \geq 0$. The elements $\mathbf{\Lambda}$ were computed using equations (16.6):

$$\mathbf{\Lambda} = \begin{pmatrix} \gamma_{11} & 0 & \gamma_{13} \\ 0 & 0 & 0 \\ \gamma_{31} & 0 & \gamma_{33} \end{pmatrix}, \tag{16.19}$$

where

$$\gamma_{11} = \gamma_{13} = L^*\left(c_{el}L^* + c_{pa}A^*\right)/(V^*)^2$$
$$\gamma_{31} = \gamma_{33} = \mu_a c_{pa}(A^*/V^*)^2.$$

Since the larvae and adults are exposed to the same habitat fluctuations, the cross-spectra will equal the spectra. The theoretical value for the spectral matrix of the noise is

$$\mathbf{S}_{\boldsymbol{\varepsilon}}(f) = \frac{\sigma^2}{1 + a^2 - 2a\cos(2\pi f)} \begin{pmatrix} 1 & 0 & 1 \\ 0 & 0 & 0 \\ 1 & 0 & 1 \end{pmatrix}, \tag{16.20}$$

where $\sigma = 1.2$. Matrix (16.18) and matrix (16.19) can be substituted into expression (16.7) to obtain the transfer function $\mathbf{T}(f)$ for the linearized stochastic process. This can be combined with the spectral matrix (16.20) for the noise to obtain, from equation (16.9), the spectral matrix $\mathbf{S}_{\mathbf{w}}(t)$ for the population.

Figure 16.8 shows the results of the spectral analyses of the data and the predictions from linear filter theory for larval fluctuations. The thin unbroken curves in Figure 16.8A show the spectral densities for the four populations subjected to white noise sequences of habitat size, which were smoothed using Daniell windows with spans of three and five time steps. The thick unbroken curve is the average for the four replicates. The dashed curve is the spectral density $\tilde{S}_{\mathbf{w},11}(f)$ computed from linear filter theory with $\sigma = 1.2$ and $a = 0$. The observed and theoretical spectra are skewed toward high frequencies (short time scales) with a spectral peak near $f = 0.5$ (period-two cycle). Figure 16.8B shows the results for red noise habitat sequences ($\sigma = 1.2$ and $a = 0.9$). Both the observed and predicted spectral densities are U-shaped, with peaks at both low and high frequencies. The reddened environmental spectra altered the time scales of the larval dynamics when compared to white environmental noise. In Figure 16.8C are plots of the spectral densities for larval numbers subjected to blue noise habitat sequences ($\sigma = 1.2$ and $a = -0.9$). The spectra are similar to white noise, with a stronger shift toward high frequencies. Considering the shortness of the experimental time series ($n = 41$) and the magnitude of the life stage fluctuations relative to the constant habitat control populations, the agreement between the data and the results for the linearized model are good.

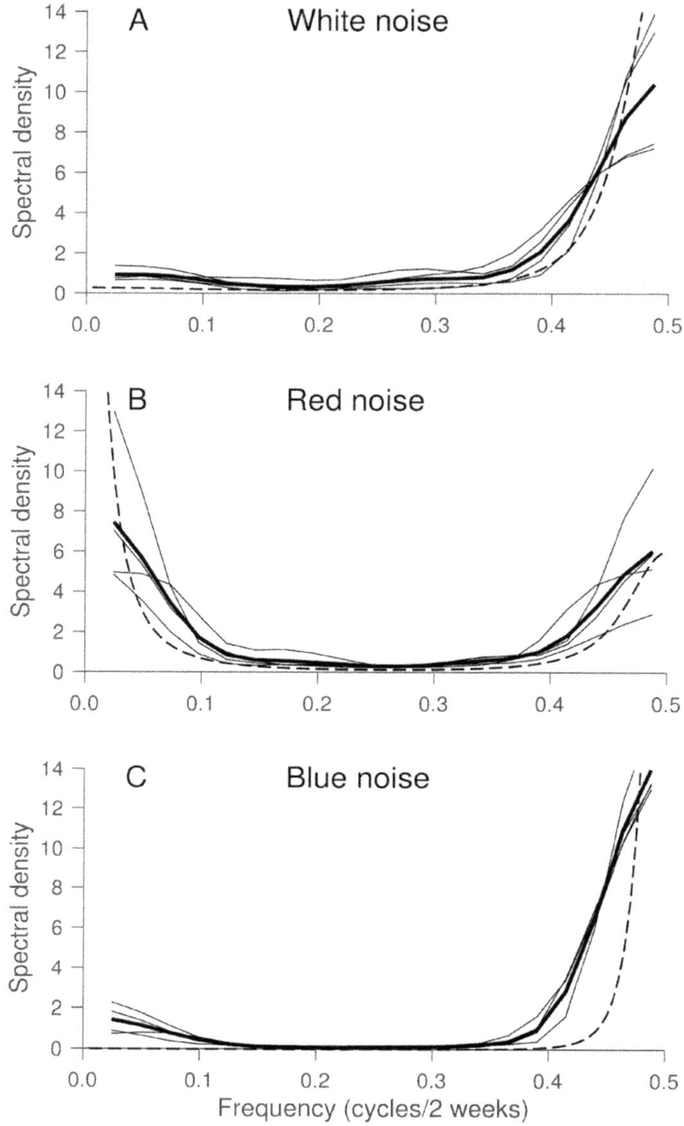

Figure 16.8 Spectral analyses for *Tribolium* experimental data with three treatment groups of colored environmental noise and four replicates per group. For all replicates, two of the LPA model parameters were manipulated to have values of $c_{pa} = 0.0842$ and $\mu_a = 0.3$. Initial numbers were $[L_0, P_0, A_0] = [166, 133, 63]$. Environmental noise was simulated using random sequences of habitat sizes $V_t = V^* + \varepsilon_t$, where $V^* = 2.7$ and ε_t is a random variable with a mean of zero and a standard deviation of 1.2. Spectral mimicry [2] was used to obtain each habitat sequence based on a first-order autoregressive model with the regression coefficient set at $a = 0$ (A, white noise), $a = 0.9$ (B, red noise), and $a = -0.9$ (C, blue noise). Demographic stochasticity was ignored. In each plot, thin lines are the larval spectra for the three individual replicates and the thick line is the mean. Two passes of the Daniell method were used to smooth the spectra, with window spans of three and five. The dashed curves are the predicted spectral densities obtained by applying linear filter theory to the random habitat size LPA model (16.4), with the remaining parameter values from Table 7.2 and the environmental spectral matrix function (16.20).

These experimental results support the general conclusions obtained using the Ricker model (Figure 16.7). Here, the predicted equilibrium dynamics for the LPA model are overcompensatory (the dominant eigenvalue is large and negative, $\lambda_1 = -0.7033$), so in the presence of white noise, the effects of the environmental fluctuations combine with the deterministic tendency for a damped two-cycle, which skews the population spectrum toward greater power at high frequencies (Figure 16.8A). When subjected to blue noise, the high-frequency environmental perturbations act synergistically with the deterministic tendency toward damped two-cycles, and the spectral power of the population fluctuations is further enhanced at high frequencies (Figure 16.8C). On the other hand, when subjected to red noise, the environmental perturbations are less frequent and of longer period. The populations track these long time scale changes while still exhibiting small high-frequency oscillations due to the influence of the deterministic forces, and the populations display peaks at both ends of the power spectra (Figure 16.8B).

16.5 CONCLUSIONS

Spectral analyses allow a different perspective on the dynamics of populations fluctuating around an attractor. Even without an explicit model, one could obtain insight into the dynamics of the populations. In the case of populations with equilibrium dynamics, one could, in principle, distinguish between undercompensatory and overcompensatory dynamics. For populations with more complex dynamics, one might be able to detect the signatures of periodic cycles or chaos. One might also learn how the dynamics of separate components of a population differ. For example, if immatures show dynamics on a short time scale and adults show dynamics on a long time scale, this might indicate the presence of nonlinear interactions among the immatures and/or an increased susceptibility of immatures to environmental fluctuations. These points were demonstrated using the Ricker model and *Tribolium* experimental data.

Linear filter theory provides an opportunity to connect the spectral properties of environmental fluctuations to the spectral properties of ecological populations fluctuating around a stable equilibrium. With an ecological model, one can get predictions for the interaction effects between the dynamical properties of the population and the time scales of the changes in environmental variables (color of the environmental noise). One can also use linear filter theory to predict the impact of changes in the values of life history parameters on the characteristics of the population power spectra, and, in particular, the time scales at which population numbers fluctuate. An experimental test using laboratory populations of flour beetles supported the predictions from linear filter theory.

Bibliography

[1] Brillinger DR (2001) Time Series: Data Analysis and Theory. Philadelphia, Society for Industrial and Applied Mathematics.

[2] Cohen, JE, Newman CM, Cohen AE, Petchey OL, Gonzalez A (1999) Spectral mimicry: A method of synthesizing matching time series with different Fourier spectra. Circuits, Systems and Signal Processing 18:431–442. https://doi.org/10.1007/BF01200792

[3] Costantino RF, Desharnais RA, Cushing JM, Dennis B (1997) Chaotic dynamics in an insect population. Science 275:389–391. https://doi.org/10.1126/science.275.5298.389

[4] Daniell, PJ (1946) Discussion on the papers by Bartlett, Foster, Cunningham and Hynd. Supplement to the Journal of the Royal Statistical Society 8:88–90. https://doi.org/10.2307/2983611

[5] Dennis B, Desharnais RA, Cushing JM, Henson SM, Costantino RF (2001) Estimating chaos and complex dynamics in an insect population. Ecological Monographs 71:277–303. https://doi.org/10.1890/0012-9615(2001)071[0277:ECACDI]2.0.CO;2

[6] Desharnais RA, Muchlinski AE, Ortiz JL, Alvidrez RI, Gatza BP (2022) Timescale analyses of fluctuations in coexisting populations of a native and invasive tree squirrel. Ecology and Evolution 12:e8779. https://doi.org/10.1002/ece3.8779

[7] Greenman JV, Benton TG (2005) The frequency spectrum of structured discrete time population models: Its properties and their ecological implications. Oikos 110:369–389. https://doi.org/10.1111/j.0030-1299.2005.13652.x

[8] Greenman JV, Benton TG (2005) The impact of environmental fluctuations on structured discrete time population models: Resonance, synchrony and threshold behaviour. Theoretical Population Biology 68:217–235. https://doi.org/10.1016/j.tpb.2005.06.007

[9] Henson SM, Costantino RF, Cushing JM, Desharnais RA, Dennis B, King AA (2001) Lattice effects observed in chaotic dynamics of experimental populations. Science 294:602–605. https://doi.org/10.1126/science.1063358

[10] King AA, Costantino RF, Cushing JM, Henson SM, Desharnais RA, Dennis B (2004) Anatomy of a chaotic attractor: Subtle model-predicted patterns revealed in population data. Proceedings of the National Academy of Sciences USA 101:408–413. https://doi.org/10.1073/pnas.2237266100

[11] Percival DB, Walden AT (1993) Spectral Analysis for Physical Applications. Cambridge, Cambridge University Press.

[12] Reinsel GC (1997) Elements of Multivariate Time Series Analysis. Second Edition. New York, Springer.

[13] Reuman DC, Costantino RF, Desharnais RA, Cohen JE (2008) Colour of environmental noise affects the nonlinear dynamics of cycling, stage-structured populations. Ecology Letters 11:820–830. https://doi.org/10.1111/j.1461-0248.2008.01194.x

[14] Schwager M, Johst K, Jeltsch F (2006) Does red noise increase or decrease extinction risk? Single extreme events versus series of unfavorable conditions. The American Naturalist 167:879–888. https://doi.org/10.1086/503609

[15] Vasseur DA, Yodzis P (2004) The color of environmental noise. Ecology 85:1146–1152. https://doi.org/10.1890/02-3122

CHAPTER 17

Time Scales of Ecological Synchrony

Ecological synchrony involves correlated fluctuations in the densities of spatially distinct populations. Population correlations can occur over different time scales and this aspect of synchrony is often overlooked. In this chapter we apply spectral analysis and linear filter theory to a metapopulation version of the LPA model and obtain predictions for the effects of dispersal and different colors of environmental noise on population synchrony. These predictions are tested experimentally using laboratory populations of *Tribolium*.

17.1 ECOLOGICAL SYNCHRONY

Populations that fluctuate in a coherent fashion exhibit *ecological synchrony*. A group of conspecific populations that are separated geographically but have the potential to exchange individuals by dispersal is referred to as a *metapopulation*. Metapopulation synchrony has been observed in insects, human pathogens, fish, amphibians, mammals and other taxa [29] and has implications for conservation biology, natural resource management, agriculture, and public health. Synchronous outbreaks of pests and pathogens can cause widespread damage [2, 17, 32, 33, 35, 41]. Synchrony in fragmented habitats can negatively affect the persistence of metapopulations [18, 21, 23] and ecological synchrony may be increasing due to climate change [16, 25, 28, 36, 41, 42]. Metapopulations are often the result of habitat fragmentation due to human activities. An increased understanding of ecological synchrony could lead to better natural resource management and conservation decisions.

There are two main factors that can lead to metapopulation synchrony: environmental correlations and dispersal. Spatial covariance in environmental variables that affect population growth can lead to synchrony, a phenomenon known as the "Moran effect" [24, 31, 38]. In this case, environmental correlations between locations lead to population correlations in those areas, a result Moran demonstrated using a pair of linear second-order difference equations with correlated stochastic noise [31]. Dispersal of individuals among populations can also lead to ecological synchrony [6, 22, 30, 37, 40]. Solé and Gamarra examined a pair of identical deterministic difference equations exhibiting chaotic dynamics and coupled them using a dispersal rate, d. They showed that, given different initial conditions, the chaotic dynamics of the two populations will synchronize whenever $d > 0.5\left(1 - e^{-\lambda}\right)$, where λ is the Lyapunov exponent[1] of each equation in the absence of dispersal [43]. These two

[1]For a chaotic deterministic system, the Lyapunov exponent λ is the exponential rate of separation of two nearby trajectories.

DOI: 10.1201/9781003467120-17

mechanisms, environmental correlations and dispersal, can operate separately or interact in combination to synchronize metapopulation fluctuations [27].

These two main mechanisms of metapopulation synchrony can be demonstrated using a pair of stochastic Ricker equations:

$$\begin{aligned} \widehat{N}_1 &= b_1 N_{1,t} e^{-c_1 N_{1,t}/V_1} e^{\varepsilon_{1,t}} \\ \widehat{N}_2 &= b_2 N_{2,t} e^{-c_2 N_{2,t}/V_2} e^{\varepsilon_{2,t}}, \end{aligned} \tag{17.1}$$

where $\widehat{N}_i$ in equation (17.1) stands for the population numbers before dispersal. After dispersal,

$$\begin{aligned} N_{1,t+1} &= (1-d)\,\widehat{N}_1 + d\widehat{N}_2 \\ N_{2,t+1} &= (1-d)\,\widehat{N}_2 + d\widehat{N}_1, \end{aligned} \tag{17.2}$$

where the dispersal rate, d, is the fraction of individuals moving between populations. The parameters b_i, c_i, and V_i are the reproductive rate, density-sensitivity, and habitat size for population i. The random variables $\boldsymbol{\epsilon}_t = [\varepsilon_{1,t},\ \varepsilon_{2,t}]$ are[2] from a bivariate normal distribution and $\boldsymbol{\epsilon}_0, \boldsymbol{\epsilon}_1, \boldsymbol{\epsilon}_2, \ldots$ are independently and identically distributed. Figure 17.1 shows a simulation for these equations with $b_1 = b_2 = 5$, $c_1 = c_2 = 0.001$, and $V_1 = V_2 = 1$. The random variables $\varepsilon_{1,t}$ and $\varepsilon_{2,t}$ have a mean of zero and a variance of 0.01. In Figure 17.1A, the environmental noise has a correlation of zero ($\rho_\varepsilon = 0$) and $d = 0$. In this simulation, the two populations fluctuate independently and have a small correlation of $r = 0.04$. Figure 17.1B shows a realization where the environmental noise has a correlation of 0.9 and there is no dispersal ($d = 0$). The two populations are strongly synchronous ($r = 0.89$). In Figure 17.1C, the environmental noise has a correlation of zero and the dispersal rate is $d = 0.3$. Dispersal synchronizes the two populations ($r = 0.88$). These examples demonstrate that both environmental covariance and dispersal are potential causal mechanisms for metapopulation synchrony.

Synchrony has been measured in several ways. In most cases, metrics based on correlation have been used to characterize the coherence between two time series [1, 4, 19]. However, this approach ignores the possibility that synchrony can vary at different time scales. Several studies have used spectral methods to measure synchrony at different frequencies [9, 10, 11, 20, 26, 41, 44]. They show that the time scales at which synchrony occurs matter for understanding the influence of environmental factors on regional-scale population dynamics.

17.2 TIME SCALE OF SYNCHRONY

In Chapter 16, we showed that the variance of a stationary time series can be decomposed into contributions at different frequencies using the power spectrum. This allows a visualization of the distribution of the fluctuations of the time series at different time scales. Similarly, for a stationary multivariate time series, the in-phase covariance between any two elements of the time series can be visualized in the frequency domain using the cospectrum. Applying these spectral methods to a metapopulation time series allows a time scale analysis of ecological synchrony.

[2] We use the square-bracket notation $[\varepsilon_{1,t},\ \varepsilon_{2,t}]$ to mean the column vector $(\varepsilon_{1,t},\ \varepsilon_{2,t})^T$, which is the transpose of the row vector $(\varepsilon_{1,t},\ \varepsilon_{2,t})$.

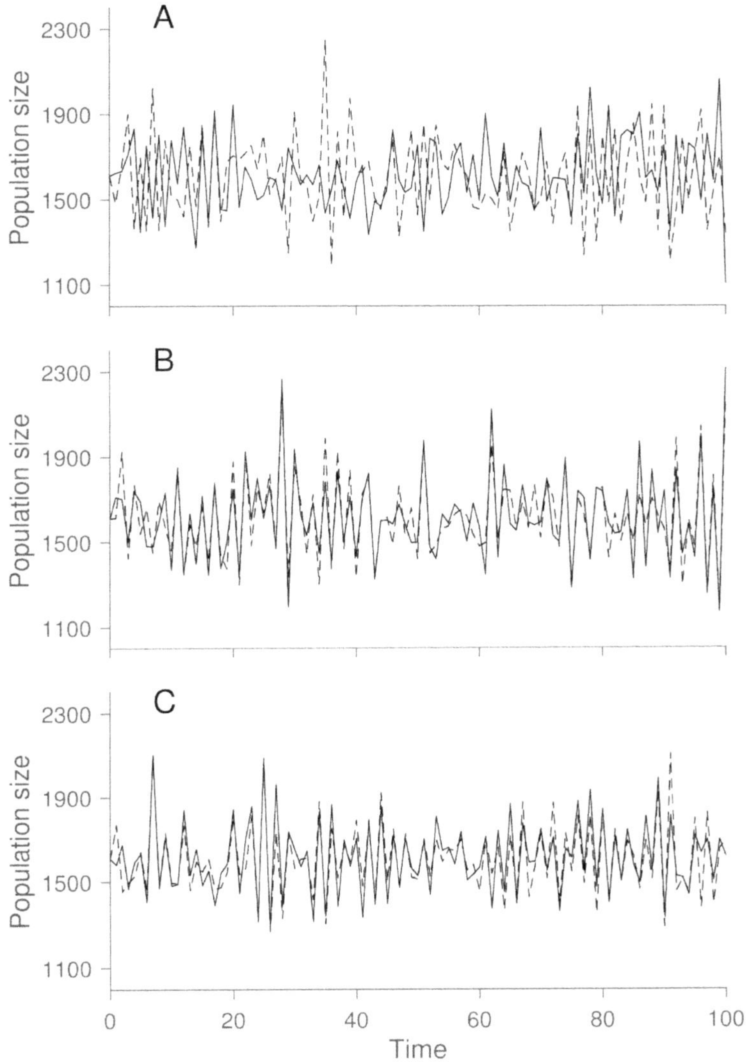

Figure 17.1 Realizations of the Ricker metapopulation model (17.1) – (17.2) demonstrating two causes of synchrony. A. No synchrony. B. Synchrony due to correlation in the environmental noise. C. Synchrony caused by dispersal. Solid and dashed lines are used to distinguish the two populations. See text for details.

17.2.1 Cospectra and Corrspectra

Let $\mathbf{X}_0, \mathbf{X}_2, \ldots, \mathbf{X}_{n-1}$ represent a sample of equally-spaced observations from a multivariate stationary time series where $\mathbf{X}_t = [x_{1t}, x_{2t}, \ldots, x_{\omega t}]$. As was shown in the previous chapter, $\mathbf{X}_t$ has a spectral matrix, $\mathbf{S}_{\mathbf{X}}(f)$ at frequency f, where the element in row p and column q is the product of the Fourier transform of x_{pt} and the complex conjugate of the Fourier transform of x_{qt}:

$$S_{\mathbf{X},pq}(f) = \Im_{\mathbf{X},p}(nf)\,\bar{\Im}_{\mathbf{X},q}(nf)/n \ , \qquad f = n^{-1}, 2n^{-1}, \ldots, mn^{-1},$$

where $m = n/2$ if n is even, or $m = (n-1)/2$ if n is odd. The diagonal elements of $\mathbf{S_X}(f)$ are the power spectra (nonnegative real-valued functions) and the off-diagonal elements are the cross-spectra. In general, a cross-spectrum is composed of a real-valued cospectrum, $C_{\mathbf{X},pq}(f)$, and an imaginary-valued quadrature spectrum, $Q_{\mathbf{X},pq}(f)$:

$$S_{\mathbf{X},pq}(f) = C_{\mathbf{X},pq}(f) + iQ_{\mathbf{X},pq}(f).$$

The cospectrum measures the in-phase frequency distribution of the covariance of x_{pt} with x_{qt}, while the quadrature spectrum measures the frequency distribution of the covariance of x_{pt} with x_{qt} lagged by a phase shift of 1/4 of a cycle at frequency f [3, 34, 39]. The cospectral power can be positive or negative. The total variance and covariance are related to the spectrum and cospectrum:

$$\begin{aligned} \mathrm{Var}(x_{p,t}) &= 2(n-1)^{-1}\sum_f S_{\mathbf{X},pp}(f) \\ \mathrm{Cov}(x_{p,t}, x_{q,t}) &= 2(n-1)^{-1}\sum_f C_{\mathbf{X},pq}(f), \end{aligned} \tag{17.3}$$

where the summation occurs over all values of the frequencies, f.[3] Similarly, the Pearson correlation coefficient, $\rho_{\mathbf{X},pq}$, of x_{pt} with x_{qt} can be obtained using

$$\rho_{\mathbf{X},pq} = \frac{2(n-1)^{-1}\sum_f C_{\mathbf{X},pq}(f)}{\sqrt{\mathrm{Var}(x_{p,t})\,\mathrm{Var}(x_{q,t})}}. \tag{17.4}$$

Equations (17.3) and (17.4) can be used to partition the variance, covariance, and correlation into time scale ranges [15, 16]. For example, if two populations were censused monthly for a long period of time, the total correlation in their fluctuations, ρ, could be partitioned into time scales based on less than a season ρ_1 for $f > 3/12$, seasonal ρ_2 for $3/12 \geq f > 1/12$, and annual or longer ρ_3 for $f \leq 1/12$, with $\rho = \rho_1 + \rho_2 + \rho_3$.

Another useful function is the *corrspectrum* for two time series [15, 16]. It is obtained by dividing the cospectrum by the square root of the product of the two spectra at each frequency:

$$\tilde{C}_{\mathbf{X},pq}(f) = \frac{C_{\mathbf{X},pq}(f)}{\sqrt{S_{\mathbf{X},pp}(f)\,S_{\mathbf{X},qq}(f)}}. \tag{17.5}$$

The corrspectrum is calculated in a manner analogous to the computation of the Pearson correlation coefficient; it represents the cospectral power as a fraction of the geometric mean of the spectral powers of the component time series. It is bounded so that $-1 \leq \tilde{C}_{\mathbf{X},pq}(f) \leq 1$. The corrspectrum reveals frequency-specific changes in synchrony of two time series independent of frequency-specific changes in the variances.

It is important to consider the time scales of ecological synchrony. A single metric, such as the correlation coefficient, may be insufficient to describe the dynamics of synchrony. As an example, consider two noisy time series, each with sinusoidal oscillations at a short and long time scale (Figure 17.2). In Figure 17.2A, the two time series are synchronous on the short time scale, but anti-synchronous on the long time scale. The opposite is true in Figure 17.2B. In both cases, the total correlation between the two time series is zero. Figures 17.2C,D show the cospectra of the two examples. The contributions from the short and long time scales cancel in their contribution to the total correlation.

---

[3]The factor of two appears because, theoretically, the spectra and cospectra are defined in the interval $0 \leq f \leq 1$ and are symmetric around $f = 0.5$. This means that, using equations (17.3), a contribution at $f = 0.5$ is counted twice. If n is even, one must adjust $S_{\mathbf{X},pp}(0.5)$ and $C_{\mathbf{X},pq}(0.5)$ by dividing by them by two.

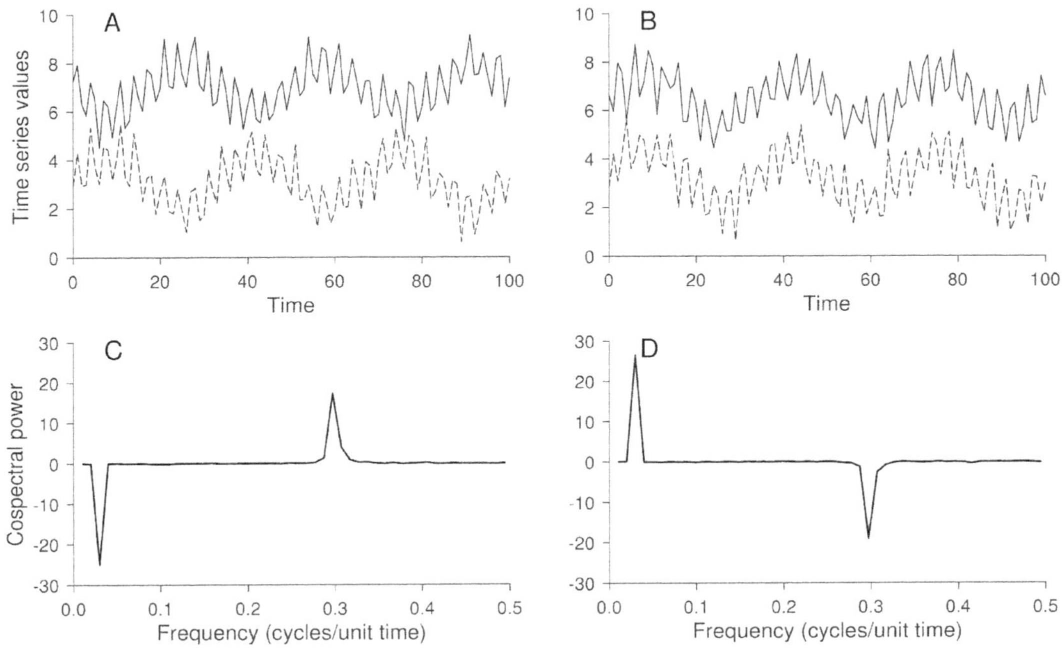

Figure 17.2 Two noisy bivariate sinusoidal time series and their cospectra. In panel A, the time series show synchrony on a short time scale and anti-synchrony on a long time scale. In panel B, the time series show synchrony on a long time scale and anti-synchrony on a short time scale. The cospectra in panels C and D show the frequency-specificity of the synchrony. In both panels A and B, the total correlation between the two time series is zero.

17.2.2 Influence of Noise and Dispersal

Metapopulation synchrony is affected by the spectral properties of the environmental noise, rates of dispersal, and dynamics of the populations. For example, consider equations (17.1) and (17.2) for the Ricker metapopulation model with $b_1 = b_2 = b$, $c_1 = c_2 = c$, and $V_1 = V_2 = 1$. The deterministic equilibrium population sizes are $N_1^* = N_2^* = N^* = (\ln b)/c$. Assume the deterministic metapopulation equilibrium is positive and stable[4] and noise is weak. We can approximate the dynamics of the stochastic metapopulation using a model linearized about the equilibrium:

$$\begin{aligned} w_{1,t+1} &= (1-d)\,\lambda w_{1,t} + d\lambda w_{2,t} + (1-d)\,\varepsilon_{1t} + d\varepsilon_{2t}, \\ w_{2,t+1} &= d\lambda w_{1,t} + (1-d)\,\lambda w_{2,t} + d\varepsilon_{1t} + (1-d)\,\varepsilon_{2t}, \end{aligned} \tag{17.6}$$

where $w_{i,t} = (N_{i,t} - N^*)/N^*$ and $\lambda = 1 - \log(b)$. We relax the assumption that the noise vectors $\boldsymbol{\epsilon}_0, \boldsymbol{\epsilon}_1, \boldsymbol{\epsilon}_2, \ldots$ are independent through time and assume that the elements of $\boldsymbol{\epsilon}_t$ are identically distributed with the same spectrum, $S_\varepsilon(f)$, and a real-valued cross-spectrum. The assumption of a real cross-spectrum implies that there are no phase differences between the noise variables; the quadrature spectrum is zero and the cross-spectrum is composed of

[4]The positive equilibrium is stable when $1 < b < e^2 \approx 7.389$.

only the cospectrum $C_\varepsilon(f)$. Defining the dispersal matrix as

$$\mathbf{D} = \begin{pmatrix} 1-d & d \\ d & 1-d \end{pmatrix}, \tag{17.7}$$

the system (17.6) can be written in matrix notation: $\mathbf{w}_{t+1} = \lambda \mathbf{D} \mathbf{w}_t + \mathbf{D} \epsilon_t$, where $w_t = [w_{1,t}, w_{2,t}]$. As discussed in Section 16.3.1, we can use linear filter theory to obtain the spectral matrix for the population fluctuations, $\mathbf{S_w}(f)$, as a function of the spectral matrix of the environmental noise, $\mathbf{S_\epsilon}(f)$. The matrix $\mathbf{S_w}(f)$ is symmetric and real, with two equal spectra:

$$\mathbf{S_w}(f) = \begin{pmatrix} S_w(f) & C_w(f) \\ C_w(f) & S_w(f) \end{pmatrix}. \tag{17.8}$$

In Appendix 17.5, we show that the spectrum and cospectrum of the population fluctuations are

$$\begin{aligned} S_w(f) &= \frac{S_\varepsilon(f)}{|1-\lambda z|^2}\left(\frac{|s|^2+d^2}{|s+d|^2}\right) + \frac{C_\varepsilon(f)}{|1-\lambda z|^2}\left(1-\frac{|s|^2+d^2}{|s+d|^2}\right) \\ C_w(f) &= \frac{S_\varepsilon(f)}{|1-\lambda z|^2}\left(1-\frac{|s|^2+d^2}{|s+d|^2}\right) + \frac{C_\varepsilon(f)}{|1-\lambda z|^2}\left(\frac{|s|^2+d^2}{|s+d|^2}\right), \end{aligned} \tag{17.9}$$

where $s = 1 - \lambda(1-2d)z - d$ and $z = e^{2\pi i f}$. The population spectra and cospectra are weighted sums of the environmental spectra and cospectra with a weighting factor $\left(|s|^2 + d^2\right)/|s+d|^2$.

We examined how the population cospectrum, $C_w(f)$, is affected by the type of population dynamics, the time scale of the environmental correlation, and the rate of dispersal, d. Two options were chosen for the populations' equilibrium dynamics: undercompensating, $\lambda = 0.50$ ($b = 1.65$), and overcompensating, $\lambda = -0.50$ ($b = 4.50$). In the first case, when noise is absent, the populations approach the equilibrium monotonically. In the second case, the populations exhibit damped oscillations near the equilibrium. Two options were chosen for the time scale of the environmental correlations. In the first case, the environmental covariance was skewed toward short time scales ("blue-tinged"), by using the function $C_\varepsilon(f) = 2f$; that is, the environmental correlations have greater power at high frequencies. In the second case, the environmental covariance was skewed toward long time scales ("red-tinged"), by using the function $C_\varepsilon(f) = 1 - 2f$; that is, the environmental correlations have greater power at low frequencies. In both cases, the spectral power for the environmental noise was held constant at $S_\varepsilon(f) \equiv 1$; there were no serial autocorrelations in the noise variables ("white noise"). For the third factor, six levels of dispersal were used: $d = 0, 0.1, \ldots, 0.5$. When $d = 0$, the two populations are isolated; when $d = 0.5$, the two populations are completely mixed and have the same dynamics as a large single population.

Figure 17.3 shows the plots of the population cospectrum, $C_w(f)$, from equations (17.9) for the combinations of the three factors described above. As shown in Figure 17.3B, the effects of dispersal are small, that is, the cospectral curves are similar, when the populations have overcompensating dynamics and the environmental correlations occur on a short time scale (blue-tinged). Population synchrony occurs on short time scales (high frequencies) for all levels of dispersal. Similarly, the effects of dispersal are small when the populations have undercompensating dynamics and the environmental correlations occur on a long time scale (red-tinged). In this case, population synchrony occurs on long time scales (low frequencies) for all levels of dispersal (17.3C). Dispersal has a strong effect on the time scales of population synchrony when the populations' dynamics are undercompensating and the environmental covariance is blue-tinged (Figure 17.3A). The effects of

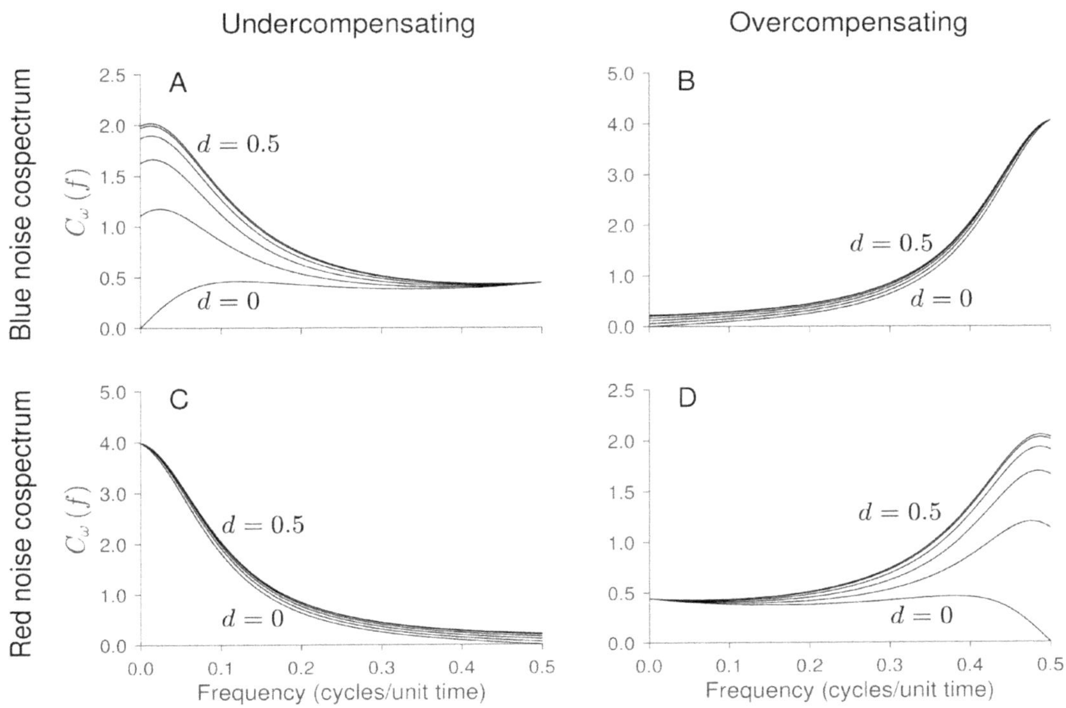

Figure 17.3 Plots of the population cospectrum (17.9) for different types of equilibrium dynamics, time scales of environmental correlation, and rates of dispersal. The environmental cospectrum is blue-tinged ($C_\varepsilon(f) = 2f$) in panels A and B and red-tinged ($C_\varepsilon(f) = 1 - 2f$) in panels C and D. The populations' dynamics are undercompensating ($\lambda = 0.5$) in panels A and C and overcompensating ($\lambda = -0.5$) in panels B and D. In all examples, the dispersal rates are $d = 0, 0.1, \ldots, 0.5$ and the environmental random variables have a white noise spectrum ($S_\varepsilon(f) \equiv 1$).

dispersal are most pronounced on long time scales (low frequencies). The opposite is true when the populations' dynamics are overcompensating and the environmental covariance is red-tinged (Figure 17.3D). The effects of dispersal are most pronounced on short time scales (high frequencies). These examples demonstrate that the time scale of synchrony depends on complex interactions among the populations' dynamics, color of the environmental correlations, and rate of dispersal.

17.2.3 Implications for Species Persistence

For metapopulation extinction to occur, the numbers of all the component populations must reach zero at the same time. This is more likely to happen if the populations are synchronous [23]. Factors that increase synchrony will increase extinction rates. For example if the population dynamics are undercompensating and the time scale of synchrony is red-tinged, the risk of extinction is increased [8]. Similarly, if population dynamics are overcompensating and the time scale of synchrony is blue-tinged, risk of extinction also increases [8]. Thus, the type of population dynamics and the time scales of both the variance and covariance of the environmental noise must be considered.

The effects of dispersal rates are more subtle. Dispersal allows surviving populations to "seed" localities where population numbers are low or zero, preventing metapopulation extinction. On the other hand, dispersal increases the overall level of population synchrony, which increases the likelihood of extinction. Furthermore, the dispersal rate interacts with the type of population dynamics and the color of the environmental noise in its effect of synchrony (Figure 17.2), making it difficult to generalize.

We examined probabilities of extinction using a version of the stochastic Ricker metapopulation model, based on equations (17.1) and (17.2), that includes demographic stochasticity (Chapter 10). The population numbers before dispersal are determined using a Poisson distribution with expected values based on equation (17.1):

$$\begin{aligned} \widehat{N}_{1,t+1} &= \text{Pois}\left(b_1 N_{1,t} e^{-c_1 N_{1,t}/V_1}\, e^{\varepsilon_{1,t}}\right) \\ \widehat{N}_{2,t+1} &= \text{Pois}\left(b_2 N_{2,t} e^{-c_2 N_{2,t}/V_2}\, e^{\varepsilon_{2,t}}\right). \end{aligned} \tag{17.10}$$

Population numbers are obtained from compound distributions where the parameter of each Poisson is a random variable based on environmental stochasticity. After dispersal,

$$\begin{aligned} N_{1,t+1} &= \left(\widehat{N}_{1,t+1} - \eta_{1,t+1}\right) + \eta_{2,t+1} \\ N_{2,t+1} &= \left(\widehat{N}_{2,t+1} - \eta_{2,t+1}\right) + \eta_{1,t+1}, \end{aligned} \tag{17.11}$$

where $\eta_{1,t+1}$ is the number of individuals migrating out of population 1 into population 2 at time $t+1$, and $\eta_{2,t+1}$ is defined similarly. Keeping with the paradigm of demographic stochasticity, binomial distributions are used to determine the number of migrants:

$$\begin{aligned} \eta_{1,t+1} &= \text{Bin}\left(\widehat{N}_{1,t+1}, d\right) \\ \eta_{2,t+1} &= \text{Bin}\left(\widehat{N}_{2,t+1}, d\right), \end{aligned} \tag{17.12}$$

where d is the probability that an individual migrates. The random variables $\eta_{1,t+1}$ and $\eta_{2,t+1}$ are determined independently.

Metapopulation extinction probabilities were estimated using realizations of equations (17.10), (17.11), and (17.12). We set $c_1 = c_2 = 0.01$, $b_1 = b_2 = 1.2$ for undercompensating dynamics, and $b_1 = b_2 = 5$ for overcompensating dynamics. Since the likelihood of extinction depends on population size, we fixed $V_1 = V_2 = 100c/\ln b$ so that the deterministic model always has $N_1^* = N_2^* = 100$ as its equilibrium. We ran simulations for two dispersal rates, $d = 0.10$ and $d = 0.25$. To examine how extinction probabilities varied with the color of the environmental spectra, we assumed a linear spectrum for each noise variable which varied from short time scale fluctuations (slope = 4, blue noise), to fluctuations with no autocorrelation (slope = 0, white noise), to long time scale fluctuations (slope = –4, red noise) and kept the cospectrum constant at $C_\epsilon(f) \equiv 0.5$. Similarly, we assumed a linear cospectrum which varied from short time scale correlations (slope = 2, blue-tinged) to long time scale fluctuations (slope = –2, red-tinged) and kept the spectra constant at $S_\epsilon(f) \equiv 1$. Figure 17.4 illustrates this protocol. The slopes were varied in 100 increments. In all cases, the total variance was one and the total covariance (and correlation) was 0.5. The random environmental fluctuations were generated using an algorithm that takes a spectral matrix function and generates time series of random vectors with the desired spectral properties [5]. For each combination of dynamics, dispersal rate, spectra, and cospectrum, we generated 2000 realizations of the metapopulation model, with each realization lasting 100 time intervals. If both populations had zero numbers by the end of a realization, it was recorded as a

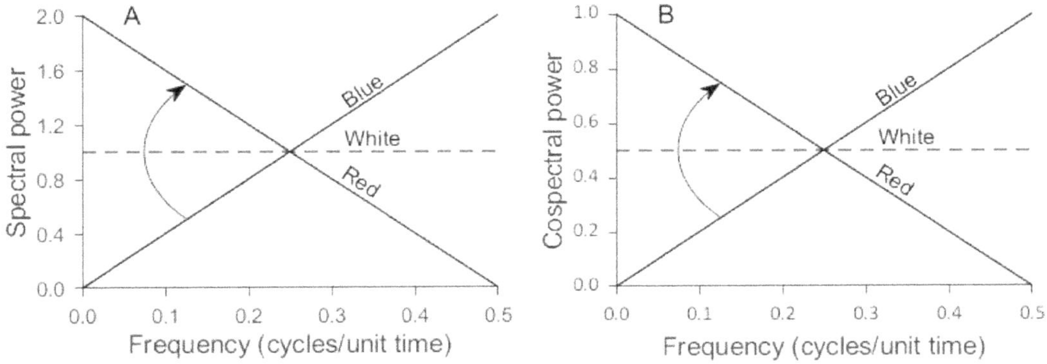

Figure 17.4 Protocol for changing smoothly the timescales of (A) the environmental spectra and (B) the cospectrum for the simulations examining metapopulation extinction. Linear spectra and cospectra were used. The blue spectrum and cospectrum are skewed toward high frequencies (short time scales). The white spectrum and cospectrum have fluctuations evenly distributed across all frequencies. The red spectrum and cospectrum are skewed toward low frequencies (long time scales). Color gradients of environmental spectra and cospectra were created by varying the slopes from blue to red across 101 evenly-spaced values (curved arrows).

metapopulation extinction event. The extinction probability was estimated as the number of extinction events divided by the number of realizations.

The results of the simulation protocol for extinction probabilities are illustrated in Figure 17.5. When the population dynamics are undercompensating, extinction probabilities increase with a shift from short (blue) to long (red) time scales of environmental noise (Figure 17.5A), and the same is true for the correlation between the noise variables (Figure 17.5C). An increase in the rate of dispersal between populations causes a slight decrease in the likelihood of extinction (Figure 17.5A,C). The opposite is true when the population dynamics are overcompensating. A shift from short to long time scale environmental noise causes a decrease in the probabilities of extinction (Figure 17.5B), and, to a lesser extent, the same is true for the correlation between the noise variables (Figure 17.5D). Higher dispersal rates significantly increase the likelihood of extinction (Figure 17.5B,D). Overall, extinction probabilities were higher when the population dynamics were overcompensating. These analyses have been extended to other types of population dynamics and metapopulations with more than two component populations [8]. The time scales of the environmental variance and covariance are an important consideration when trying to assess the risk to biodiversity in fragmented populations.

17.3 LPA MODEL PREDICTIONS AND RESULTS

We designed and carried out an experiment to evaluate how metapopulation synchrony is affected by changes in the time scale of environmental stochasticity and dispersal [15]. As with the experiments described in Chapter 16, laboratory populations of *Tribolium* were used and habitat size was manipulated to vary environmental noise. Pairs of flour beetle populations were linked through changes in habitat size and dispersal. Linear filter theory,

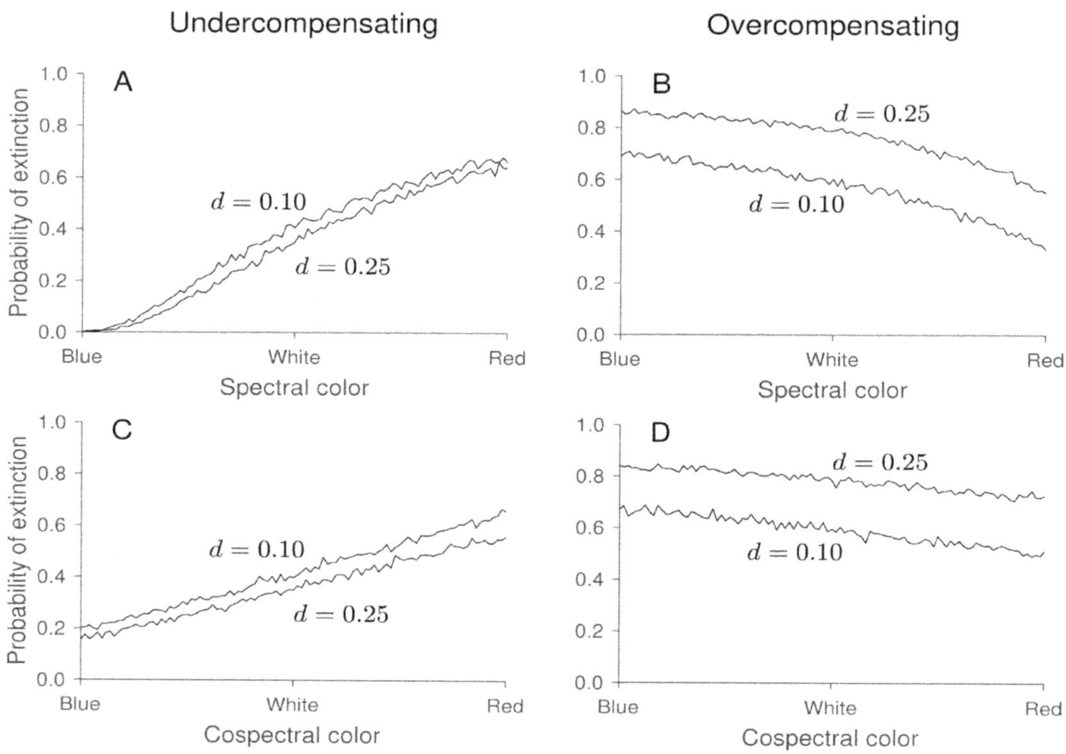

Figure 17.5 Plots of the estimated probabilities of metapopulation extinction for changing color gradients of environmental noise (Figure 17.4). In panels A and B the time scales of the environmental noise spectra were altered from blue to white to red while the cospectrum was held constant at $C_\epsilon(f) \equiv 0.5$. In panels C and D, the time scales of the environmental noise cospectra were altered from blue to white to red while the spectra were held constant at $S_\epsilon(f) \equiv 1$. Two types of population dynamics were investigated: undercompensating ($b = 1.2$) in panels A and C, and overcompensating ($b = 5$) in panels B and D. Two dispersal rates were used for each combination of population dynamics and spectral properties: $d = 0.10$ and $d = 0.25$.

applied to a metapopulation version of the LPA model, was used to predict the effects on synchrony. Model predictions were evaluated using experimental data.

17.3.1 LPA Metapopulation Model

For our experimental studies of synchrony, we considered a metapopulation version of the lattice LPA model with demographic noise. In order to place the populations in a region of parameter space where there is a stable equilibrium, an experimental protocol was used where "harvesting mortality" was imposed on the larva, m_l, and adults, m_a, at the time of census. Prior to manipulation, the LPA equations for populations $j = 1, 2$ are

$$\begin{aligned}
\widehat{L}_{j,t+1} &= \mathrm{rd}\left(\left[\sqrt{bA_{j,t}\exp\left(-\frac{c_{el}L_{j,t}}{V_{j,t}} - \frac{c_{ea}A_{j,t}}{V_{j,t}}\right)} + \varepsilon_{lj,t}\right]^2\right) \\
\widehat{P}_{j,t+1} &= \mathrm{rd}\left(\left[\sqrt{L_{j,t}\left(1-\mu_l\right)} + \varepsilon_{pj,t}\right]^2\right) \\
\widehat{A}_{j,t+1} &= \mathrm{rd}\left(\left[\sqrt{P_{j,t}\exp\left(-\frac{c_{pa}A_{j,t}}{V_{j,t}}\right) + A_{j,t}\left(1-\mu_a\right)} + \varepsilon_{aj,t}\right]^2\right),
\end{aligned} \tag{17.13}$$

where the quantities to be squared are taken to be zero if negative and $\mathrm{rd}(x)$ means rounding to the nearest integer. $\widehat{L}_{j,t+1}$, $\widehat{P}_{j,t+1}$, and $\widehat{A}_{j,t+1}$ are the life stage numbers for population j at time $t+1$ prior to harvesting and dispersal. Parameters b, μ_l, μ_a c_{el}, c_{ea}, and c_{pa} have their usual definitions and common values for both populations. Habitat sizes, $V_{1,t}$ and $V_{2,t}$, were varied randomly in a manner described below; they represent the environmental noise. The random variables $\varepsilon_{lj,t}$, $\varepsilon_{pj,t}$, and $\varepsilon_{aj,t}$ simulate demographic noise added on a square root scale and are chosen from a trivariate normal distribution with a diagonal variance-covariance matrix $\mathbf{\Sigma}$ common to both populations. The vectors $\boldsymbol{\epsilon}_{j,t} = [\varepsilon_{lj,t}, \varepsilon_{pj,t}, \varepsilon_{aj,t}]$ are generated independently for each population ($j = 1, 2$) and are assumed to have no serial autocorrelation. At the time of census, harvesting and dispersal were simulated by removing beetles and exchanging a fraction d of the adults between population pairs:

$$\begin{aligned}
L_{j,t+1} &= \mathrm{rd}\left(\left(1-m_l\right)\widehat{L}_{j,t+1}\right) \\
A_{1,t+1} &= \mathrm{rd}\left(\left(1-m_a\right)\left(1-d\right)\widehat{A}_{1,t+1}\right) + \mathrm{rd}\left(\left(1-m_a\right)d\widehat{A}_{2,t+1}\right) \\
A_{2,t+1} &= \mathrm{rd}\left(\left(1-m_a\right)d\widehat{A}_{1,t+1}\right) + \mathrm{rd}\left(\left(1-m_a\right)\left(1-d\right)\widehat{A}_{2,t+1}\right).
\end{aligned} \tag{17.14}$$

Pupae were not manipulated, so $P_{j,t+1} = \widehat{P}_{j,t+1}$. The component populations of each experimental metapopulation are linked by the rate of dispersal, d, and the covariance between the habitat sizes, $V_{1,t}$ and $V_{2,t}$.

17.3.2 Experimental Design

The *Tribolium* metapopulation experiment was designed to examine the effects of environmental noise frequencies and dispersal rate on synchrony [15]. The experimental unit was a metapopulation composed of two vials of flour beetles. Different pairs of vials were manipulated in different ways according to two crossed experimental factors: environmental noise fluctuations and adult dispersal. There were two levels for each factor. Environmental noise was simulated by altering the habitat size. Each metapopulation replicate received a pair of random flour volume sequences, $V_{1,t}$ and $V_{2,t}$, that were either positively correlated on short time scales (high frequencies) and negatively correlated on long time scales (low frequencies), denoted as Blue$^+$/Red$^-$, or negatively correlated on short time scales and positively correlated on long time scales, Blue$^-$/Red$^+$. How the habitat sequences were generated is described below. The two levels of adult dispersal were no dispersal, $d = 0$, or high dispersal, $d = 0.25$. There were three replicates for each of the four treatment groups to yield 12 experimental metapopulations containing 24 paired laboratory cultures.

A census was made for each population every two weeks. At each census, an additional "harvesting mortality" was imposed on larvae, $m_l = 0.70$, and adults, $m_a = 0.05$, to yield predicted population dynamics with a stable equilibrium, based on previous parameter estimates (Table 7.2). For metapopulations in the treatments with dispersal, 25% of the remaining unharvested adults were exchanged between populations within a replicate. All populations were established with the initial condition $[L_{j,0}, P_{j,0}, A_{j,0}] = [67,\ 82,\ 486]$, which is close to the predicted stable equilibrium. The experiment was terminated after 80 weeks ($t = 40$).

A habitat size sequence was generated for each replicate metapopulation. First, we simulated an autoregressive first-order time series (AR1), $y_{t+1} = ay_t + \xi_t$, of length 40, where y_0 and ξ_t are independently identically distributed standard normal variables. We used an autocorrelation coefficient of $a = -0.9$ for blue noise and $a = 0.9$ for red noise. These were generated independently for each replicate. Then we created a time series of flour volumes with 20 values of –23 g and 20 values of 23 g. We used spectral mimicry [7] to permute the time series of flour volumes to have the same rank order as $y_0, y_1, \ldots, y_{39}$. Mimicry ensured an equal number of 23s and –23s. Let ν_b represent the series of flour volumes that mimics the blue noise autoregressive process and define ν_r similarly for the red noise process. For the Blue$^+$/Red$^-$ habitat regime, we used $\nu_b + \nu_r + 54$g for the habitat sequence of one population and $\nu_b - \nu_r + 54$g for the habitat sequence of the other population. For the Blue$^-$/Red$^+$ habitat regime, we used $\nu_b + \nu_r + 54$g for the habitat sequence of one population and $-\nu_b + \nu_r + 54$g for the habitat sequence of the other population. Thus, the only amounts of medium used were 8 g, 54 g, and 100 g. In terms of habitat sizes, $V_{j,t}$, the only possible values are 0.4, 2.7, and 5.0. For the Blue$^+$/Red$^-$ treatments, the bivariate time series used for the habitat sizes were synchronized at short time scales and anti-synchronized at long time scales. The opposite was true for the Blue$^-$/Red$^+$ treatments. The mean value of every habitat time series was exactly 2.7 and the total correlation of the bivariate habitat time series for each replicate was exactly zero: $\rho(V_{1,t}, V_{2,t}) = 0$. Under traditional definitions of synchrony in terms of total correlation, all experimental replicates experienced unsynchronized noise. Correlation does not reveal the time scale (or, equivalently, frequency-specific) structure of synchrony in the habitat time series.

17.3.3 Parameter Estimates

Parameters for the LPA metapopulation model were estimated using the model (17.13) prior to harvesting and dispersal. We used the method of conditional least squares (CLS; see Chapter 5) which involves minimizing the sum of the squared differences between the square roots of the observed life stage numbers $L_{j,t+1}$, $P_{j,t+1}$, and $A_{j,t+1}$ prior to manipulation and the one-time-step values based on equation (17.13), conditional on the observed data and habitat sizes at time step t:

$$\begin{aligned}
\sigma_l^2\left(b, c_{el}, c_{ea}\right) &= \sum_{\text{all reps}} \sum_{t=0}^{39} \sum_{j=1}^{2} \left(\sqrt{L_{j,t+1}} - \sqrt{\widehat{L}_{j,t+1}}\right)^2 \\
\sigma_p^2\left(\mu_l\right) &= \sum_{\text{all reps}} \sum_{t=0}^{39} \sum_{j=1}^{2} \left(\sqrt{P_{j,t+1}} - \sqrt{\widehat{P}_{j,t+1}}\right)^2 \qquad (17.15) \\
\sigma_a^2\left(c_{pa}, \mu_a\right) &= \sum_{\text{all reps}} \sum_{t=0}^{39} \sum_{j=1}^{2} \left(\sqrt{A_{j,t+1}} - \sqrt{\widehat{A}_{j,t+1}}\right)^2 .
\end{aligned}$$

The outer summation is taken over all metapopulation replicates. The experiment ran for 80 weeks, so the final census occurred at $t = 40$. The minimized values in equations (17.15) are CLS estimates of the diagonal elements of the matrix $\mathbf{\Sigma}$. The parameter estimates appear in

TABLE 17.1 Conditional least squares (CLS) parameter estimates for the LPA metapopulation model. The estimated diagonal elements of the variance-covariance matrix Σ are $\sigma_l^2 = 3.625$, $\sigma_p^2 = 0.9719$, and $\sigma_a^2 = 0.07799$.

Parameter	CLS estimate	95% confidence interval
b	9.125	(8.764, 9.463)
μ_l	0.1735	(0.1630, 0.1838)
μ_a	0.03664	(0.03457, 0.03868)
c_{el}	0.01478	(0.01420, 0.01533)
c_{ea}	0.01295	(0.01264, 0.01324)
c_{pa}	0.005116	(0.004996, 0.005239)

Table 17.1. They agree well with previous estimates from independent experiments utilizing the LPA single population model [12, 13].

We produced confidence intervals for the parameter estimates by using the model equations (17.13) and (17.14) and CLS parameter estimates (Table 17.1) to simulate the entire experimental protocol 2000 times. We employed the same experimental bivariate habitat size sequences for each replicate, but we generated independent realizations of the random variables $\varepsilon_{lj,t}$, $\varepsilon_{pj,t}$, and $\varepsilon_{aj,t}$ using the CLS estimates for the variance-covariance matrix $\mathbf{\Sigma}$. We obtained a vector of CLS parameter estimates for each simulated experiment. The 2.5th and 97.5th percentiles of these estimates yielded the 95% confidence intervals in Table 17.1.

17.3.4 Spectra for Environmental Noise

We obtained the spectral properties of the habitat size sequences by using the algorithm for computing flour volumes described above to generate bivariate sequences of length 2^{15} for the Blue$^+$/Red$^-$ and Blue$^-$/Red$^+$ habitat regimes. We computed the periodograms and smoothed them using the modified Daniell kernel with spans of 301 and 303. Figure 17.6 shows the estimated spectral power, corrspectra, and pointwise normalized quadrature spectra. The corrspectra and point-wise normalized quadrature spectra are obtained by dividing the cross-spectrum (a complex number) by the square root of the product of the two spectral powers at each frequency. The corrspectrum is given by equation (17.4). The pointwise normalized quadrature spectrum $\tilde{Q}_{\mathbf{X},pq}(f)$ is defined similarly by using $Q_{\mathbf{X},pq}(f)$ on the right-hand-side of equation (17.4) in place of $C_{\mathbf{X},pq}(f)$; it measures the frequency-specific phase shifts between the two time series. The four spectral densities were nearly identical for the two components of the time series and two habitat regimes, so only one of them is shown (Figure 17.6A). Similarly, the two pointwise normalized quadrature spectra were nearly identical for the two habitat regimes (Figure 17.6D).

Since linear filter theory requires frequency-specific functions for the spectral properties of the environmental noise to obtain spectral predictions for metapopulation numbers, we fitted curves to approximate the shape of the normalized spectra and corrspectra (broken lines in Figure 17.6). The algorithm used to generate the habitat sizes adds together random sequences obtained by applying spectral mimicry to two first-order autoregressive processes with coefficients 0.9 and −0.9, so we approximated the spectra using

$$S_\epsilon(f) = \beta_0 + \beta_1 \left(\frac{1}{1 + a^2 + 2a\cos(2\pi f)} + \frac{1}{1 + a^2 - 2a\cos(2\pi f)} \right), \tag{17.16}$$

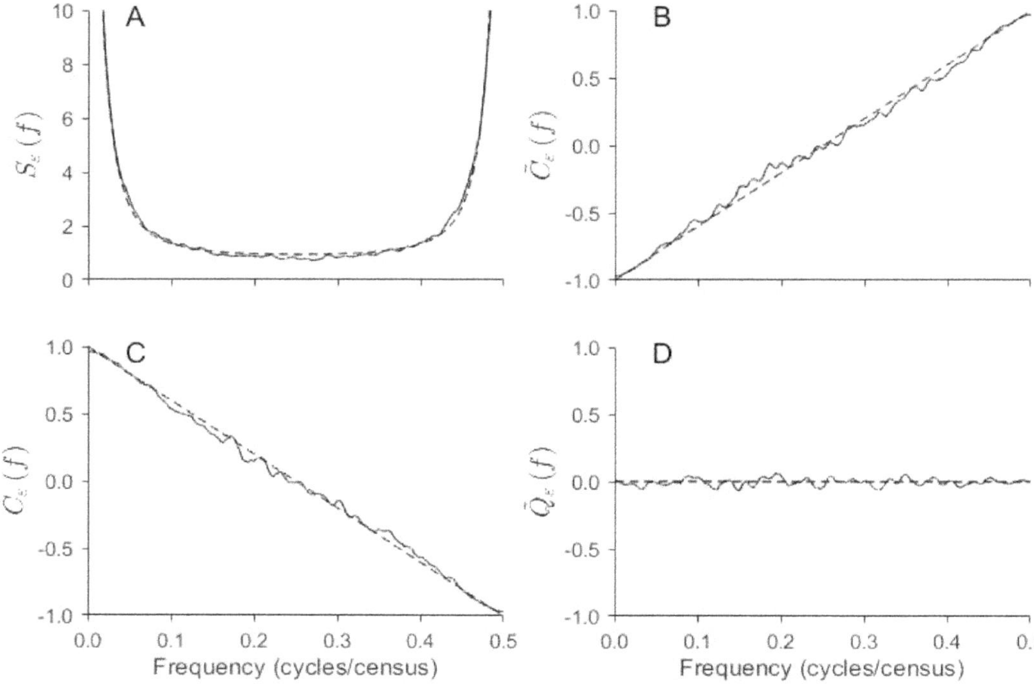

Figure 17.6 Spectral properties of the experimental habitat regimes. The solid curves are the estimated power spectrum (A), corrspectra (B,C), and point-wise normalized quadrature spectrum (D) for random bivariate habitat size sequences of length 2^{15}. All the power spectra were nearly identical, so only one example appears in panel A. Panels B and C show the corrspectra (17.4) for the Blue$^+$/Red$^-$ and Blue$^-$/Red$^+$ habitat regimes, respectively. For both habitat regimes, the pointwise normalized quadrature spectra fluctuated near zero; only one example is shown in panel D. Broken lines are the functional forms for the spectrum (17.16) and corrspectra (17.17) with $\tilde{Q}_\epsilon(f) \equiv 0$.

with $a = 0.9$. The expression in parentheses is the sum of the spectral density functions for two independent first-order autoregressive processes with coefficients a and $-a$. Using nonlinear least squares regression, we obtained values of $\beta_0 = 0.7373$ and $\beta_1 = 0.1918$ for the coefficients in equation (17.16). For the corrspectral functions, we used

$$\tilde{C}_\epsilon(f) = \begin{cases} 4f - 1, & \text{for Blue}^+/\text{Red}^- \text{ habitat regimes,} \\ 1 - 4f, & \text{for Blue}^-/\text{Red}^+ \text{ habitat regimes,} \end{cases} \tag{17.17}$$

with $0 \leq f \leq 0.5$. Since there was no evidence of frequency-specific phase shifts (Figure 17.6D), we set the pointwise normalized quadrature spectra to zero. These functions are excellent fits to the estimated habitat spectra and cross-spectra (Figure 17.6) and are used in the next section to obtain predictions for the population spectral densities and corrspectra.

17.3.5 Model Predictions

Model predictions are based on a non-lattice version of the LPA metapopulation model without demographic stochasticity:

$$\begin{aligned}
\widehat{L}_{j,t+1} &= bA_{j,t} \exp\left(-\frac{c_{el}L_{j,t}}{V_{j,t}} - \frac{c_{ea}A_{j,t}}{V_{j,t}}\right) \\
\widehat{P}_{j,t+1} &= L_{j,t}\left(1-\mu_l\right) \\
\widehat{A}_{j,t+1} &= P_{j,t} \exp\left(-\frac{c_{pa}A_{j,t}}{V_{j,t}}\right) + A_{j,t}\left(1-\mu_a\right),
\end{aligned} \tag{17.18}$$

with

$$\begin{aligned}
L_{j,t+1} &= \left(1-m_l\right)\widehat{L}_{j,t+1} \\
A_{1,t+1} &= \left(1-m_a\right)\left(1-d\right)\widehat{A}_{1,t+1} + \left(1-m_a\right)d\widehat{A}_{2,t+1} \\
A_{2,t+1} &= \left(1-m_a\right)d\widehat{A}_{1,t+1} + \left(1-m_a\right)\left(1-d\right)\widehat{A}_{2,t+1},
\end{aligned} \tag{17.19}$$

and $P_{j,t+1} = \widehat{P}_{j,t+1}$, $j = 1, 2$. We simulated the experimental design using equations (17.18) and (17.19), the estimated parameter values (Table 17.1), and the same initial conditions and habitat sequences as employed in the experiment. Pooling across all populations and time intervals, the magnitudes of the life stage variances in the simulated data compared to the those in the observed data was 97.3% for larvae, 93.9% for pupa, and 91.1% for adults. We concluded that lattice effects and demographic stochasticity represented a relatively small contribution to the population fluctuations and that most of the variation was due to environmental stochasticity.

Figure 17.7 shows representative time series predicted by the metapopulation model (17.18) and (17.19) using the estimated parameter values (Table 17.1) and same initial conditions and habitat sequences as employed in the experiment. In each plot, the solid and broken lines are for the two beetle cultures that make up a metapopulation. There are four groupings of figure panels—one for each experimental treatment. The two groups on the left are for the Blue$^+$/Red$^-$ habitat regimes and the two groups on the right are for the Blue$^-$/Red$^+$ habitat regimes. The two groups at the top are for the replicates with no dispersal ($d = 0$). The adult dispersal rate was $d = 0.25$ for the groups at the bottom. Within each group, the top panels show the random sequences of habitat sizes. These illustrate the habitat fluctuations on the short and long time time scales and the environmental synchrony and anti-synchrony on those different time scales. The middle and bottom panels are the predicted time series of larval and adult numbers, respectively, based on the metapopulation model and habitat sequences in the top panel. Figure 17.7 shows only one simulated replicate from each of the four experimental treatments; however, all the time series data and simulations are available in the supporting information that accompanies the original report [15].

Linear filter theory can be used to make predictions for the spectral properties of the LPA metapopulation model (17.18) – (17.19). We can view the random habitat sizes, $V_{1,t}$ and $V_{2,t}$, as environmental noise fluctuating around a fixed value of $V^* = 2.7$, that is, $V_{j,t} = V^* + \varepsilon_{j,t}$ for $j = 1, 2$. Applying the parameter estimates from Table 17.1, $m_l = 0.70$, and $m_a = 0.05$, the deterministic version of the model ($\varepsilon_{j,t} \equiv 0$) predicts a stable equilibrium with $L_1^* = L_2^* = 94.53$, $P_1^* = P_2^* = 78.13$, and $A_1^* = A_2^* = 405.72$ for both $d = 0$ and $d = 0.25$. Linearization around this equilibrium produces a model of the form $\mathbf{w}_{t+1} = \mathbf{M}\mathbf{w}_t + \boldsymbol{\Lambda}\boldsymbol{\epsilon}_t$, where $\mathbf{w}_t$ is a 6×1 vector of deviations of the life stage numbers from their equilibrium values, $\boldsymbol{\epsilon}_t$ is a 6×1 vector of random deviations of habitat sizes from V^*

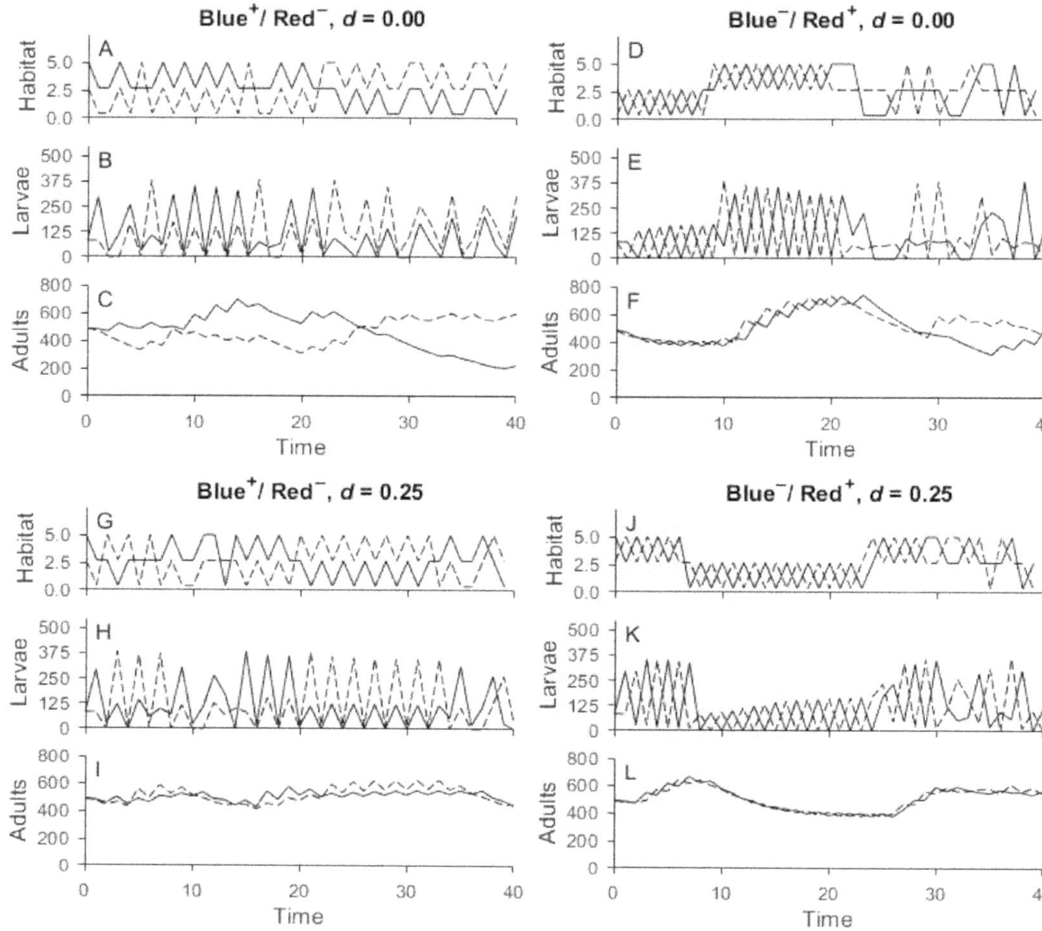

Figure 17.7 Habitat sizes and predicted life stage numbers for the LPA metapopulation model (17.18) – (17.19) with the random habitat sequences used in the experiment. See text for details.

corresponding to each life stage, $\mathbf{M}$ is a 6×6 matrix of linearization coefficients for the elements of $\mathbf{w}_t$, and $\mathbf{\Lambda}$ is a 6×6 matrix of linearization coefficients for the elements of $\boldsymbol{\epsilon}_t$. Using this linearized stochastic process, we obtained the spectral matrix $\mathbf{S}_{\mathbf{w}}(f)$ for the life stage fluctuations. Details of these analyses are provided in Appendix 17.6.

The spectral predictions for the LPA metapopulation model (17.18) and (17.19) appear in Figure 17.8. The results are grouped into predictions for each of the four experimental treatments based on the habitat regime and the level of adult dispersal. Each group contains four panels, two plots of the spectral densities for larvae and adults,[5] and two plots of the corrspectra (17.5) between larvae and between adults in the two populations.[6] The spectra

[5]Given the assumption that the two populations within each metapopulation have identical parameter values, the two power spectra corresponding to the same life stage are equal, and only one of them is presented. The spectral density $\tilde{S}_{\mathbf{w},pp}(f)$ is the power spectrum normalized to have a total area of one.

[6]There are six corrspectra, one for each pairwise combination of life stages between the two populations. We limited the presentation to the larva-larva and adult-adult pairs.

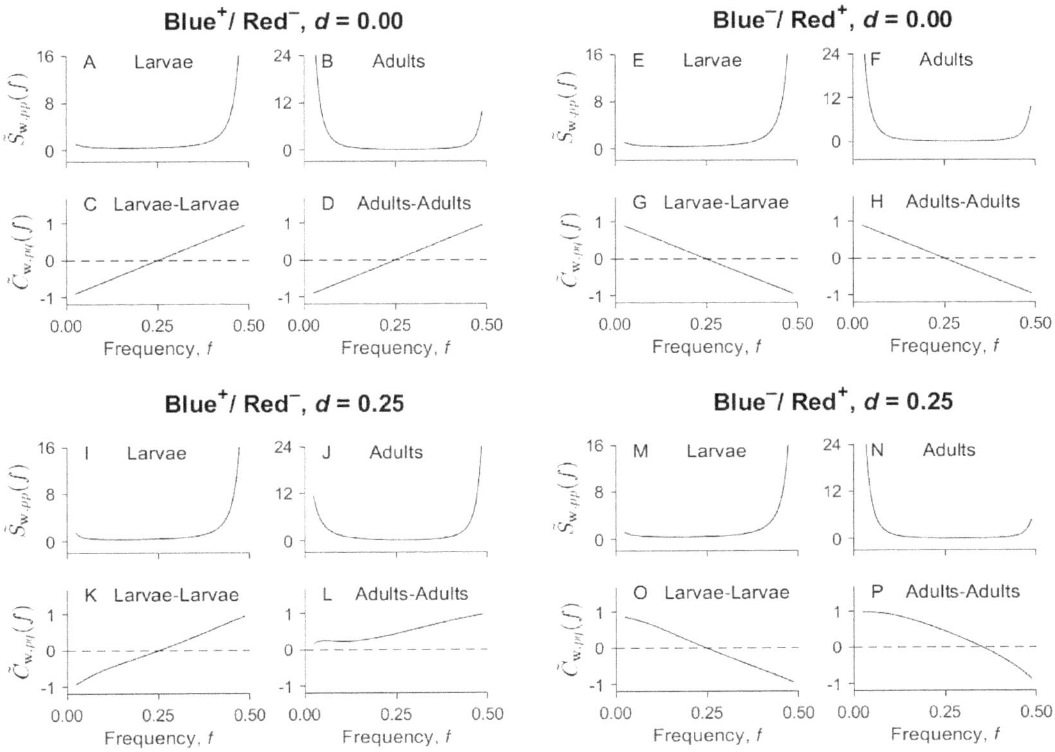

Figure 17.8 Spectral predictions for the LPA metapopulation model (17.18) – (17.19) from linear filter theory. See text for details.

for larvae, both with and without adult dispersal, are skewed toward short time scales (high frequencies) for both types of habitat regimes (Figure 17.8A,E,I,M). When adult dispersal is absent ($d = 0$), the opposite is true for adults; their spectra are skewed toward long time scales (Figure 17.8B,F). The corrspectra for larvae (Figure 17.8C,G,K,O) mimic the corrspectra for habitat size (Figure 17.6B,C). When $d = 0$, the same is true for the corrspectra of adults (17.8D,H). The greatest amount of interaction between environmental noise and dispersal occurs in the adult spectra and corrspectra. For the Blue$^+$/Red$^-$ habitat regime with $d = 0.25$, the adult spectrum is skewed toward short time scales (Figure 17.8J), while the opposite is true for the Blue$^-$/Red$^+$ habitat regime with $d = 0.25$ (Figure 17.8N). For the Blue$^+$/Red$^-$ habitat regime with $d = 0.25$, the adult corrspectrum shows a big shift; it has a shallower slope and is positive for all frequencies (Figure 17.8L). Lastly, for the Blue$^-$/Red$^+$ habitat regime with $d = 0.25$, the adult corrspectrum shows a negative slope, but with an upward curvature (Figure 17.8P). These predictions are confronted with experimental data in the next section.

17.3.6 Experimental Results

Figure 17.9 shows the observed census data for larvae and adults. One replicate is shown from each experimental treatment. The layout is identical to Figure 17.7. There is a remarkable agreement between the model-predicted values (Figure 17.7) and the observed

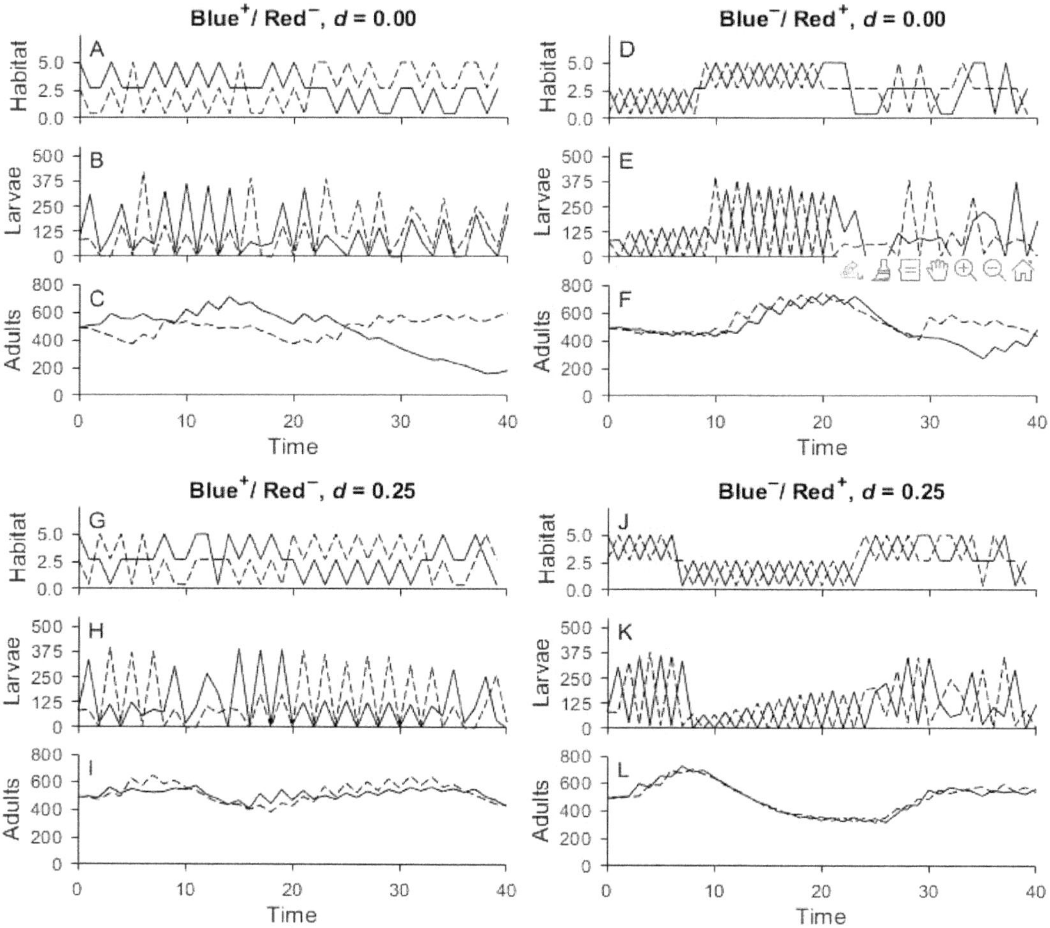

Figure 17.9 Habitat sizes and observed life stage numbers for the synchrony experiment. This figure is so similar to the predictions in Figure 17.7 that one is tempted to think it is an error, but that is not the case. See text for details.

values (Figure 17.9) given the same habitat sequences.[7] Even subtle changes in population numbers are well predicted by the model.[8] Despite its simplicity, the LPA metapopulation model predicts well the population trajectories when habitat size is altered. This suggests two characteristics of the system: (a) that demographic stochasticity makes relatively small contributions to the variability of life-stage fluctuations, and (b) that changes in habitat size, and their impact on rates of cannibalism, are a major factor in the population dynamics of these insect cultures.

The time scale patterns of synchrony and anti-synchrony are apparent in the bivariate plots of habitat size and population numbers (Figure 17.9). For the Blue$^+$/Red$^-$ regime,

[7]The predicted time series are generated by plugging the experimental initial conditions and the habitat sequences into equations (17.18) and (17.19); they are not step-by-step predictions conditional on the data.

[8]The agreement is so close that one might be tempted to assume that an error was made and the same data were plotted in both Figure 17.7 and Figure 17.9, but that is not the case. See Figures S17–S19 of the Supporting Information in [15] for plots of the life stage residuals.

habitat sizes are positively correlated above or below the mean value of 2.7 for short periods of time, but on a long time scale they are negatively correlated (Figure 17.9A,G). The opposite is true for the Blue$^-$/Red$^+$ treatments (Figure 17.9D,J). The effects of these different patterns of time scale-specific habitat changes can be seen in the way the two population time series for each metapopulation co-vary. In the Blue$^+$/Red$^-$ habitat regime, larvae numbers vary synchronously over short time scales, but anti-synchronously over long time scales (Figure 17.9B,H). The same is true for adults when dispersal is absent (Figure 17.9C). The converse is true for the larvae and adults in the Blue$^-$/Red$^+$ habitat regime (Figure 17.9D,E,J,K). The one exception occurs in the Blue$^+$/Red$^-$ habitat regime when dispersal is present ($d = 0.25$); adult numbers are synchronous on both the short and long time scales (Figure 17.9I).

There are major differences between the immature and adult life stages in the variation of population numbers. Larval numbers oscillate widely on a short time scale while the magnitudes of those oscillations vary over a long time scale. Numbers of adults vary largely over long time scales, while changes in habitat size induced much smaller oscillations over short time scales. The difference arises because cohorts of immatures develop quickly into adults, whereas adults are long-lived. Egg cannibalism by larvae and adults is an important mechanism driving demographic oscillations in the immature life stages of *Tribolium* [14], and so high-frequency changes in habitat size driving similar changes in rates of egg-cannibalism can cause rapid changes in immature numbers. Since adults are not subject to cannibalism, the magnitude of changes is limited by the rate of adult mortality, which is small relative to the intercensal period ($\mu_a + m_a - \mu_a m_a = 0.08481$).[9] The result is that immatures are sensitive and adults are insensitive to high-frequency changes in habitat size.

Figure 17.10 shows the observed spectral densities and corrspectra for the synchrony experiment. The spectra and cospectra were smoothed using two passes of the modified Daniell kernel with window spans of five and five. The individual replicates are plotted as dashed lines and the heavy solid line is the mean of the replicates.[10] Most of the variation in larval numbers occurs at high frequencies whereas most variation in adult numbers occurs at low frequencies for all treatment groups (plots of $\tilde{S}_{w,pp}(f)$ vs. f in Figure 17.10). The corrspectra reflect the properties of experimentally imposed habitat sequences. For the Blue$^+$/Red$^-$ habitat regime, the corrspectra of both the larvae and adults shift gradually from strong anti-synchrony at low frequencies to strong synchrony at high frequencies. The opposite trend is seen in the Blue$^-$/Red$^+$ habitat regime, except for adults with dispersal (Figure 17.10L). Dispersal transformed anti-synchrony at low frequencies into synchrony, flattening the corrspectra. This finding shows that dispersal and the time scale of environmental correlation can interact in their effects on synchrony.

Linear filter theory predicted well the different responses of the immatures and adults to habitat fluctuations. (Compare Figure 17.8 to Figure 17.10.) The larvae were predicted to be most sensitive to high-frequency fluctuations as indicated by the concentration of the spectral power at the highest frequencies. On the other hand, the adults effectively filtered out high-frequency fluctuations in habitat size, so the spectral power was concentrated at low frequencies. These predicted patterns (Figure 17.8) match well overall with the observed spectral densities (Figure 17.10). The one exception is in Figure 17.10J, where the increase in spectral density at high frequencies is much smaller than expected (Figure 17.8J). Also, as predicted by linear filter theory, larval corrspectra show a gradual shift from anti-synchrony to synchrony in the Blue$^+$/Red$^-$ habitat regime, with an

[9] In model (17.18)–(17.19), the adult survivorship is $1 - (1 - \mu_a)(1 - m_a) = \mu_a + m_a - \mu_a m_a$.

[10] Since each metapopulation has two spectra—one for each population—and one cospectrum, there are six replicate curves for the spectral densities and three for the corrspectra.

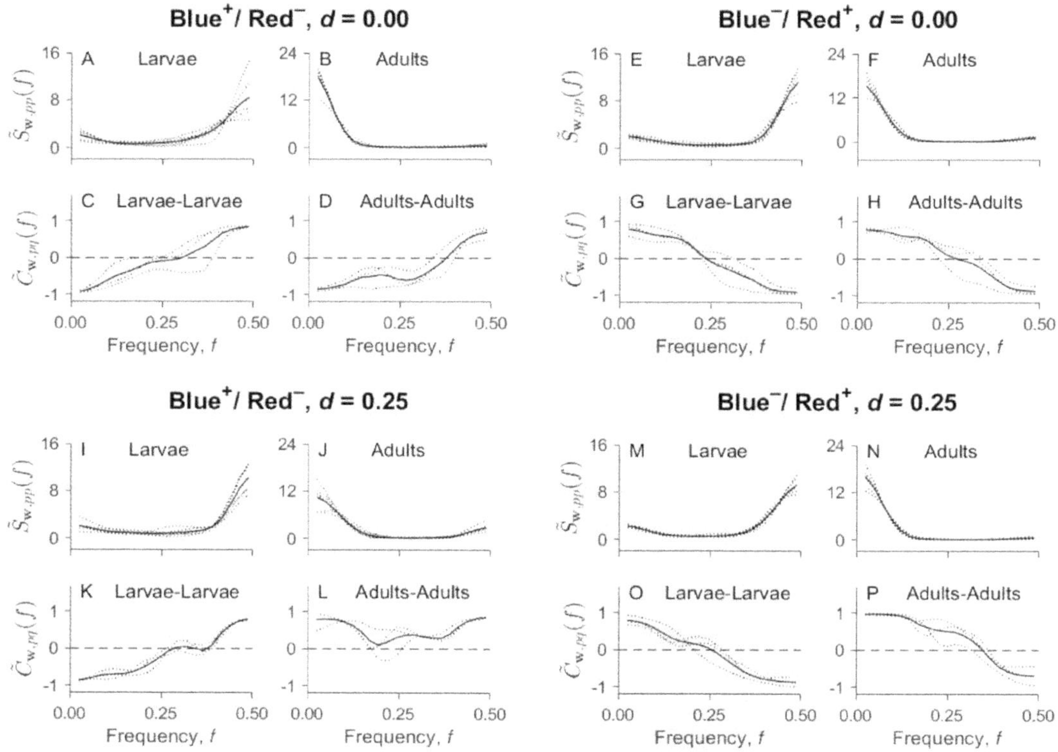

Figure 17.10 Observed spectral densities and corrspectra for the larval and adult life stages of the synchrony experiment. See text for details.

opposite tend in the Blue$^-$/Red$^+$ treatments. This predicted trend also holds for adults in the corrspectra of the $d = 0$ treatments (Figure 17.10D,H). With adult dispersal present, the adult corrspectrum in the Blue$^+$/Red$^-$ habitat regime is predicted to be positive and relatively flat at all frequencies (Figure 17.8L), which is similar to the observed corrspectrum (Figure 17.10L).

Table 17.2 gives the total synchrony and its decomposition into low- and high-frequency ranges for the larvae and adults. The low- and high-frequency ranges correspond to $0 \leq f < 0.25$ and $0.25 \leq f < 0.50$, respectively. The Pearson correlation coefficient between populations within a metapopulation was computed for larvae and adults using equation (17.4). Shown are the means and standard errors among replicates within a treatment group. In all cases except one, the total synchrony is the sum of a positive correlation at low frequencies and a negative correlation at high frequencies or vice versa. The exception is for the adults when the habitat regime is Blue$^+$/Red$^-$ and dispersal links the two populations ($d = 0.25$). In this case, the two populations have positive correlations for both the low- and high-frequency ranges, again indicating that the time scale of environmental correlations and dispersal interact in their effects on total synchrony. In every case, there is significant total synchrony or anti-synchrony (relative to the standard errors), despite zero overall correlation in habitat sizes.

TABLE 17.2 Synchrony between habitats within metapopulations.

Habitat regime	Dispersal	Frequencies	Larvae	Adults
$Blue^+/Red^-$	$d = 0$	Low	-0.114 ± 0.027	-0.726 ± 0.049
		High	0.522 ± 0.014	0.017 ± 0.006
		Total	0.409 ± 0.041	-0.708 ± 0.043
$Blue^-/Red^+$	$d = 0$	Low	0.138 ± 0.025	0.678 ± 0.047
		High	-0.648 ± 0.056	-0.065 ± 0.018
		Total	-0.510 ± 0.042	0.612 ± 0.063
$Blue^+/Red^-$	$d = 0.25$	Low	-0.166 ± 0.008	0.598 ± 0.035
		High	0.414 ± 0.012	0.136 ± 0.010
		Total	0.248 ± 0.016	0.734 ± 0.030
$Blue^-/Red^+$	$d = 0.25$	Low	0.133 ± 0.013	0.919 ± 0.028
		High	-0.617 ± 0.051	-0.019 ± 0.002
		Total	-0.483 ± 0.038	0.900 ± 0.030

17.4 CONCLUSIONS

There are two main causes of ecological synchrony in population dynamics, correlations in environmental noise and dispersal. Theoretical and experimental results show that changes in the time scale of the environmental covariance can induce changes in both the total population synchrony and the time scale structure of population synchrony. This finding is true even when the spectrum of the environmental fluctuations in each location and the total correlation between environmental time series are unchanged. In short, the spectral properties of environmental variance and covariance influence synchrony.

Dispersal can affect the time scale structure of synchrony. Our theoretical analysis showed that, at each frequency, the population cospectrum is a weighted average of the spectrum and cospectrum of environmental variation and that the balance of this weighting depends on the level of dispersal. The time scale-specificity of this weighting varies with the dynamic stability properties of the population, resulting in a three-way interaction among dispersal, the time scale of environmental stochasticity, and population stability in their effects on the time scale of synchrony (Figure 17.3). The synchrony experiment provides an empirical example. The time scale of the synchrony of adult numbers is strongly affected by dispersal, but only in the metapopulations with a $Blue^+/Red^-$ habitat regime (Figure 17.10).

There are practical implications for conservation biology and natural resource management. The time scale of environmental noise and rates of dispersal have implications for the risk of metapopulation extinction (Figure 17.5). Extinction probabilities are increased when environmental spectra and cospectra vary on long time scales and the population dynamics are undercompensating, whereas extinction probabilities are decreased when environmental spectra and cospectra vary on long time scales and the population dynamics are overcompensating. Increased dispersal rates increase the likelihood of extinction when the population dynamics are overcompensating (Figure 17.5b,d). The time scale of synchrony should not be ignored.

17.5 APPENDIX: SPECTRAL MATRIX FOR THE RICKER MODEL

The linearized Ricker metapopulation model is $\mathbf{w}_{t+1} = \lambda \mathbf{D} \mathbf{w}_t + \mathbf{D} \boldsymbol{\epsilon}_t$, where the dispersal matrix $\mathbf{D}$ is given by equation (17.7). Since, by assumption, the deterministic system $\mathbf{w}_{t+1} = \lambda \mathbf{D} \mathbf{w}_t$ is stable, the stochastic process has a stationary distribution (see Section 16.3.1).

The transfer function for this linear stochastic process is $\mathbf{T}(f) = \mathbf{D}^{-1}(\mathbf{I} - \lambda \mathbf{D}z)$, where $z = e^{2\pi i f}$ [3, 39]. The spectral matrix for the population fluctuations is

$$\begin{aligned}\mathbf{S_w}(f) &= \mathbf{T}(f)^{-1}\mathbf{S}_\epsilon(f)\left(\bar{\mathbf{T}}(f)^{\mathrm{T}}\right)^{-1} \\ &= \left(\mathbf{D}^{-1}(\mathbf{I} - \lambda\mathbf{D}z)\right)^{-1}\mathbf{S}_\epsilon(f)\,\overline{\left(\mathbf{D}^{-1}(\mathbf{I} - \lambda\mathbf{D}z)\right)}^{-1} \\ &= (\mathbf{I} - \lambda\mathbf{D}z)^{-1}\mathbf{D}\mathbf{S}_\epsilon(f)\,\mathbf{D}(\mathbf{I} - \lambda\mathbf{D}\bar{z})^{-1}. \end{aligned} \tag{17.20}$$

Here, we use the fact that, if $\mathbf{A}$ and $\mathbf{B}$ are square nonsingular matrices, $\overline{\mathbf{AB}} = \bar{\mathbf{B}}\bar{\mathbf{A}}$ and $(\mathbf{AB})^{-1} = \mathbf{B}^{-1}\mathbf{A}^{-1}$. For an explicit solution in terms of λ and d, we can begin by computing

$$\begin{aligned}(\mathbf{I} - \lambda\mathbf{D}z)^{-1} &= \begin{pmatrix} 1 - \lambda(1-d)z & -\lambda dz \\ -\lambda dz & 1 - \lambda(1-d)z \end{pmatrix}^{-1}, \\ &= \frac{\begin{pmatrix} 1 - \lambda(1-d)z & \lambda dz \\ \lambda dz & 1 - \lambda(1-d)z \end{pmatrix}}{(1 - \lambda(1-d)z)^2 - (\lambda dz)^2}, \\ &= \frac{\begin{pmatrix} 1 - \lambda(1-d)z & \lambda dz \\ \lambda dz & 1 - \lambda(1-d)z \end{pmatrix}}{(1 - \lambda z)(1 - \lambda z + 2\lambda dz)}. \end{aligned}$$

Postmultiplying by the dispersal matrix gives

$$\begin{aligned}(\mathbf{I} - \lambda\mathbf{D}z)^{-1}\mathbf{D} &= \frac{\begin{pmatrix} 1 - \lambda(1-d)z & \lambda dz \\ \lambda dz & 1 - \lambda(1-d)z \end{pmatrix}\begin{pmatrix} 1-d & d \\ d & 1-d \end{pmatrix}}{(1 - \lambda z)(1 - \lambda z + 2\lambda dz)}, \\ &= \frac{\begin{pmatrix} s & d \\ d & s \end{pmatrix}}{(1 - \lambda z)(s + d)}, \end{aligned} \tag{17.21}$$

where $s = 1 - \lambda z + 2\lambda dz - d$.

We can use equations (17.20), (17.8), and (17.21) to solve for the spectral matrix of the population fluctuations:

$$\begin{aligned} S_{\mathbf{w}}(f) &= \frac{|s|^2 S_\epsilon(f) + sdC_\epsilon(f) + \bar{s}dC_\epsilon(f) + d^2 S_\epsilon(f)}{|1 - \lambda z|^2 |s + d|^2} \\ &= \frac{\left(|s|^2 + d^2\right) S_\epsilon(f) + (s + \bar{s})\, dC_\epsilon(f)}{|1 - \lambda z|^2 |s + d|^2} \\ &= \frac{\left(|s|^2 + d^2\right) S_\epsilon(f) + 2\,\mathrm{Re}(s)\, dC_\epsilon(f)}{|1 - \lambda z|^2 |s + d|^2} \\ &= \frac{\left(|s|^2 + d^2\right)(S_\epsilon(f) - C_\epsilon(f) + C_\epsilon(f)) + 2\,\mathrm{Re}(s)\, dC_\epsilon(f)}{|1 - \lambda z|^2 |s + d|^2} \\ &= \frac{\left(|s|^2 + d^2\right)(S_\epsilon(f) - C_\epsilon(f)) + \left(|s|^2 + 2\,\mathrm{Re}(s)\, d + d^2\right) C_\epsilon(f)}{|1 - \lambda z|^2 |s + d|^2} \\ &= \frac{\left(|s|^2 + d^2\right)(S_\epsilon(f) - C_\epsilon(f)) + |s + d|^2 C_\epsilon(f)}{|1 - \lambda z|^2 |s + d|^2} \\ &= \frac{S_\epsilon(f)}{|1 - \lambda z|^2}\left(\frac{|s|^2 + d^2}{|s + d|^2}\right) + \frac{C_\epsilon(f)}{|1 - \lambda z|^2}\left(1 - \frac{|s|^2 + d^2}{|s + d|^2}\right). \end{aligned}$$

where $\mathrm{Re}(s)$ represents the real part of the complex number s. The cospectra are the off-diagonal elements:

$$
\begin{aligned}
C_{\mathbf{w}}(f) &= \frac{sdS_{\epsilon}(f) + d^2 C_{\epsilon}(f) + |s|^2 C_{\epsilon}(f) + \bar{s}dS_{\epsilon}(f)}{|1-\lambda z|^2 |s+d|^2} \\
&= \frac{\left(|s|^2 + d^2\right) C_{\epsilon}(f) + 2\,\mathrm{Re}(s)\, dS_{\epsilon}(f)}{|1-\lambda z|^2 |s+d|^2} \\
&= \frac{C_{\epsilon}(f)}{|1-\lambda z|^2}\left(\frac{|s|^2+d^2}{|s+d|^2}\right) + \frac{S_{\epsilon}(f)}{|1-\lambda z|^2}\left(1 - \frac{|s|^2+d^2}{|s+d|^2}\right).
\end{aligned}
$$

17.6 APPENDIX: SPECTRA FOR LPA METAPOPULATION MODEL

To obtain predictions for the spectral matrix of the linearized LPA metapopulation model $\mathbf{w}_{t+1} = \mathbf{M}\mathbf{w}_t + \boldsymbol{\Lambda}\boldsymbol{\epsilon}_t$, we used the equation

$$
\mathbf{S}_{\mathbf{w}}(f) = \mathbf{T}(f)^{-1}\mathbf{S}_{\epsilon}(f)\left(\bar{\mathbf{T}}(f)^{\mathrm{T}}\right)^{-1}, \tag{17.22}
$$

where $\mathbf{T}(f) = \boldsymbol{\Lambda}^{-1}\left(\mathbf{I} - \mathbf{M}e^{2\pi f i}\right)$ is the transfer function for the linear stochastic process. In this appendix, we provide expressions for the matrices $\mathbf{M}$ and $\boldsymbol{\Lambda}$, and the matrix function $\mathbf{S}_{\epsilon}(f)$.

The 6×6 matrix $\mathbf{M} = \{M_{ij}\}$ contains the linearization coefficients for the life stage variables in the metapopulation model (17.18) – (17.19), with $V_{j,t} = V^* + \varepsilon_{j,t}$. These were obtained by taking the partial derivatives of the six model equations with respect to the life stage variables, setting the life stage numbers to their equilibrium values, and letting $\boldsymbol{\epsilon}_t = \mathbf{0}$:

$$
\begin{aligned}
M_{11} &= M_{44} = -c_{el}L^*/V^* \\
M_{13} &= M_{46} = (L^*/A^*)\left(1 - c_{ea}A^*/V^*\right) \\
M_{21} &= M_{54} = 1 - \mu_l \\
M_{32} &= M_{65} = e^{-c_{ea}A^*/V^*}\left(1-m_a\right)\left(1-d\right) \\
M_{33} &= M_{66} = \left(1 - \left(c_{pa}/V^*\right)P^* e^{-c_{pa}A^*/V^*}\right)\left(1-m_a\right)\left(1-d\right) \\
M_{35} &= M_{62} = e^{-c_{ea}A^*/V^*}\left(1-m_a\right)d \\
M_{36} &= M_{63} = \left(1 - \left(c_{pa}/V^*\right)P^* e^{-c_{pa}A^*/V^*}\right)\left(1-m_a\right)d.
\end{aligned} \tag{17.23}
$$

All other M_{ij} values are zero.

The 6×6 matrix $\boldsymbol{\Lambda}$ contains the linearization coefficients for the habitat random variables in the vector $\boldsymbol{\epsilon}_t$. There are no random variables for the pupal life stage. The coefficients were obtained by taking the partial derivatives of the six model equations with respect to the random variables, setting the life stage numbers to their equilibrium values, and letting $\boldsymbol{\epsilon}_t = \mathbf{0}$:

$$
\boldsymbol{\Lambda} = \begin{pmatrix}
\vartheta_l & 0 & 0 & 0 & 0 & 0 \\
0 & 0 & 0 & 0 & 0 & 0 \\
0 & 0 & \vartheta_a(1-d) & 0 & 0 & \vartheta_a d \\
0 & 0 & 0 & \vartheta_l & 0 & 1 \\
0 & 0 & 0 & 0 & 0 & 0 \\
0 & 0 & \vartheta_a d & 0 & 0 & \vartheta_a(1-d)
\end{pmatrix}, \tag{17.24}
$$

where

$$\vartheta_l = L^* \left(c_{el}L^* + c_{ea}A^*\right)\left(V^*\right)^{-2}$$
$$\vartheta_a = \left(\mu_a + m_a - \mu_a m_a\right) c_{pa}(A^*/V^*)^2.$$

The matrix $\mathbf{S}_{\boldsymbol{\epsilon}}(f)$ gives the spectral properties of the random vector $\boldsymbol{\epsilon}_t$. The two spectra for the bivariate sequence of habitat fluctuations in a metapopulation are equal and given by $S_{\boldsymbol{\epsilon}}(f)$ in equation (17.16). Larva and adults in the same population are exposed to the same habitat fluctuations, therefore the elements of the noise vector have the same spectrum $S_{\boldsymbol{\epsilon}}(f)$ and the cospectrum between them is equal to the spectrum. The dynamics of the pupal life stage does not depend on habitat size, so its spectral function is zero. The cospectrum for the bivariate sequence of habitat fluctuations in a metapopulation is $C_\varepsilon(f) = S_\varepsilon(f)\tilde{C}_\varepsilon(f)$, where the corrspectrum $\tilde{C}_\varepsilon(f)$ is given by (17.17).[11] The spectral matrix for $\boldsymbol{\epsilon}_t$ is

$$\mathbf{S}_{\boldsymbol{\epsilon}}(f) = S_\varepsilon(f)\begin{pmatrix} 1 & 0 & 1 & \tilde{C}_\varepsilon(f) & 0 & \tilde{C}_\varepsilon(f) \\ 0 & 0 & 0 & 0 & 0 & 0 \\ 1 & 0 & 1 & \tilde{C}_\varepsilon(f) & 0 & \tilde{C}_\varepsilon(f) \\ \tilde{C}_\varepsilon(f) & 0 & \tilde{C}_\varepsilon(f) & 1 & 0 & 1 \\ 0 & 0 & 0 & 0 & 0 & 0 \\ \tilde{C}_\varepsilon(f) & 0 & \tilde{C}_\varepsilon(f) & 1 & 0 & 1 \end{pmatrix}.$$

We used the parameter estimates (Table 17.1), the equilibrium life stage values $L^* = 94.53$, $P^* = 78.13$, and $A^* = 405.72$, the harvesting values $m_l = 0.70$ and $m_a = 0.05$, and the appropriate dispersal value d to obtain the coefficient matrices (17.23) and (17.24). We then substituted $\mathbf{M}$, $\boldsymbol{\Lambda}$, and $\mathbf{S}_{\boldsymbol{\epsilon}}(f)$ into equation (17.22) to obtain $\mathbf{S}_{\mathbf{w}}(f)$, the spectral matrix for the population fluctuations. These are the spectral densities and corrspectra plotted in Figure 17.8.

Bibliography

[1] Björnstad ON, Falck W (2001) Nonparametric spatial covariance functions: Estimation and testing. Environmental and Ecological Statistics 8:53–70. https://doi.org/10.1023/A:1009601932481

[2] Björnstad ON, Liebhold AM, Johnson DM (2008) Transient synchronization following invasion: Revisiting Moran's model and a case study. Population Ecology 50:379–389. https://doi.org/10.1007/s10144-008-0105-5

[3] Brillinger DR (2001) Time Series: Data Analysis and Theory. Philadelphia, Society for Industrial and Applied Mathematics.

[4] Buonaccorsi JP, Elkinton JS, Evans SR, Liebhold AM (2001) Measuring and testing for spatial synchrony. Ecology 82:1668–1679. https://doi.org/10.1890/0012-9658(2001)082[1668:MATFSS]2.0.CO;2

[5] Chambers MJ (1995) The simulation of random vector time series with given spectrum. Mathematical and Computer Modelling 22:1–6. https://doi.org/10.1016/0895-7177(95)00106-C

[11]Since $S_{\boldsymbol{\epsilon},pp}(f) \equiv S_{\boldsymbol{\epsilon},qq}(f)$ and $\tilde{C}_{\boldsymbol{\epsilon},pq}(f) = C_{\boldsymbol{\epsilon},pq}(f)/\sqrt{S_{\boldsymbol{\epsilon},pp}(f)\,S_{\boldsymbol{\epsilon},qq}(f)}$, we can write $C_{\boldsymbol{\epsilon},pq}(f) = S_{\boldsymbol{\epsilon},pp}(f)\,\tilde{C}_{\boldsymbol{\epsilon},pq}(f)$.

[6] Chevalier M, Laffaille P, Grenouillet G (2014) Spatial synchrony in stream fish populations: Influence of species traits. Ecography 37:960–968. https://doi.org/10.1111/ecog.00662

[7] Cohen, JE, Newman CM, Cohen AE, Petchey OL, Gonzalez A (1999) Spectral mimicry: A method of synthesizing matching time series with different Fourier spectra. Circuits, Systems and Signal Processing 18:431–442. https://doi.org/10.1007/BF01200792

[8] Danielian S (2016) The Effects of Environmental Noise and Dispersal on Ecological Extinction and the Spectral Properties of Metapopulations. MS Thesis, California State University, Los Angeles. https://scholarworks.calstate.edu/handle/10211.3/173814

[9] Defriez EJ, Sheppard, LW, Reid PC, Reuman DC (2016) Climate change-related regime shifts have altered spatial synchrony of plankton dynamics in the North Sea. Global Change Biology 22:2069–2080. https://doi.org/10.1111/gcb.13229

[10] Defriez, EJ, Reuman DC (2017) A global geography of synchrony for marine phytoplankton. Global Ecology and Biogeography 26:867–877. https://doi.org/10.1111/geb.12594

[11] Defriez EJ, Reuman DC (2017) A global geography of synchrony for terrestrial vegetation. Global Ecology and Biogeography 26:878–888. https://doi.org/10.1111/geb.12595

[12] Dennis B, Desharnais RA, Cushing JM, Costantino RF (1995) Nonlinear demographic dynamics: Mathematical models, statistical methods, and biological experiments. Ecological Monographs 65:261–282. https://doi.org/10.2307/2937060

[13] Dennis B, Desharnais RA, Cushing JM, Henson SM, Costantino RF (2001) Estimating chaos and complex dynamics in an insect population. Ecological Monographs 71:277–303. https://doi.org/10.1890/0012-9615(2001)071[0277:ECACDI]2.0.CO;2

[14] Desharnais RA, Liu L (1987) Stable demographic limit cycles in laboratory populations of *Tribolium castaneum*. Journal of Animal Ecology 56:885–906. https://doi.org/10.2307/4955

[15] Desharnais RA, Reuman DC, Costantino RF, Cohen JE (2018) Temporal scale of environmental correlations affects ecological synchrony. Ecology Letters 21:1800–1811. https://doi.org/10.1111/ele.13155

[16] Desharnais RA, Muchlinski AE, Ortiz JL, Alvidrez RI, Gatza BP (2022) Timescale analyses of fluctuations in coexisting populations of a native and invasive tree squirrel. Ecology and Evolution 12:e8779. https://doi.org/10.1002/ece3.8779

[17] Earn DJD, Rohani P, Grenfell BT (1998) Persistence, chaos and synchrony in ecology and epidemiology. Proceedings of the Royal Society of London Series B Biological Sciences 265:7–10. https://doi.org/10.1098/rspb.1998.0256

[18] Earn DJD, Levin SA, Rohani P (2000) Coherence and conservation. Science 290:1360–1364. https://doi.org/10.1126/science.290.5495.1360

[19] Gouhier TC, Guichard F, Dray S (2014) Synchrony: Quantifying variability in space and time. Methods in Ecology and Evolution 5:524–533. https://doi.org/10.1111/2041-210X.12188

[20] Grenfell BT, Björnstad ON, Kappey J (2001) Travelling waves and spatial hierarchies in measles epidemics. Nature 414:716–723. https://doi.org/10.1038/414716a

[21] Harrison S, Quinn JF (1989) Correlated environments and the persistence of metapopulations. Oikos 56:293–298. https://doi.org/10.2307/3565613

[22] Haydon D, Steen H (1997) The effects of large—and small—scale random events on the synchrony of metapopulation dynamics: A theoretical analysis. Proceedings of the Royal Society of London Series B Biological Sciences 264:1375–1381. https://doi.org/10.1098/rspb.1997.0191

[23] Heino M, Kaitala V, Ranta E, Lindstrom J (1997) Synchronous dynamics and rates of extinction in spatially structured populations. Proceedings of the Royal Society of London Series B Biological Sciences 264:481–486. https://doi.org/10.1098/rspb.1997.0069

[24] Hudson PJ, Cattadori IM (1999) The Moran effect: A cause of population synchrony. Trends in Ecology and Evolution 14:1–2. https://doi.org/10.1016/S0169-5347(98)01498-0

[25] Kahilainen A, van Nouhuys S, Schulz T, Saastamoinen M (2018) Metapopulation dynamics in a changing climate: Increasing spatial synchrony in weather conditions drives metapopulation synchrony of a butterfly inhabiting a fragmented landscape. Global Change Biology 24:4316–4329. https://doi.org/10.1111/gcb.14280

[26] Keitt TH (2008) Coherent ecological dynamics induced by large-scale disturbance. Nature 454:331–334. https://doi.org/10.1038/nature06935

[27] Kendall BE, Björnstad ON, Bascompte J, Keitt TH, Fagan WF (2000) Dispersal, environmental correlation, and spatial synchrony in population dynamics. American Naturalist 155:628–636. https://doi.org/10.1086/303350

[28] Koenig WD, Liebhold AM (2016) Temporally increasing spatial synchrony of North American temperature and bird populations. Nature Climate Change 6:614–617. https://doi.org/10.1038/nclimate2933

[29] Liebhold A, Koenig WD, Björnstad ON (2004) Spatial synchrony in population dynamics. Annual Review of Ecology Evolution and Systematics 35:467–490. https://doi.org/10.1146/annurev.ecolsys.34.011802.132516

[30] Martin AE, Pearce-Higgins JW, Fahrig L (2017) The spatial scale of time-lagged population synchrony increases with species dispersal distance. Global Ecology and Biogeography 26:1201–1210. https://doi.org/10.1111/geb.12630

[31] Moran PAP (1953) The statistical analysis of the Canadian lynx cycle II: Synchronization and meteorology. Australian Journal of Zoology 1:291–298. https://doi.org/10.1071/ZO9530291

[32] Myers JH (1990) Population-cycles of western tent caterpillars—experimental introductions and synchrony of fluctuations. Ecology 71:986–995. https://doi.org/10.2307/1937367

[33] Myers JH (1998) Synchrony in outbreaks of forest lepidoptera: A possible example of the Moran effect. Ecology 79:1111–1117. https://doi.org/10.1890/0012-9658(1998)079[1111:SIOOFL]2.0.CO;2

[34] Percival DB, Walden AT (1993) Spectral Analysis for Physical Applications. Cambridge, Cambridge University Press.

[35] Porter JH, Parry ML, Carter TR (1991) The potential effects of climatic change on agricultural insect pests. Agricultural and Forest Meteorology 57:221–240. https://doi.org/10.1016/0168-1923(91)90088-8

[36] Post E, Forchhammer MC (2004) Spatial synchrony of local populations has increased in association with the recent Northern Hemisphere climate trend. Proceedings of the National Academy of Sciences USA 101:9286–9290. https://doi.org/10.1073/pnas.0305029101

[37] Ranta E, Kaitala V, Lindstrom J, Linden H (1995) Synchrony in population dynamics. Proceedings of the Royal Society of London Series B Biological Sciences 262:113–118. https://doi.org/10.1098/rspb.1995.0184

[38] Ranta E, Kaitala V, Lindstrom J, Helle E (1997) The Moran effect and synchrony in population dynamics. Oikos 78:136–142. https://doi.org/10.2307/3545809

[39] Reinsel GC (1997) Elements of Multivariate Time Series Analysis. Second Edition. New York, Springer.

[40] Ripa J (2000) Analysing the Moran effect and dispersal: Their significance and interaction in synchronous population dynamics. Oikos 89:175–187. https://doi.org/10.1034/j.1600-0706.2000.890119.x

[41] Sheppard LW, Bell JR, Harrington R, Reuman DC (2016) Changes in large-scale climate alter spatial synchrony of aphid pests. Nature Climate Change 6:610–613. https://doi.org/10.1038/nclimate2881

[42] Shestakova, TA, Gutiérrez E, Kirdyanov AV, Camarero JJ, Génova M, Knorre AA, Linares JC, de Dios VR, Sánchez-Salguero R, Voltas J (2016) Forests synchronize their growth in contrasting Eurasian regions in response to climate warming. Proceedings of the National Academy of Sciences USA 113:662–667. https://doi.org/10.1073/pnas.1514717113

[43] Solé RV, Gamarra JGP (1998) Chaos, dispersal and extinction in coupled ecosystems. Journal of Theoretical Biology 193:539–541. https://doi.org/10.1006/jtbi.1998.0716

[44] Vasseur DA, Gaedke U (2007) Spectral analysis unmasks synchronous and compensatory dynamics in plankton communities. Ecology 88:2058–2071. https://doi.org/10.1890/06-1899.1

CHAPTER 18

Dynamics of Natural Selection

Genetically variable populations can evolve through time. Natural selection favors alleles which increase fitness traits like survival probabilities and reproductive rates. These same traits affect population dynamics. In this chapter, we couple models of natural selection and population growth. We present a genetic version of the LPA model, make predictions regarding the effects of selection on population dynamics, and test these predictions using experimental data from laboratory populations of *Tribolium*.

18.1 ECOLOGICAL GENETICS

The field of population genetics began in the 1920s and 1930s with the pioneering work of R.A. Fisher, J.B.S. Haldane, and Sewall Wright [13, 18, 24]. The focus was on mathematical models for the frequencies of alternate alleles segregating at one or two genetic loci. These researchers and others addressed a variety of phenomena including genotype frequencies, inbreeding, natural selection, linkage disequilibrium, mutation pressure, mutation-selection balance, assortative mating, genetic drift, and probability distributions of allele frequencies. This work provided a theoretical foundation for the study of evolution.

Three decades later, an ecological perspective was added to the theory of natural selection [1, 2, 19, 22]. This seemed like a natural progression since variation in genotypic fitness is defined in terms of heritable differences in rates of reproduction and mortality—the same factors that determine population growth. These ecological-genetic models examined the impact of natural selection on population dynamics.

18.1.1 Single-Locus Model of Natural Selection

A simple population genetic model of natural selection involves multiple alleles segregating at single autosomal locus. Mating is random and genotypes differ in their expected contributions to the next generation. Let $\mathcal{A}_1, \mathcal{A}_2, \ldots, \mathcal{A}_n$ represent the n alleles with population frequencies $p_{1,t}, p_{2,t}, \ldots, p_{n,t}$ at time t. With a population size N_t, the total number of alleles in a diploid population is $2N_t$. After random mating, alleles from the parental gametes recombine to form new genotypes $\mathcal{A}_i\mathcal{A}_j$ with frequencies $p_{i,t}p_{j,t}$. We will assume the per capita fertility is β and genotypes differ in their viabilities to reproductive age, s_{ij}. The total number of $\mathcal{A}_i\mathcal{A}_j$ genotypes in the next generation is $N_t\beta s_{ij}p_{i,t}p_{j,t}$. Let $W_{ij} = \beta s_{ij}$ represent the reproductive output or "fitness" of genotype $\mathcal{A}_i\mathcal{A}_j$.[1] The average fitness in

[1] For autosomal loci, fitness rarely depends on which parents contribute each allele, so $W_{ij} = W_{ji}$.

DOI: 10.1201/9781003467120-18

generation t is

$$W_t = \sum_{i=1}^{n} \sum_{j=1}^{n} p_{i,t} p_{j,t} W_{ij},$$

and the population size in the next generation is the sum of the number of genotypes, $N_{t+1} = N_t W_t$. The frequency of $\mathcal{A}_i$ alleles in the next generation is obtained by summing together the number of genotypes carrying one $\mathcal{A}_i$ allele and dividing by the total number of genotypes:

$$p_{i,t+1} = \frac{N_t \sum_{j=1}^{n} p_i p_j W_{ij}}{N_{t+1}} = \frac{N_t p_{i,t} W_{i,t}}{N_t W_t} = \frac{p_{i,t} W_{i,t}}{W_t}, \qquad i = 1, \ldots, n, \tag{18.1}$$

where $W_{i,t} = \sum_{j=1}^{n} p_{j,t} W_{ij}$ is defined as the fitness for allele $\mathcal{A}_i$. An often used version of this model looks at the change in allele frequency, $\Delta p_{i,t} = p_{i,t+1} - p_{i,t}$, at each generation:

$$\Delta p_{i,t} = p_{i,t} \left(\frac{W_{i,t} - W_t}{W_t} \right). \tag{18.2}$$

Equation (18.1) or (18.2) can be used to study the evolution of the population allele frequencies through time.

Many population genetic studies use a continuous-time version of (18.2). When the differences between the genotypic fitnesses are small (weak selection), one can approximate the allele frequency and population dynamics using

$$\begin{aligned} \frac{dp_i(t)}{dt} &= p_i(t)\left(m_i(t) - m(t)\right) \qquad i = 1, 2, \ldots, n \\ \frac{dN(t)}{dt} &= N(t)\, m(t), \end{aligned} \tag{18.3}$$

where $m_i(t)$ and $m(t)$ are the mean allelic and population fitnesses, respectively, expressed using Malthusian versions of the genotypic fitnesses, $W_{ij} = e^{m_{ij}}$. Derivations of equation (18.3) using continuous-time models with overlapping generations and weak selection are available in books by Nagylaki [21] and Charlesworth [3].

Figure 18.1 illustrates the possible dynamics of allele frequency for the simplest case of two alleles ($n = 2$). Let $p = p_1$ and $1 - p = p_2$. The rate of change in allele frequency (18.3) is plotted as a function of allele frequency and the arrows indicate the direction of change. Directional selection occurs when $m_{11} \geq m_{12} \geq m_{22}$ (Figure 18.1A) or $m_{11} \leq m_{12} \leq m_{22}$ (Figure 18.1B).[2] In these cases one of the alleles becomes fixed in the population and the other is eliminated. A stable polymorphic equilibrium p^* exists when $m_{11} < m_{12} > m_{22}$, that is, the heterozygote has a higher fitness than either homozygote (Figure 18.1C). The equilibrium value is

$$p^* = \frac{m_{12} - m_{22}}{2m_{12} - m_{11} - m_{22}}. \tag{18.4}$$

If the heterozygote is inferior to both homozygotes, $m_{11} > m_{12} < m_{22}$, either $\mathcal{A}_1$ or $\mathcal{A}_2$ becomes fixed, where the surviving allele depends on the initial allele frequency (Figure 18.1D). If $p(0) > p^*$, then $\mathcal{A}_1$ is selected to fixation; if $p(0) < p^*$, $\mathcal{A}_2$ becomes fixed, where p^* is given by equation (18.4). This is sometimes referred to as the "founder effect." The discrete-generation model (18.2) has the same qualitative dynamics.

---

[2]For directional selection, at least one strict inequality must exist among the fitness values, that is, $m_{11} \neq m_{12}$ and/or $m_{12} \neq m_{22}$.

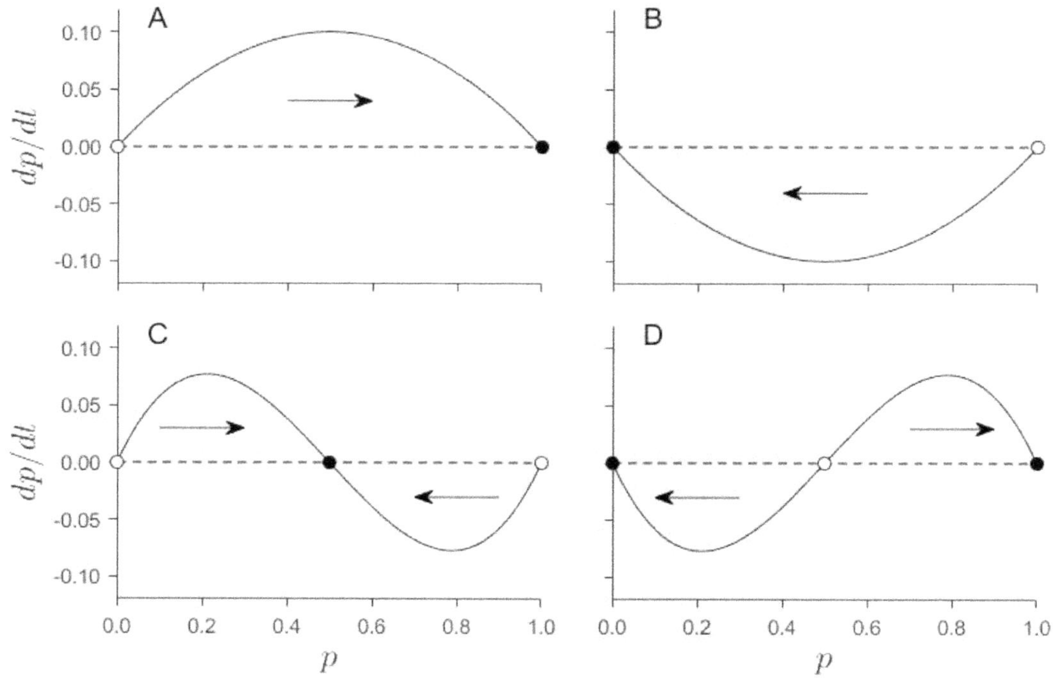

Figure 18.1 Plots of the rate of change in allele frequency dp/dt versus allele frequency p for the two-allele continuous-time model of natural selection (18.3). The solid circles locate stable equilibria and the open circles are for unstable equilibria. The arrows indicate the direction of change in allele frequency. Represented are directional selection (A,B), balancing selection (C), and disruptive selection (D). See text for more details.

An important result in population genetics theory is the Fundamental Theorem of Natural Selection. Derived by R.A. Fisher [13], it relates the change in the mean population fitness to the variance in allelic fitnesses, also called the "additive genetic variance":

$$\frac{dm(t)}{dt} = 2\sum_{i=1}^{n} p_i(m_i(t) - m(t))^2 = \mathcal{V}_g(t). \tag{18.5}$$

Crow and Kimura [6] provide a straightforward derivation of this result based on the continuous-time model (18.3). The factor of two in the variance term comes from the fact that each genotype carries two alleles. The quantity $\mathcal{V}_g \geq 0$, and the equality only holds at genetic equilibrium.[3] The expression (18.5) implies that natural selection causes the mean population fitness to increase until it reaches a local maximum at genetic equilibrium. The rate of increase is related to the amount of genetic variation present in the population. The fundamental theorem also can be obtained from the discrete-time model (18.2), although its derivation is more complex [21].

[3]This can be seen by examining model (18.3). For alleles lost to selection, $p_i^* = 0$. For alleles present at equilibrium, $p_i^* > 0$ and $m_i^* = m^*$.

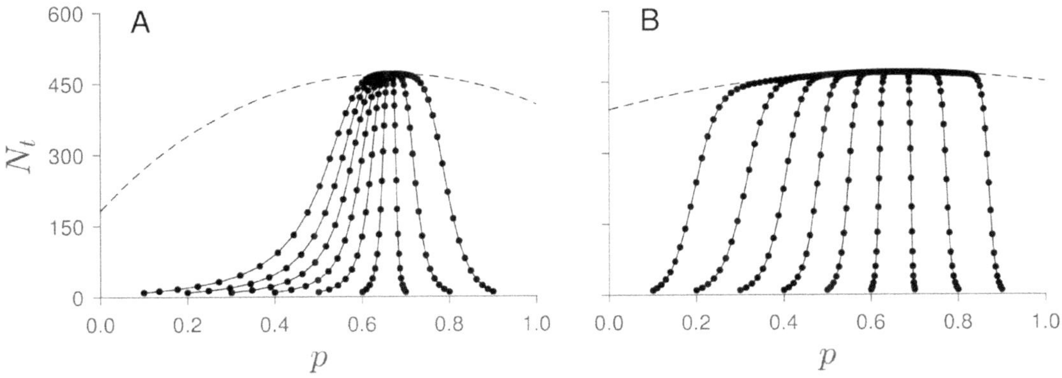

Figure 18.2 State space plots of simulations of the two-allele discrete-time Ricker model of density-dependent selection (18.6). A. Strong selection. B. Moderate selection. The dashed curves separate regions where the population growth rate is positive (below) and negative (above). See text for more details.

18.1.2 Coupling Genetics With Population Growth

Changes in population numbers and allele frequencies can be linked by allowing the genotypic fitness values to be functions of population density: $W_{ij} = W_{ij}(N)$. For example, let β represent the per capita fertility and assume genotypes differ with respect to their probabilities of survival to reproductive age, s_{ij}. Let $b_{ij} = \beta s_{ij}$ be the reproductive output of genotype $\mathcal{A}_i\mathcal{A}_j$ and assume the overall fitness is limited by population numbers, such that $W_{ij}(N) = b_{ij}e^{-cN}$. In this case, the model equations are

$$\begin{aligned} p_{i,t+1} &= \frac{p_{i,t}b_{i,t}}{b_t} \qquad i = 1, 2, \ldots, n \\ N_{t+1} &= b_t N_t e^{-cN_t}. \end{aligned} \tag{18.6}$$

Here, $b_{i,t} = \sum_j p_{j,t}b_{ij}$ and $b_t = \sum_{ij} p_{i,t}p_{j,t}b_{ij}$ are the allelic and population mean values for the b_{ij}. This is a Ricker model for density-dependent selection.

Figure 18.2 shows some simulation examples for the Ricker selection model (18.6) for two alleles ($n = 2$). Let $p = p_1$ and $1 - p = p_2$. The three genotypic values for the reproductive output parameter are $b_{11} = 1.6 - 0.2s$, $b_{12} = 1.6 + 0.1s$, and $b_{22} = 1.6 - 0.4s$, where s represents the selection intensity. The density-regulation parameter is $c = 0.001$. At equilibrium, $p^* = 0.667$, $b^* = 1.57$, and $N^* = (\ln b^*)/c = 451.1$ for all values of $s > 0$. Initial allele frequencies were $p_0 = 0.1, 0.2, \ldots, 0.9$ and the initial population density was $N_0 = 10$. For strong selection ($s = 1$, Figure 18.2A), allele frequency and population size quickly converge to their equilibrium values. For moderate selection ($s = 0.3$, Figure 18.2B), allele frequencies change more slowly. Population numbers quickly approach a "pseudo-equilibrium" value $\phi(p_t) = (\ln b_t)/c$, where b_t is the population mean value that depends on the allele frequency. The dashed curves in Figure 18.2 show the manifold $\phi(p)$ for $0 \le p \le 1$. As allele frequency slowly converges to p^*, population size tracks these changes, hugging the curve $\phi(p)$. This demonstrates that for weak selection, there are two time scales for the system: a "fast ecological time scale" and a "slow evolutionary time scale."

When selection is density dependent, at equilibrium, $W^* = 1$, or, equivalently, $m^* = 0$. The principal that natural selection maximizes the mean population fitness no longer holds. However, one can show that under some mild conditions for the density-dependent genotypic fitness functions, natural selection will maximize the equilibrium population size. More specifically, let $\mathbf{p}$ denote an arbitrary vector of n allele frequencies. If $\mathbf{p}$ is held constant, we can assume that population numbers N approach a pseudo-equilibrium, $\phi(\mathbf{p})$. These pseudo-equilibria $\phi(\mathbf{p})$ represent a manifold in state space where population growth is zero. As $t \to \infty$, $\mathbf{p} \to \mathbf{p}^*$ and $N \to \phi(\mathbf{p}^*)$, where $\phi(\mathbf{p}^*)$ is a local maximum of the manifold $\phi(\mathbf{p})$. In other words, natural selection results in a maximization of the equilibrium population density N^*. Derivations of this principal exist for multiple alleles [14] and discrete-time models with stable equilibria [23]. The simulations of Ricker model (18.6) in Figure 18.2 provide an example. The dashed curves represent the manifold of pseudo population equilibria, $\phi(\mathbf{p})$. Population numbers converge to the stable equilibrium N^*, which is a local (in these examples global) maximum of $\phi(\mathbf{p})$.

18.2 COMBINING GENETICS WITH THE LPA MODEL

In this section we present a single locus population genetic version of the LPA model with multiple alleles [4]. Unlike the simple discrete generation model (18.1), it is necessary to include equations for the genotypes of each life stage, since, in general, genotype frequencies cannot be expressed as the product of allele frequencies. We describe an empirical example where the LPA genetic model predicts the dynamics of selection for pesticide resistance [4, 8].

18.2.1 The gLPA Model

Let $p_{i,t}$ denote the frequency of $\mathcal{A}_i$ alleles, $i = 1, 2, \ldots, n$, among adults at time t. Assume all females have the same fecundity β and that males and females have the same viability and rates of cannibalism. Let s_{ij} denote the probability that a newly oviposited egg with genotype $\mathcal{A}_i\mathcal{A}_j$ survives to the larval stage in the absence of cannibalism. Given an equal sex ratio, the same allele frequencies in the two sexes, and random mating, the number of larva with genotype $\mathcal{A}_i\mathcal{A}_j$ produced per adult at time t is $p_{i,t}p_{j,t}b_{ij}$, where $b_{ij} = 0.5\beta s_{ij}$. The genotype-specific LPA equations are

$$\begin{aligned} L_{ij,t+1} &= p_{i,t}p_{j,t}b_{ij}A_t e^{-c_{el,t}L_t - c_{ea,t}A_t} \\ P_{ij,t+1} &= L_{ij,t}\left(1-\mu_{l,ij}\right) \\ A_{ij,t+1} &= P_{ij,t}e^{-c_{pa,t}A_t} + A_{ij,t}\left(1-\mu_{a,ij}\right). \end{aligned} \tag{18.7}$$

The total populations of each of the life stages L_t, P_t, and A_t are the sums of the populations of each genotype:

$$L_t = \sum_{i=1}^{n}\sum_{j=1}^{n} L_{ij,t}, \quad P_t = \sum_{i=1}^{n}\sum_{j=1}^{n} P_{ij,t}, \quad A_t = \sum_{i=1}^{n}\sum_{j=1}^{n} A_{ij,t}. \tag{18.8}$$

The cannibalism parameters $c_{el,t}$, $c_{ea,t}$, and $c_{pa,t}$ are averages based on the frequencies of each genotype:

$$\begin{aligned} c_{el,t} &= \frac{\sum_{i=1}^{n}\sum_{j=1}^{n} c_{el,ij}L_{ij,t}}{L_t} \\ c_{ea,t} &= \frac{\sum_{i=1}^{n}\sum_{j=1}^{n} c_{ea,ij}A_{ij,t}}{A_t} \\ c_{pa,t} &= \frac{\sum_{i=1}^{n}\sum_{j=1}^{n} c_{pa,ij}A_{ij,t}}{A_t}. \end{aligned} \tag{18.9}$$

The mortality parameters in model (18.7) are genotypic-specific versions of the same parameters found in the non-genetic LPA model. The allele frequency $p_{i,t}$ is obtained by tallying up the numbers of $\mathcal{A}_i$ alleles among adults:

$$p_{i,t} = \frac{2A_{ii,t} + \sum\limits_{j \neq i}^{n} A_{ij,t}}{2A_t}, \qquad i = 1, 2, \ldots, n. \tag{18.10}$$

We call equations (18.7)–(18.10) the gLPA model. The gLPA model can be used to predict the ecological and evolutionary dynamics of *Tribolium* populations with heritable genetic variation in life-history parameters.

18.2.2 An Example of Evolutionary Rescue

We present an application of the gLPA model (18.7) to experimental data [4, 8]. Flour beetle populations were initiated in medium containing the pesticide malathion at concentrations of 0 ppm, 1.5 ppm, and 3.0 ppm. The experimental cultures were kept in 120 ml Wheaton vials containing 20 g of media (92.5% bleached white flour, 5% dry brewer's yeast, 1.5% ground fumagillin, and 1% sunflower oil.) The malathion was diluted into the sunflower oil before adding to the flour. Each population was initiated with 250 larvae, 5 pupae, and 250 adults. All cultures were kept in the same darkened incubator at a temperature of 31°C, and a relative humidity of 55%.

We will focus on the populations subjected to 3.0 ppm of malathion. Three treatment groups were established, each with four replicate populations. The three treatment groups varied with respect to the initial frequency of a dominant malathion resistant allele, denoted using the symbol R. The recessive wild type malathion susceptible allele is denoted using the symbol S. We will let p represent the frequency of the R allele among adults. The genotypes RR and RS can tolerate concentrations of the pesticide which are toxic to the SS homozygotes. The three treatment groups differed with respect to the initial frequency of the R allele among adults. In the susceptible treatment group, all insects were SS genotypes ($p_0 = 0$). In the "evolving" treatment group, populations were initiated with 250 SS larvae, 5 SS pupae, 245 SS adults, and 5 RS adults, for an initial allele frequency of $p_0 = 0.01$. In the resistant treatment group, all insects were RR genotypes ($p_0 = 1$). Life stages were counted every two weeks and added back to fresh medium. The experiment was maintained for 80 weeks ($t = 0, 1, 2, \ldots, 40$). Adult mortality was set at 50% by removing half the number of adults counted in the previous census minus the number of dead adults found in the current census. One of the replicate populations in the evolving treatment group was lost to disease at week 56 ($t = 28$).

Maximum likelihood parameter estimates were obtained using a Poisson-binomial version of the LPA model [4] which is similar to the stochastic model described in Chapter 10. The estimation procedure was applied to data from the susceptible (SS) and resistant (RR) treatment groups. These appear in Table 18.1.[4] Since the R allele is dominant, these estimates can be used as parameter values for the RR, RS, and SS genotypes in the gLPA model (18.7)–(18.10).

Figure 18.3 shows the census results and model predictions for the three treatment groups. As predicted by the model, the susceptible populations quickly become extinct (Figure 18.3A). The resistant populations (Figure 18.3C) appeared healthy and showed low amplitude oscillations. While the deterministic gLPA model predicted damped oscillations, the oscillations in the experimental data decrease in amplitude, but persist, which is likely a result of stochasticity. Adult numbers for the populations in the evolving treatment group

[4]The value of $\mu_a = 0.5$ was fixed by the experimental protocol.

TABLE 18.1 Maximum likelihood parameter estimates for the gLPA model.

Parameter	RR estimate	SS estimate
b	9.650	0.6564
c_{el}	0.006730	0.005727
c_{ea}	0.009901	0.0
c_{pa}	0.01051	0.01330
μ_l	0.1115	0.6586
μ_a	0.5	0.5

($p_0 = 0.01$) initially drop towards zero, then around week 10 they rebound and approach a steady-state density that is similar to the resistant populations (Figure 18.3B). The latter is an example of "evolutionary rescue" [7, 16], in which natural selection allows a population to adapt to environmental conditions that would otherwise result in extinction. In all three cases, the gLPA model excels at predicting the observed population dynamics.

Allele frequencies were estimated at the end of the experiment ($t = 40$) in the three surviving populations of the evolving treatment group by sampling and isolating 45–50 large larva or pupa, mating them with *SS* genotypes, and testing the survival of their offspring in malathion media to determine the genotypes of the sampled insects. This yielded estimates of the frequency of the R allele of 0.857, 0.948, and 0.889, with a mean and standard error of $p_{40} = 0.898 \pm 0.027$ [8]. In comparison, at $t = 40$, the model predicts a frequency of the R allele of $p_{40} = 0.894$, which is very close to the mean experimental value.

18.3 EFFECTS OF GENETIC DISEQUILIBRIUM

Early in the development of population genetics theory, J.B.S. Haldane introduced the concept of the *cost of natural selection* [17]. The cost was defined in terms of the total number of "selective deaths" that must occur for an allele to become established in a population. Selective deaths were defined abstractly as fractions of the total population. Population dynamics were not considered explicitly in Haldane's analyses. There was no obvious way to test Haldane's theory. In this section, we will present research that provides testable hypotheses regarding the "cost" of establishing rare alleles into a population and present experimental evidence from *Tribolium* experiments that support the theoretical predictions.

18.3.1 Fitness Entropy and Life-Not-Lived

Ginzburg introduced a function, that he called *fitness entropy* [15], to represent the deviation of a vector of n arbitrary allele frequencies $\mathbf{p}$ from their final equilibrium values $\mathbf{p}^*$:

$$H(\mathbf{p}, \mathbf{p}^*) = -\sum_{i=1}^{n} p_i^* \ln \left(\frac{p_i}{p_i^*} \right). \qquad (18.11)$$

Here, we assume that the system has a polymorphic equilibrium where $0 > p_i^* > 1$ for $i = 1, 2, \ldots, n$. The function (18.11) has the following properties: (a) $H(\mathbf{p}^*, \mathbf{p}^*) = 0$, (b) $H(\mathbf{p}, \mathbf{p}^*) > 0$ if $\mathbf{p} \neq \mathbf{p}^*$, and $H(\mathbf{p}, \mathbf{p}^*) \to \infty$ as any $p_i \to 0$. Plots of the fitness entropy appear in Figure 18.4 for two alleles ($n = 2$). The function (18.11) is a measure of genetic disequilibrium.

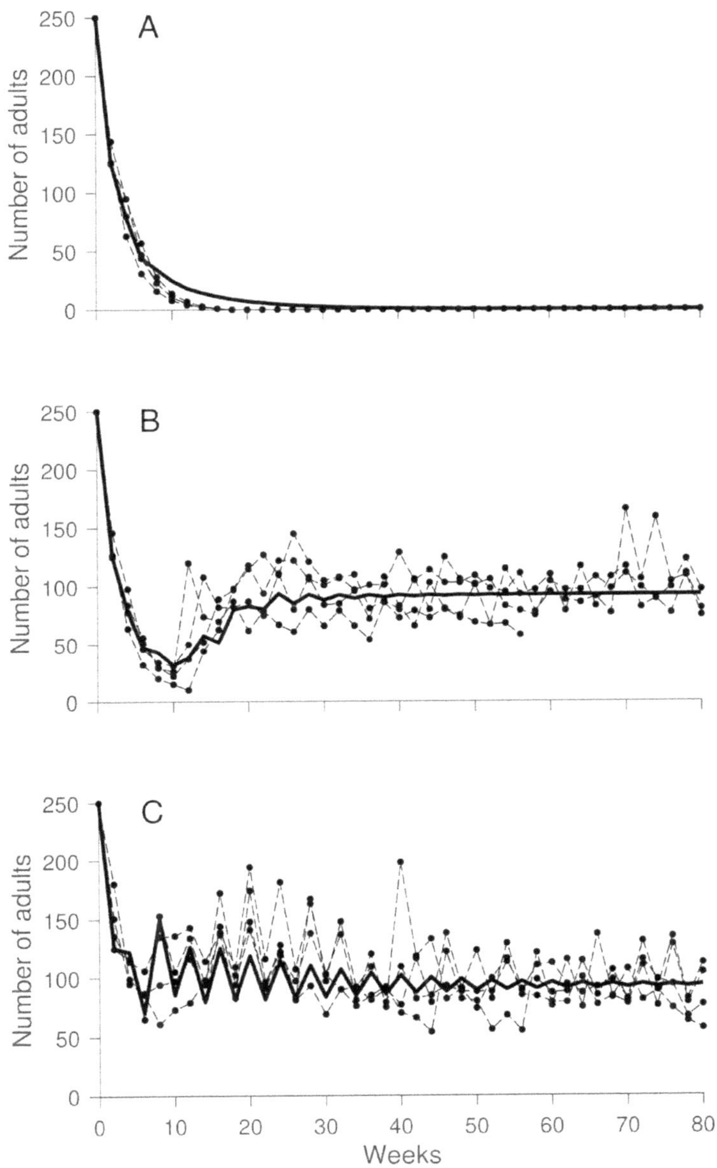

Figure 18.3 Time series plots of adult numbers for the three treatment groups of the malathion experiment. The solid circles and dashed lines are for the four replicate populations. The heavy lines show the predictions from the gLPA model (18.7)–(18.10) using the maximum likelihood parameter estimates from Table 18.1. The three panels show the results for the treatments. A. Susceptible. B. Evolving. C. Resistant. See text for more details.

In essence, the cost of natural selection represents the cost to the population of having to make the journey from $\mathbf{p}(0) \rightarrow \mathbf{p}^*$. This can be conceptualized by comparing the dynamics of an *evolving population* $N(t)$ with initial allele frequencies $\mathbf{p}(0) \neq \mathbf{p}^*$, to a *reference population* $\tilde{N}(t)$ that is identical to the evolving population, except that $\mathbf{p}(0) = \mathbf{p}^*$. Using the model (18.3), when the fitness values are positive constants, $m_{ij} > 0$, population numbers will increase without bound. Since the mean population fitness, $m(t)$, is maximized by

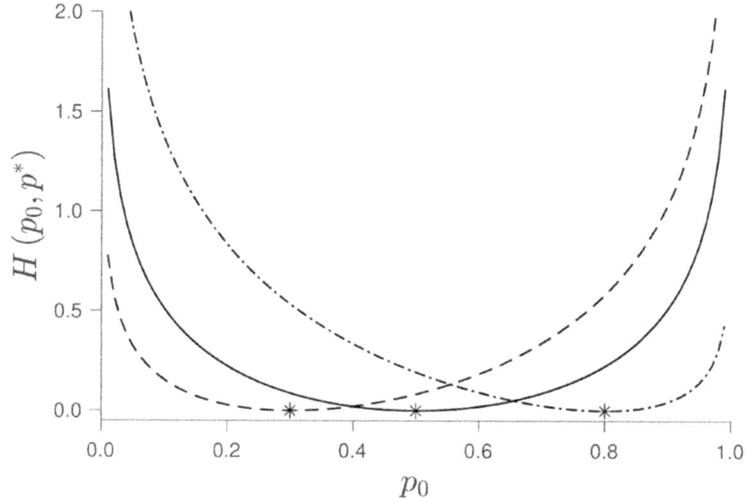

Figure 18.4 Plots of the fitness entropy function (18.11) for two alleles ($n = 2$) and $p = p_1$. Examples for three equilibrium values are shown: $p^* = 0.3$ (dashed line), $p^* = 0.5$ (solid line), and $p^* = 0.8$ (dot-dashed line).

selection, the reference population has a maximal growth rate $\tilde{N}(t) = N_0 e^{m^* t}$. It can be shown that the evolving population will, eventually, as $t \to \infty$, grow at the same maximal rate, but with a time lag relative to the reference population: $N(t)/\tilde{N}(t-\tau) \to 1$. Ginzburg referred to the time lag τ as the "selective delay." If the genetic equilibrium is polymorphic ($p_i^* > 0$, $i = 1, 2, \ldots, n$), then

$$H_0 = m^* \tau, \tag{18.12}$$

where $H_0 = H(\mathbf{p}(0), \mathbf{p}^*)$. In this case, the "cost" to the evolving population is the time lost while reaching genetic equilibrium, and it is directly proportional to the fitness entropy measure of the initial genetic disequilibrium [15]. The derivation of this result can be found in Appendix 18.5. A graphical depiction of the selective delay appears in Figure 18.5A.

The approach used by Ginzburg was extended to the case of density-dependent selection [9]. Let $m_{ij}(N) = r_{ij} - aN$, where r_{ij} is the intrinsic rate of increase for genotype $\mathcal{A}_i\mathcal{A}_j$, and a is a density-sensitivity parameter. The growth rate of the evolving population is $dN(t)/dt = N(t)(r(t) - aN(t))$, where $r(t)$ is the population mean for the r_{ij}'s, which depend on the changing allele frequencies. For the reference population, $d\tilde{N}(t)/dt = \tilde{N}(t)\left(r^* - a\tilde{N}(t)\right)$, where r^* is the maximum mean intrinsic rate of increase that is achieved at genetic equilibrium. It can be shown that

$$H_0 = a\mathfrak{L}_N, \tag{18.13}$$

where

$$\mathfrak{L}_N = \int_0^\infty \left(\tilde{N}(t) - N(t)\right) dt$$

is the "life-not-lived" due to genetic disequilibrium. Since allele frequencies and population numbers are directly measurable entities, this hypothesis of the cost of natural selection can be challenged by experimentation. The result (18.13) has been extended to models of coevolution in interacting species [11] and age-structured populations [12].

An interesting property of the relationship between fitness entropy and the life-not-lived is that it is dependent only on the initial and equilibrium allele frequencies. It is independent of the initial population density and the strength of selection. This is illustrated in

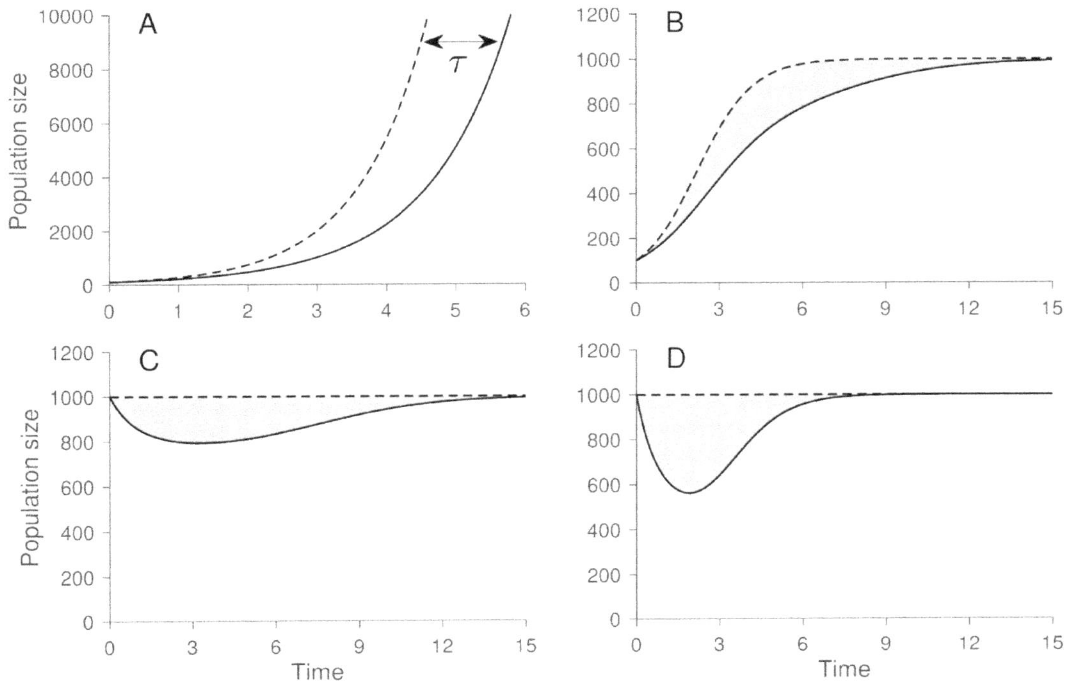

Figure 18.5 Demonstration of the concepts of selective delay τ and life-not-lived $\mathfrak{L}_N$ for a single-locus continuous-time model of natural selection with two alleles (18.3). Each panel compares the trajectories of an evolving population with $p_0 = 0.01$ (solid curves) with a reference population with $p_0 = p^* = 0.5$ (dashed curves). In panel A, the fitness values are constants, $m_{11} = m_{22} = 1 - s$ and $m_{12} = 1 + s$, with $s = 0.25$, and population growth is exponential. The double arrow shows the selective delay τ for the evolving population as it lags behind the reference population. The remaining panels are for density-dependent selection, $m_{ij}(N) = r_{ij} - aN$, $r_{11} = r_{22} = 1 - s$, $r_{12} = 1 + s$, and $a = 0.001$. The density of the evolving population remains below the reference population as they converge to the equilibrium $N^* = 1000$. The total shaded area bounded by the two trajectories is the life-not-lived, $\mathfrak{L}_N$. In panel B, $N_0 = 100$ and the selection intensity is moderate, $s = 0.25$. In panel C, $N_0 = N^*$ and the selection intensity is moderate, $s = 0.25$. In panel D, $N_0 = N^*$ and the selection intensity is strong, $s = 0.75$.

Figure 18.5. The intensity of selection is the same in Figures 18.5B and 18.5C, but the initial population density differs. The initial population density is the same in Figures 18.5C and 18.5D, but the strength of selection differs. The value of $\mathfrak{L}_N$ (shaded areas) is the same in all three cases.

18.3.2 Robustness of the Theory

The predicted relationship (18.13) between fitness entropy and life-not-lived was obtained using a continuous-time approximation of the discrete-generation model (18.2) for weak selection and logistic density-dependence. This was done for mathematical convenience as calculus can be used to derive the results. This raises the question of whether or not the prediction in (18.13) is robust to more complex models and non-equilibrial population dynamics. In this section, we use simulations of the Ricker (18.6) and gLPA (18.7) models to test the robustness of prediction (18.13).

Simulations were run for the discrete-time Ricker model (18.6) with two alleles. The polymorphic equilibrium value was $p^* = 0.5$. Initial allele frequencies p_0 for the evolving population N_t were chosen to give fitness entropy values of $H_0 = 0.5, 1.0, \ldots, 4.0$.[5] For the reference population, $\tilde{N}_t$, $\tilde{p}_0 = p^* = 0.5$. The initial population density was $N_0 = 10$ and $c = 0.001$ for both populations. The initial genotype numbers were in Hardy-Weinberg equilibrium. The genotype-specific reproductive rates were $b_{11} = b_{22} = b(1-s)$ and $b_{12} = b(1+s)$. The parameter b was used to vary the type of population dynamics and parameter s was used to vary the intensity of selection. To accommodate the possibility of cycles as attractors, we computed the live-not-lived using

$$\mathfrak{L}_N = \sum_{t=0}^{\infty} \sum_{k=0}^{\theta-1} \left(\tilde{N}_{t\theta+k} - N_{t\theta+k} \right), \tag{18.14}$$

where θ is the integer period of the discrete cycle. This allows for the possibility that the evolving and reference populations may, asymptotically, be out of phase as they converge to the cyclic attractor. In cases like these, the sequence of inner sums will converge to zero and $\mathfrak{L}_N$ will converge to a constant value.[6] In practice, one will compute $\mathfrak{L}_N$ using a finite number of terms for the outer sum in equation (18.14). To check that one has included enough terms, one can examine how quickly the values of the inner sums converge to zero. When the attractor is an equilibrium, $\theta = 1$. From the fitness definition $W_{ij}(N) = b_{ij}e^{-cN}$, it is easy to see, using the Ricker model (18.6), that the allele frequency dynamics are independent of the population dynamics, so $p_t \to p^*$.

Figure 18.6 shows the plots of the life-not-lived $\mathfrak{L}_N$ versus the fitness entropy H_0 for the Ricker model simulations. Four types of population dynamics were investigated: an undercompensatory equilibrium ($b = 1.5$), an overcompensatory equilibrium ($b = 5$), a two-cycle ($b = 10$), and a four-cycle ($b = 14$). For each dynamic, three selection intensities were used: weak selection ($s = 0.1$, circles), moderate selection ($s = 0.2$, triangles), and strong selection ($s = 0.3$, squares). For each value of s, the dashed lines are linear least-squares fits with a zero intercept. As predicted by theory, a strong linear relationship exists between $\mathfrak{L}_N$ and H_0 for all four types of population dynamics with only slight changes in the slopes as the selection intensity increases (Figure 18.6). For weak selection intensities, the slope is nearly equal to c^{-1}, which is the equivalent of the predicted slope a for the logistic model (18.13). The continuous time approximation uses $m_{ij} = \ln W_{ij}$ (Appendix 18.5.1), so r_{ij} is the equivalent of $\log b_{ij}$ and a is the equivalent of c^{-1}. It appears that the theoretical prediction (18.13) can be applied to discrete-time systems and is insensitive to the type of population dynamics and the intensity of selection.

Simulations of the two-allele version of the gLPA model (18.7) were also conducted to explore the robustness of the life-not-lived prediction (18.13). We used parameter values from Table 7.2, and, as for the Ricker model, genotype-specific reproductive rates $b_{11} = b_{22} = b(1-s)$ and $b_{12} = b(1+s)$, where s is the intensity of selection. This results in an equilibrium allele frequency of $p^* = 0.5$ for all values of s which does not depend on the type of population dynamics [4]. The same initial allele frequencies were also used resulting in the same values of H_0. Each simulation was initiated with a density of 20 adults

[5] This corresponds to $p_0 = 0.1025, 0.03506, 0.01261, 0.004600, 0.001687, 0.0006201, 0.0002280$, and 0.00008373.

[6] In general, it is not necessarily the case that the infinite sum of a sequence tending to zero is finite. For example, the limit of the harmonic sequence $\{1/t\}_{t=1}^{\infty}$ is zero, but the sum of the harmonic sequence $\Sigma_{t=1}^{\infty} 1/t = \infty$, and this is the case for any sequence that declines to zero more slowly than $1/t$. We know from linearization theory, however, that in the neighborhood of a stable equilibrium or cycle, the solutions of a dynamical system approach the attractor approximately exponentially. Hence, the sequence of inner sums in equation (18.14) will approach zero exponentially, rendering the outer sum finite.

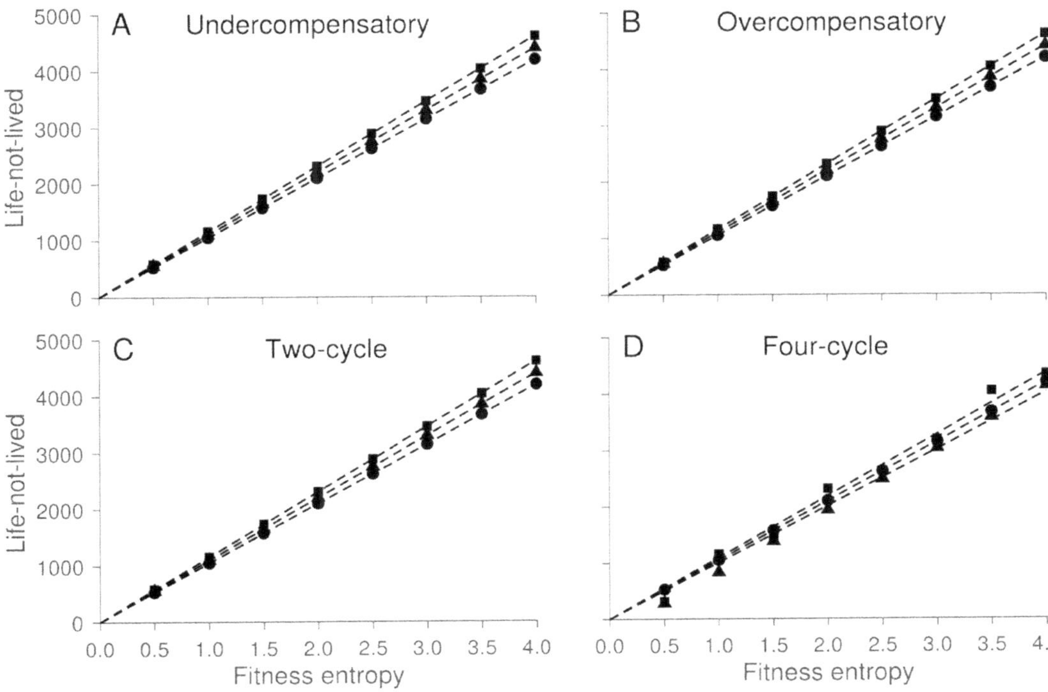

Figure 18.6 Life-not-lived $\mathfrak{L}_N$ versus fitness entropy H_0 for simulations of the discrete-time Ricker model (18.6) with two alleles. Each panel shows the results for three selection intensities: weak (circles), moderate (triangles), and strong (squares). Dashed lines were fit using a least-squares linear model with no intercept. Simulations were conducted for four different types of asymptotic population dynamics (panels A–D). See text for details.

whose genotypes were in Hardy-Weinberg equilibrium. The type of asymptotic population dynamics was chosen by varying the parameters c_{pa} and μ_a. Life-not-lived was computed using equation (18.14) with A_t and $\tilde{A}_t$ used in place of N_t and $\tilde{N}_t$ for the evolving and reference populations.

Figure 18.7 shows the plots of the life-not-lived for adults $\mathfrak{L}_A$ versus the fitness entropy H_0 for the gLPA model simulations. Four types of population dynamics were investigated: a stable equilibrium ($c_{pa} = 0.004619$, $\mu_a = 0.007629$), a two-cycle ($c_{pa} = 0.001$, $\mu_a = 0.96$), a three-cycle ($c_{pa} = 0.5$, $\mu_a = 0.96$), and a five-cycle ($c_{pa} = 0.02$, $\mu_a = 0.96$). For each dynamic, three selection intensities were used: weak selection ($s = 0.1$, circles), moderate selection ($s = 0.2$, triangles), and strong selection ($s = 0.3$, squares). For each value of s, the dashed lines are linear least-squares fits with a zero intercept. As with the Ricker model, a strong linear relationship exists between $\mathfrak{L}_A$ and H_0 for all four types of population dynamics with only slight to moderate changes in the slopes as the selection intensity increases (Figure 18.7). These results, together with the simulation results from the genetic Ricker model (Figure 18.6) suggest that the linear relationship between life-not-lived and fitness entropy predicted by equation (18.13) is robust.

18.3.3 Empirical Support

An experimental *Tribolium* study using the corn oil sensitive (*cos*) mutation [20], conducted prior to the publication of the fitness entropy prediction (18.13), produced data that can

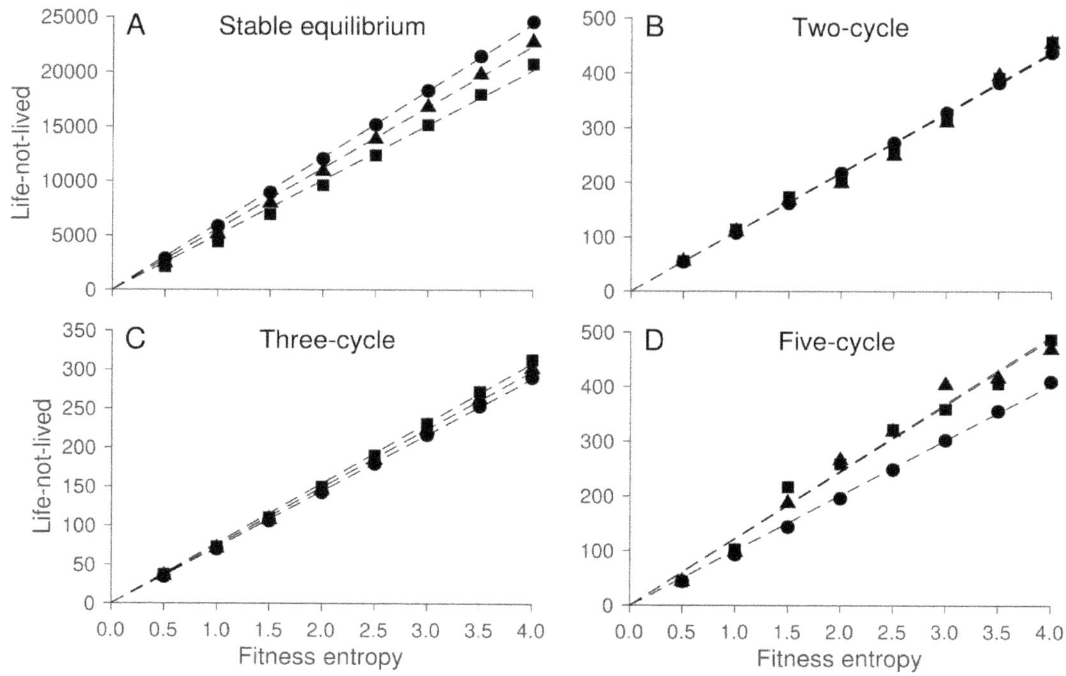

Figure 18.7 Life-not-lived for adults $\mathfrak{L}_A$ versus fitness entropy H_0 for simulations of the gLPA model (18.7) with two alleles. Each panel shows the results for three selection intensities: weak (circles), moderate (triangles), and strong (squares). Dashed lines were fit using a least-squares linear model with no intercept. Simulations were conducted for four different types of asymptotic population dynamics (panels A–D). See text for details.

be used to test the theory. When cultured in flour medium containing 5% liquid corn oil, *cos*/*cos* homozygotes have lower larval viability and reduced egg production [20]. The wild type +/+ and heterozygous +/*cos* genotypes do better, with the +/*cos* genotypes having a small life history advantage over the +/+ homozygotes [20].

A 68-week experiment was initiated where the treatment effect was the initial frequency of the *cos* and + alleles. Denoting p as the frequency of the *cos* allele, the initial allele frequencies were $p_0 = 0.0, 0.1, 0.2, \ldots, 1.0$ among the 20 founding adults in each population. Cultures were maintained on 20 g of flour medium (90% flour, 5% yeast, and 5% liquid corn oil) in darkened incubators at 33°C and 42% relative humidity. Life stages were counted every two weeks and returned to fresh medium.

In a separate experiment, populations were initiated with *cos* allele frequencies of $p_0 =$ 0.1, 0.3, 0.5. 0.7 and 0.9. There were 12 replicates for each treatment. At weeks 2, 6, 12 and 16, three of the 12 replicates were sacrificed and allele frequencies were estimated. At week 16, the populations had approached a polymorphic equilibrium with a mean *cos* allele frequency of 0.252 and with a median frequency of 0.28.

In order to evaluate the theoretical prediction (18.13), we used the population with $p_0 = 0.3$ as the "reference population," since its initial allele frequency was closest to the estimated polymorphic equilibrium. The cultures with $p_0 = 0.1, 0.2, 0.4, 0.5, \ldots, 0.9$ were used as "evolving populations." Two versions of the life-not-lived were computed, one for adults, $\mathfrak{L}_A$, and one for the total of the life stages, $\mathfrak{L}_N$, by summing over time the differences

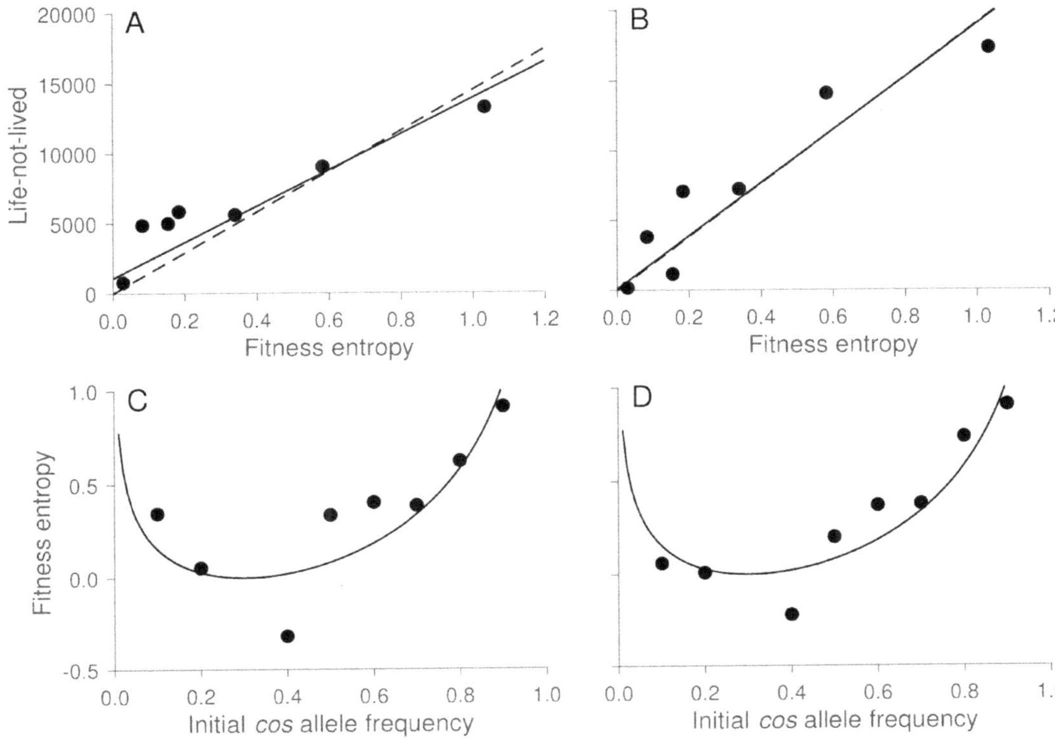

Figure 18.8 Test of the prediction of a linear relationship between life-not-lived and fitness entropy using *Tribolium* experimental data [20]. See text for details.

between the mean beetle numbers for the reference populations minus the mean for the evolving populations from each treatment. The fitness entropy, H_0, was computed using $p^* = 0.3$ and the value of p_0 for each of the evolving population treatments.

The values of $\mathfrak{L}_A$ and $\mathfrak{L}_N$ versus H_0 are plotted in Figures 18.8A,B (solid circles). Linear least-squares regressions were fit to each set of data (solid lines). In both cases, the intercept is not significantly different from zero (p-values[7] of 0.485 and 0.948, for adults and totals, respectively), so a second regression line was fit that goes through origin (dashed lines). The linear relationships are strongly significant (p-values of 0.0075 and 0.0013 for adults and totals, respectively), suggesting that the results support the predictions. Figures 18.8C,D show an alternative view of the data, with the function $H_0 = H(p_0, p^*)$ (solid curve) and the experimental values of $\mathfrak{L}_A/a_1$ and $\mathfrak{L}_N/a_2$ (solid circles) plotted as a function of p_0, where a_1 and a_2 are the no-intercept regression slopes for the adult and total numbers, respectively. Overall, there is a good agreement between theory and data.

18.3.4 Cost of a Gene Substitution

The definition of fitness entropy (18.11) is based on the assumption that all n alleles are present in a polymorphic equilibrium. However, the entropy function can be redefined so that the summation in equation (18.11) only includes those alleles for which $p_i^* > 0$ [9]. While, in general, the prediction (18.13) no longer holds when alleles are lost during selection, there is a biologically important exception: when a single dominant allele, $\mathcal{A}_1$, is selected

[7] In this chapter, we use p for allele frequency. To avoid confusion, when speaking of statistical significance, we write out the word "p-value."

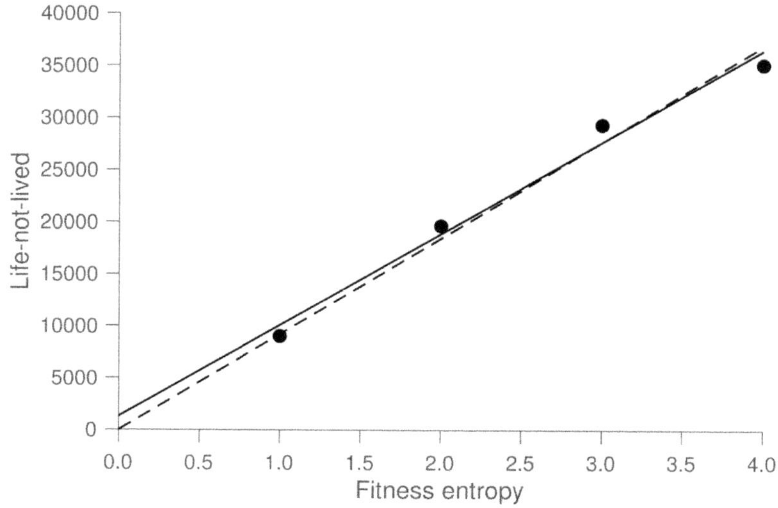

Figure 18.9 Cost of a gene substitution for a malathion-resistant allele in flour beetles. The prediction of a linear relationship between life-not-lived and fitness entropy is supported using *Tribolium* experimental data [5]. See text for details.

to fixation. In this case, $H_0 = -\ln p_1(0)$, which is exactly the value obtained by Haldane for the cost of a gene substitution [17]. In this case, because $r_{1j} = r_{11}$ for $j = 2, 3, \ldots, n$, it is easy to see from equation (18.18) in Appendix 18.5.4 that $dH/dt = r(t) - r_{11}$, and that $H_0 = a\mathfrak{L}_N$. Haldane's concept of "cost" can be restated in terms of the experimentally accessible computation of "life-not-lived."

A study was designed to test the prediction for the cost of a gene substitution [5]. The experiment used the same protocols for the malathion resistant and susceptible alleles R and S described in Section 18.2.2. Populations were initiated with 50 adults, 14 pupae, and 16 larvae. There were five initial frequencies of the R allele among adults: $p_0 = 0.02$, 0.05, 0.14, 0.37, and 1.00, with genotype frequencies in Hardy-Weinberg equilibrium. These correspond to fitness entropy values of $H_0 = 4$, 3, 2, 1, and 0. Each treatment had three replicates. The cultures were maintained in flour medium containing 1.5 ppm malathion and kept in a dark incubator at 33°C and 60% relative humidity. Adults were counted every two weeks and all life stages were returned to fresh medium. We used the treatment with $p_0 = 1$ as the reference population and the other treatments as evolving populations. The life-not-lived values for adults, $\mathfrak{L}_A$, were computed by subtracting the means for the evolving populations from the means for the reference populations and summing the differences over time. The study ran for 32 weeks. By that time the R allele had reached a high frequency in all the populations.[8]

Figure 18.9 plots the values of life-not-lived for adults $\mathfrak{L}_A$ against the values for fitness entropy H_0 (solid circles). A least-squares linear regression was fit to the data (solid line). The slope of the linear relationship is significantly different from zero (p-value 0.0082) and the intercept is not significantly different from zero (p-value 0.608). A second regression line was fit that goes through origin (dashed line). The results support the use of life-not-lived as proxy for Haldane's cost of a gene substitution.

---

[8]Since the resistant allele R is dominant, the susceptible allele S is protected from selection in RS heterozygotes. In this situation it takes a long period of time before the allele S is completely eliminated.

18.4 CONCLUSIONS

The union of population genetics and population ecology can benefit both disciplines. For population genetics, concepts such as the maximization of mean fitness and the cost of natural selection can be recast in more concrete terms such as the maximization of population numbers and life-not-lived due to genetic disequilibrium. For population ecology, natural selection can be a force for moving populations across bifurcation points, as is the case with evolutionary rescue. This idea could be expanded to include transitions into other dynamic regimes such as cycles and chaos. With the addition of the gLPA model and the availability of mutants, the *Tribolium* experimental system is an apt tool for investigating these and other phenomena in the area of ecological genetics.

18.5 APPENDIX: FITNESS ENTROPY, SELECTIVE DELAY, AND LIFE-NOT-LIVED

Let fitness entropy and allele frequencies be continuous functions of time:

$$H(t) = -\sum_{i=1}^{n} p_i^* \ln\left(\frac{p_i(t)}{p_i^*}\right).$$

Using the selection model (18.3), the time derivative of the entropy function is

$$\begin{aligned}\frac{dH(t)}{dt} &= -\sum_{i=1}^{n}\left(\frac{p_i^*}{p_i(t)}\right)\left(\frac{dp_i(t)}{dt}\right)\\ &= -\sum_{i=1}^{n} p_i^*\left(m_i(t) - m(t)\right)\\ &= m(t) - \sum_{i=1}^{n}\sum_{j=1}^{n} p_i^* p_j(t)\, m_{ij}.\end{aligned}$$

Since $m_{ij} = m_{ji}$ and, at equilibrium, $m_i^* = m^*$ for $i = 1, \ldots, n$, we have

$$\begin{aligned}\frac{dH(t)}{dt} &= m(t) - \sum_{j=1}^{n} p_j(t) \sum_{i=1}^{n} p_i^* m_{ji}\\ &= m(t) - \sum_{j=1}^{n} p_j(t)\, m_j^*\\ &= m(t) - m^*.\end{aligned} \tag{18.15}$$

Consider the dynamics of a *reference population* $\tilde{N}(t)$ that is identical to the *evolving population* $N(t)$, except that its initial allele frequencies are at the equilibrium values, $\tilde{\mathbf{p}}(0) = \mathbf{p}^*$. Its exponential growth rate will remain at the maximum value, that is, $\tilde{N}(t) = N_0 e^{-m^* t}$. From the second equation of (18.3), we have $m = N^{-1}dN/dt = d\ln(N)/dt$. Substituting into equation (18.15) and integrating gives

$$\begin{aligned}H(t) - H(0) &= \ln N(t) - \ln N_0 - m^* t\\ &= \ln N(t) - \ln N_0 - \ln \tilde{N}(t) + \ln N_0\\ &= \ln\left(\frac{N(t)}{\tilde{N}(t)}\right).\end{aligned} \tag{18.16}$$

As $t \to \infty$, $\mathbf{p}(t) \to \mathbf{p}^*$, which implies that $H(t) \to 0$. This also implies that $\tilde{N}(t)/N(t)$ approaches a constant. From the fundamental theorem, we know that $m(t) \leq m^*$, with the equality holding only at equilibrium. This means that the evolving population lags behind the reference population, but, asymptotically, they grow at the same exponential rate m^*. In other words,

$$\lim_{t\to\infty}\left(\frac{\tilde{N}(t-\tau)}{N(t)}\right) = \lim_{t\to\infty}\left(\frac{\tilde{N}(t)\,e^{-m^*\tau}}{N(t)}\right) = 1. \tag{18.17}$$

Substituting into equation (18.16) gives the result (18.12) as $t \to \infty$.

Now consider a logistic version of the density-dependent model of selection where $m_{ij}(N) = r_{ij} - aN$. The equations for allele frequencies in (18.3) will remain the same with $r_i(t) = \sum_{j=1}^{n} p_j(t)\, r_{ij}$ and $r(t) = \sum_{i=1}^{n}\sum_{j=1}^{n} p_j(t)\, r_{ij}$ replacing $m_i(t)$ and $m(t)$. The mean intrinsic rate of increase evolves to a maximum value r^*. For evolving and reference populations,

$$\begin{aligned}\frac{d\ln N(t)}{dt} &= r(t) - aN(t)\\ \frac{d\ln \tilde{N}(t)}{dt} &= r^* - a\tilde{N}(t).\end{aligned}$$

Since the r_{ij} values play the role of genotypic fitnesses, from equation (18.15) we have

$$\begin{aligned}\frac{dH(t)}{dt} &= r(t) - r^*\\ &= \frac{d\ln N(t)}{dt} + aN(t) - \frac{d\ln \tilde{N}(t)}{dt} - a\tilde{N}(t)\\ &= \frac{d\ln\left(N(t)/\tilde{N}(t)\right)}{dt} + a\left(N(t) - \tilde{N}(t)\right).\end{aligned} \tag{18.18}$$

Integrating equation (18.18) with respect to time gives

$$H(t) - H(0) = \ln\left(\frac{N(t)}{\tilde{N}(t)}\right) - \ln\left(\frac{N(0)}{\tilde{N}(0)}\right) + a\int_0^t \left(N(\xi) - \tilde{N}(\xi)\right) d\xi.$$

By definition, $N(0) = \tilde{N}(0) = N_0$. In the limit as $t \to \infty$, $H(t) \to 0$, $N(t) \to N^*$, and $\tilde{N}(t) \to N^*$, where $N^* = r^*/a$. This gives the result (18.13).

Bibliography

[1] Anderson WW (1971) Genetic equilibrium and population growth under density-regulated selection. American Naturalist 105:489–498. https://doi.org/10.1086/282741

[2] Charlesworth B (1971) Selection in density-regulated populations. Ecology 52:469–474. https://doi.org/10.2307/1937629

[3] Charlesworth B (1994) Evolution in Age-Structured Populations. Second Edition. Cambridge, Cambridge University Press.

[4] Cheung W (2002) The Effects of Natural Selection on Nonlinear Population Dynamics. MS Thesis, California State University, Los Angeles.

[5] Costantino RF, Desharnais RA (1991) Population Dynamics and the *Tribolium* Model: Genetics and Demography. Monographs on Theoretical and Applied Genetics 13. New York, Springer-Verlag.

[6] Crow JF, Kimura M (1970) An Introduction to Population Genetics Theory. New York, Harper and Row.

[7] de Vries C, Caswell H (2019) Stage-structured evolutionary demography: Linking life histories, population genetics, and ecological dynamics. American Naturalist 193:545–559. https://doi.org/10.1086/701857

[8] de Vries C, Desharnais RA, Caswell H (2020) A matrix model for density-dependent selection in stage-classified populations, with application to pesticide resistance in *Tribolium*. Ecological Modelling 416:108875. https://doi.org/10.1016/j.ecolmodel.2019.108875

[9] Desharnais RA, Costantino RF (1982) Natural selection and fitness entropy in a density-regulated population. Genetics 101:317–329. https://doi.org/10.1093/genetics/101.2.317

[10] Desharnais RA, Costantino RF (1983) Natural selection and density-dependent population growth. Genetics 105:1029–1040. https://doi.org/10.1093/genetics/105.4.1029

[11] Desharnais RA (1986) Natural selection, fitness entropy, and the dynamics of coevolution. Theoretical Population Biology 30:309–340. https://doi.org/10.1016/0040-5809(86)90039-0

[12] Desharnais RA, Cohen JE (1986) Life not lived due to disequilibrium in heterogeneous age-structured populations. Theoretical Population Biology 29:385–406. https://doi.org/10.1016/0040-5809(86)90016-X

[13] Fisher RA (1930) The Genetical Theory of Natural Selection. Oxford, Clarendon Press.

[14] Ginzburg LR (1977) The equilibrium and stability for n alleles under the density-dependent selection. Journal of Theoretical Biology 68:545–550. https://doi.org/10.1016/0022-5193(77)90104-7

[15] Ginzburg LR (1977) A macro-equation of natural selection. Journal of Theoretical Biology 67:677–686. https://doi.org/10.1016/0022-5193(77)90254-5

[16] Gomulkiewicz R, Holt RD (1995) When does evolution by natural selection prevent extinction? Evolution 49:201–207. https://doi.org/10.2307/2410305

[17] Haldane JBS (1957) The cost of natural selection. Journal of Genetics 55:511–524. https://doi.org/10.1007/BF02984069

[18] Haldane JBS (1932) The Causes of Evolution. New York, Harper Brothers.

[19] MacArthur RH (1962) Some generalized theorems of natural selection. Proceedings of the National Academy of Sciences USA 48:1893–1897. https://doi.org/10.1073/pnas.48.11.1893

[20] Moffa AM, Costantino RF (1977) Genetic analysis of a population of *Tribolium*. VI. Polymorphism and demographic equilibrium. Genetics 87:785–805. https://doi.org/10.1093/genetics/87.4.785

[21] Nagylaki T (1977) Selection in One- and Two-Locus Systems. New York, Springer Verlag.

[22] Roughgarden, J (1971) Density-dependent natural selection. Ecology 52:453–468. https://doi.org/10.2307/1937628

[23] Roughgarden J (1976) Resource partitioning among competing species—a coevolutionary approach. Theoretical Population Biology 9:388–424. https://doi.org/10.1016/0040-5809(76)90054-X

[24] Wright S (1931) Evolution in Mendelian populations. Genetics 16:97–159. https://doi.org/10.1093/genetics/16.2.97

CHAPTER 19

Continuous-time LPA Model

Because of the long tradition of continuous-time modeling in population ecology, we decided to include a brief discussion of continuous-time approaches in this book. In fact, a number of continuous-time models for *Tribolium* populations exist in the literature [1, 5, 6, 7, 9], and there is no question that continuous-time models have certain advantages. One important advantage afforded by continuous models in the present context is the ability to study dynamics as a function of varying developmental rates, which is not possible in the LPA model with its fixed life-stage durations. For example, Hastings and Costantino used continuous-time *Tribolium* models to hypothesize that, for fixed larval duration, a decrease in egg stage duration promotes equilibrium stability; in particular, the birth rate parameter b must increase an order of magnitude in order to destabilize the equilibrium if the egg duration is decreased from 9 days to 4 days [6, 7].

Despite such theoretical advantages, major obstacles challenge the continuous-time approach. These include difficulties of mathematical analysis, simulation, stochastic formulation, dense data collection, parameter estimation, and experimental design to test predictions. To date, none of the continuous-time flour beetle models have been connected to data in the rigorous way that the discrete-time LPA model has been.

In this chapter we review some of our efforts toward continuous-time modeling of the *Tribolium* system [8, 9]. We were not as successful with the continuous-time approach as we were with the LPA model, partially because of the inherent difficulties in interfacing a continuous-time model with discrete-time data and discrete life stages, but also possibly because our continuous-time approach needed more development. This chapter gives a flavor of the continuous-time approach in order to stimulate interest and further work.

At the outset, it is important to note that *a confusion exists in the literature about how to construct continuous-time structured models for single populations when intraspecific interactions are stage-dependent*, as they are in the *Tribolium* system. Some authors have taken a compartmental approach with ordinary differential equations (ODEs), analogous to Lotka-Volterra-type multiple-species models, but in which each life stage is a compartment with inflow and outflow rates specified at instantaneous time t. *In general, this approach is not correct for a structured population*, because the outflow rate from a stage class at time t typically depends on the inflow rate at time $t - \tau$, where τ is the duration of the stage. For example, in a continuous-time compartmental juvenile-adult model, the transition from the juvenile class to the adult class typically should have a delay equal to the maturation period of the organism. In the literature, many structured compartmental models, posed as ODEs, omit this delay. For a comprehensive review of this problematic issue, see [14]. One of our goals in this chapter is to demonstrate a correct way to derive a continuous-time structured analog of the LPA model, which is by integrating a partial differential equation (PDE)

DOI: 10.1201/9781003467120-19

over the durations of each life stage. This yields a system of functional delay differential equations (FDEs) rather than ODEs [15].

In this chapter we pose two PDE models. The first PDE includes an egg stage of duration $\varepsilon > 0$. We denote this the "continuous-time ELPA model," or ctELPA model. The second PDE arises when the egg stage duration ε formally approaches zero. We call this the ctLPA model; it is of interest because it is analogous to the discrete-time LPA model, which does not include an explicit egg stage.

19.1 ctELPA MODEL

We begin with an age-structured McKendrick-von Foerster-type PDE [2, 4, 10]

$$\begin{aligned} \frac{\partial \rho_\varepsilon}{\partial t} + \frac{\partial \rho_\varepsilon}{\partial a} &= -\delta_\varepsilon\left(a, \rho_\varepsilon\right)\rho_\varepsilon \\ \rho_\varepsilon\left(t, 0\right) &= \beta A_\varepsilon(t) \\ \rho_\varepsilon\left(0, a\right) &= \varphi(a), \end{aligned} \tag{19.1}$$

where $\varepsilon > 0$ is the duration of the egg stage, and $\rho_\varepsilon\left(t, a\right)$ is the per unit age density of animals at time t, that is, $\rho_\varepsilon\left(t, a\right)\Delta a$ is the number of animals at time t within the age interval $[a, a + \Delta a]$. The function $\varphi(a) \geq 0$ is the initial age distribution, and $\delta_\varepsilon\left(a, \rho_\varepsilon\right)$ denotes the death rate. The birth rate $\rho_\varepsilon\left(t, 0\right)$ at time t is proportional to the number of sexually mature adults $A_\varepsilon(t)$, where $\beta > 0$ is the number of eggs produced per adult per unit time, and the number of sexually mature adults is given by

$$A_\varepsilon(t) = \int_{a_a}^{\infty} \rho_\varepsilon(t, a)da, \tag{19.2}$$

where is a_a is the age of maturation. Note that we assume β does not depend on the age of the adult.

Similarly, the numbers of eggs $E_\varepsilon(t)$, feeding larvae $L_\varepsilon(t)$, as well as non-feeding larvae, pupae, and callow adults $P_\varepsilon(t)$ are given by

$$\begin{aligned} E_\varepsilon\left(t\right) &= \int_0^{\varepsilon} \rho_\varepsilon(t, a)da \\ L_\varepsilon(t) &= \int_{\varepsilon}^{a_p} \rho_\varepsilon(t, a)da \\ P_\varepsilon(t) &= \int_{a_p}^{a_a} \rho_\varepsilon(t, a)da, \end{aligned} \tag{19.3}$$

where $a_p > 0$ is the age at which larvae cease to feed in preparation for pupation.

We take the death rate[1] to be

$$\delta_\varepsilon\left(a, \rho_\varepsilon(t, \cdot)\right) = \begin{cases} D_{el}L_\varepsilon(t) + D_{ea}A_\varepsilon(t) & a \leq \varepsilon \\ \nu_l & \varepsilon < a \leq a_p \\ C_{pa}A_\varepsilon(t) & a_p < a \leq a_a \\ \nu_a & a_a < a, \end{cases} \tag{19.4}$$

where the parameters $D_{el}, D_{ea}, C_{pa} \geq 0$ are per capita cannibalism rates (units per individual per unit time) for larval-on-egg, adult-on-egg, and adult-on-pupa cannibalism, respectively, and $\nu_l, \nu_a > 0$ are non-density-dependent death rates. We refer to model (19.1)–(19.4)

[1] In equation (19.4), the "dot" in $\rho_\varepsilon(t, \cdot)$ is a placeholder that indicates the age input without specifying a particular age value. In this case, δ_ε depends on ρ_ε via $L_\varepsilon(t)$ and $A_\varepsilon(t)$, which are defined in equations (19.2)–(19.3) and depend on the ages in the intervals $[\varepsilon, a_p]$ and $[a_a, \infty)$. We could, in fact, write $\delta_\varepsilon(a, L_\varepsilon(t), A_\varepsilon(t))$.

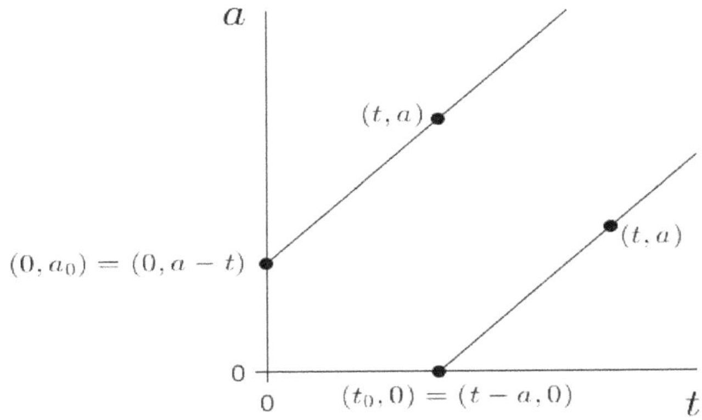

Figure 19.1 Cohort curves in the t-a plane. Cohorts that did not yet exist at $t = 0$ (lower curve) are solved for $t \geq a$. Cohorts that already existed at $t = 0$ (upper curve) are solved for $t < a$.

as the "continuous-time ELPA model," or ctELPA model [9]. The ctELPA model is a first-order nonlinear hyperbolic PDE with a nonstandard boundary condition. For basic theory of these kinds of equations, see [2, 4, 10, 11, 12, 17].

We can think of the population as a collection of cohorts of two types: cohorts that did not yet exist (i.e., whose members were not yet born) when the clock began ticking at $t = 0$ (Figure 19.1, lower curve), and cohorts that already existed (i.e., whose members were already born) when the clock began ticking at $t = 0$ (Fig 19.1, upper curve). Along cohorts of the first type, we have $t \geq a$. Along those of the second type, we have $t < a$. We can parameterize these cohort curves with a "cohort age" s, as follows.

For a cohort of the first type, if the individuals in that cohort are currently of age a at time $t \geq a$ and the cohort began at time $t_0 \geq 0$, then the "cohort age" is $s = a$, and, along the cohort, we have $t = a + t_0 = s + t_0$. Thus, along the cohort, the Chain Rule gives

$$\begin{aligned}\frac{d\rho_\varepsilon}{ds} &= \frac{\partial \rho_\varepsilon}{\partial t}\frac{dt}{ds} + \frac{\partial \rho_\varepsilon}{\partial a}\frac{da}{ds}\\ &= \frac{\partial \rho_\varepsilon}{\partial t}(1) + \frac{\partial \rho_\varepsilon}{\partial a}(1)\\ &= -\delta_\varepsilon\left(s, \rho_\varepsilon(s+t_0,\cdot)\right)\rho_\varepsilon(s+t_0,s).\end{aligned}$$

Integrating along the cohort from $s = 0$ to $s = a$, we have

$$\int_0^a \frac{1}{\rho_\varepsilon(s+t_0,s)}\frac{d\rho_\varepsilon}{ds}ds = -\int_0^a \delta_\varepsilon\left(s, \rho_\varepsilon(s+t_0,\cdot)\right)ds,$$

which yields, by the Fundamental Theorem of Calculus and the positivity of ρ_ε,

$$\ln \rho_\varepsilon(t,a) - \ln \rho_\varepsilon(t_0,0) = -\int_0^a \delta_\varepsilon\left(s, \rho_\varepsilon(s+t_0,\cdot)\right)ds.$$

Given that $t_0 = t - a$, application of the boundary condition $\rho_\varepsilon(t,0) = \beta A_\varepsilon(t)$ yields, after some algebraic manipulation,

$$\rho_\varepsilon(t,a) = \beta A_\varepsilon(t-a)\exp\left[-\int_0^a \delta_\varepsilon\left(s, \rho_\varepsilon(t-a+s,\cdot)\right)ds\right], \quad \text{for } t \geq a > 0.$$

For a cohort of the second type, if the individuals in that cohort are currently of age $a > t$ at time t and the cohort began at time $t = 0$ and age $a_0 > 0$, then the "cohort age" is $s = t$, and, along the cohort, $a = t + a_0 = s + a_0$. In this case, the Chain Rule yields

$$\frac{d\rho_\varepsilon}{ds} = -\delta_\varepsilon \left(s + a_0, \rho_\varepsilon(s, \cdot)\right) \rho_\varepsilon(s, s + a_0),$$

and integration from $s = 0$ to $s = t$ obtains

$$\ln \rho_\varepsilon \left(t, a\right) - \ln \rho_\varepsilon \left(0, a_0\right) = -\int_0^t \delta_\varepsilon \left(s + a_0, \rho_\varepsilon(s, \cdot)\right) ds.$$

Hence, given that $a_0 = a - t$, application of the initial condition $\rho_\varepsilon \left(0, a\right) = \varphi(a)$ yields, after some algebra,

$$\rho_\varepsilon \left(t, a\right) = \varphi(a - t) \exp \left[-\int_0^t \delta_\varepsilon \left(s + a_0, \rho_\varepsilon(s, \cdot)\right) ds\right].$$

A change of variables $\tau = s + a_0$ produces

$$\rho_\varepsilon \left(t, a\right) = \varphi(a - t) \exp \left[-\int_{a-t}^a \delta_\varepsilon \left(\tau, \rho_\varepsilon(\tau - a + t, \cdot)\right) d\tau\right], \text{ for } t < a. \tag{19.5}$$

Thus, after renaming the variable of integration in equation (19.5), the ctELPA model can be expressed in integral form as

$$\rho_\varepsilon \left(t, a\right) = \left\{ \begin{array}{ll} \beta A_\varepsilon(t - a) \exp \left[-\int_0^a \delta_\varepsilon \left(s, \rho_\varepsilon(t - a + s, \cdot)\right) ds\right] & \text{if } t \geq a > 0 \\ \varphi(a - t) \exp \left[-\int_{a-t}^a \delta_\varepsilon \left(s, \rho_\varepsilon(t - a + s, \cdot)\right) ds\right] & \text{if } t < a \end{array} \right. . \tag{19.6}$$

Intuitively, for each cohort with $t \geq a$, the right-hand side of equation (19.6) is the density of individuals born at time $t - a$, multiplied by an exponential that represents the probability of surviving until age a. For $t < a$, the right-hand side of equation (19.6) represents the initial density of the population of age $a - t$, multiplied by the probability of surviving from age $a - t$ until age a.

19.2 ctLPA MODEL

The discrete-time LPA model does not explicitly include an egg stage. Our goal here is to formally shrink the duration of the egg stage ε to zero in model (19.1)–(19.4). In the integral form (19.6) of that model, if $t \geq a > \varepsilon$, then

$$\rho_\varepsilon \left(t, a\right) = \beta A_\varepsilon(t - a) S_\varepsilon(t - a) \exp \left[-\int_\varepsilon^a \delta_\varepsilon \left(s, \rho_\varepsilon(t - a + s, \cdot)\right) ds\right],$$

where

$$S_\varepsilon(t - a) = \exp \left[-\int_0^\varepsilon \left(D_{ei} L_\varepsilon(t - a + s) + D_{ea} A_\varepsilon(t - a + s)\right) ds\right] \tag{19.7}$$

is the probability that an egg survives during the entire egg stage $[0, \varepsilon]$. And thus a problem arises: If L_ε and A_ε remain bounded as the egg stage duration $\varepsilon \to 0$, then egg survivorship (19.7) approaches one. Simply stated, egg cannibalism is eliminated because there is no time to cannibalize eggs. This non-biological situation is completely different from that in the discrete-time LPA model, in which cannibalism of eggs is included implicitly in the larval recruitment rate.

In order to eliminate the egg stage in model (19.1)–(19.4) while still implicitly including cannibalism of eggs, we take

$$D_{el} = \frac{C_{el}a_l}{\varepsilon}, \qquad D_{ea} = \frac{C_{ea}a_l}{\varepsilon},$$

where $a_l > 0$ is the true biological length of the egg stage, and $C_{el}, C_{ea} > 0$. Applying the change of variables $\sigma = s/\varepsilon$ in equation (19.7), we can take the formal limit:

$$\begin{aligned} S_0(t-a) &= \lim_{\varepsilon \to 0} \exp\left[-\int_0^1 \left(C_{el}a_l L_\varepsilon(t-a+\varepsilon\sigma) + C_{ea}a_l A_\varepsilon(t-a+\varepsilon\sigma)\right) d\sigma\right] \\ &= \exp\left[-C_{el}a_l L_0(t-a) - C_{ea}a_l A_0(t-a)\right]. \end{aligned}$$

We thus obtain, after dropping the subscripts, the integral equation for the formal limiting system:

$$\rho(t,a) = \begin{cases} \beta A(t-a)S(t-a)\exp\left[-\int_0^a \delta\left(s, \rho(t-a+s,\cdot)\right) ds\right] & \text{if } t \geq a > 0 \\ \varphi(a-t)\exp\left[-\int_{a-t}^a \delta\left(s, \rho(t-a+s,\cdot)\right) ds\right] & \text{if } t < a \end{cases}, \tag{19.8}$$

where

$$S(t) = \exp\left[-C_{el}a_l L(t) - C_{ea}a_l A(t)\right]$$

$$\delta\left(a, \rho(t,\cdot)\right) = \begin{cases} \nu_l & 0 < a \leq a_p \\ C_{pa}A(t) & a_p < a \leq a_a \\ \nu_a & a_a < a, \end{cases}. \tag{19.9}$$

The limiting system (19.8)–(19.9) is equivalent to the McKendrick PDE

$$\begin{aligned} &\frac{\partial \rho}{\partial t} + \frac{\partial \rho}{\partial a} = -\delta\left(a, \rho\right)\rho \\ &\rho\left(t, 0\right) = \beta A(t)\exp\left[-C_{el}a_l L(t) - C_{ea}a_l A(t)\right] \\ &\rho\left(0, a\right) = \varphi(a), \end{aligned} \tag{19.10}$$

which we call the "continuous-time LPA model" or ctLPA.

Now we can obtain equations for a compartmental model with state variables L, P, and A. For the L-stage, we find

$$\begin{aligned} \frac{dL}{dt} &= \frac{d}{dt}\int_0^{a_p} \rho\left(t,a\right) da \\ &= -\int_0^{a_p} \left(\frac{\partial \rho}{\partial a} + \nu_l \rho(t,a)\right) da \\ &= -\rho\left(t,a\right)\Big|_0^{a_p} - \nu_l \int_0^{a_p} \rho(t,a) da. \end{aligned}$$

Thus,

$$\frac{dL}{dt} = \rho(t,0) - \rho\left(t, a_p\right) - \nu_l L(t).$$

The three terms on the right-hand side describe an inflow and two outflow rates from the L compartment. The first term, $\rho(t,0)$, is the inflow rate due to larval recruitment. The second term, $\rho(t, a_p)$, is the outflow rate of larval survivors as they age into the next class. The third term, $\nu_l L(t)$, is the outflow rate due to larval deaths. For the P- and A-stages, similar computations obtain

$$\frac{dP}{dt} = \rho(t, a_p) - \rho(t, a_a) - C_{pa}A(t)P(t)$$
$$\frac{dA}{dt} = \rho(t, a_a) - \nu_a A(t).$$

Note that the differential equation for A contains only two terms because there is no transition out of the A class except through deaths, which is equivalent here to assuming $\lim_{\alpha\to\infty}\rho(t,\alpha) = 0$.

Using equations (19.8)–(19.10), we can write these differential equations in terms of L, P, and A. For $t \geq a_p$, for example, we obtain

$$\begin{aligned}\frac{dL}{dt} &= \beta A(t)\exp[-C_{el}a_l L(t) - C_{ea}a_l A(t)] \\ &\quad - \beta A(t-a_p)\exp[-C_{el}a_l L(t-a_p) - C_{ea}a_l A(t-a_p) - \nu_l a_p] \\ &\quad - \nu_l L(t).\end{aligned} \tag{19.11}$$

Here, we see that the second term, which is the rate at which surviving animals age out of the L class and enter the P-class, is equal to the recruitment rate into the L class a_p units of time ago, multiplied by the survivorship associated with the inherent death rate ν_l.

In a similar manner, for $t \geq a_a$, we obtain for the P-stage

$$\begin{aligned}\frac{dP}{dt} &= \beta A(t-a_p)\exp[-C_{el}a_l L(t-a_p) - C_{ea}a_l A(t-a_p) - \nu_l a_p] \\ &\quad - \beta A(t-a_a)\exp[-C_{el}a_l L(t-a_a) - C_{ea}a_l A(t-a_a) - \nu_l a_p \\ &\quad - \int_{a_p}^{a_a} C_{pa}A(t-a_a+s)ds] \\ &\quad - C_{pa}A(t)P(t).\end{aligned} \tag{19.12}$$

For $t \geq a_a$, we have for the A-stage

$$\begin{aligned}\frac{dA}{dt} &= \beta A(t-a_a)\exp[-C_{el}a_l L(t-a_a) - C_{ea}a_l A(t-a_a) - \nu_l a_p \\ &\quad - \int_{a_p}^{a_a} C_{pa}A(t-a_a+s)ds] \\ &\quad - \nu_a A(t).\end{aligned} \tag{19.13}$$

Equations (19.11)–(19.13) comprise the functional differential equation (FDE) model for $t \geq a_a$ [13]. These equations are not as easy to integrate numerically as ODEs, and require "initial functions" for the time interval $[0, a_a]$. When constructing these initial functions, one must pay attention to the cases in equations (19.8)–(19.9). In fact, the initial functions are restricted to solutions of the following system:

$$\frac{dL}{dt} = \beta A(t)\exp[-C_{el}a_l L(t) - C_{ea}a_l A(t)] - L_{out} - \nu_l L(t)$$
$$\frac{dP}{dt} = L_{out} - P_{out} - C_{pa}A(t)P(t)$$
$$\frac{dA}{dt} = P_{out} - \nu_a A(t),$$

TABLE 19.1 Relationships between the ctLPA model parameters and the LPA parameters.

LPA	ctLPA
$1-\mu_l$	$e^{-\nu_l a_p}$
$1-\mu_a$	$e^{-\nu_a(a_a-a_p)}$
c_{el}	$C_{el}a_l$
c_{ea}	$C_{ea}a_l$
c_{pa}	$C_{pa}\left(a_a-a_p\right)$

where

$$
\begin{aligned}
&L_{out} = \\
&\begin{cases} \beta A(t-a_p)\exp\left[-C_{el}a_lL(t-a_p)-C_{ea}a_lA(t-a_p)-\nu_la_p\right], & t\geq a_p \\ \varphi(a_p-t)\exp\left[-\nu_lt\right], & t<a_p \end{cases}, \\
&P_{out} = \\
&\begin{cases} \varphi(a_a-t)\exp\left[-\nu_l\left(a_p-a_a+t\right)-C_{pa}\int_{a_p}^{a_a}A\left(t-a_a+\sigma\right)d\sigma\right] \\ \quad \text{for } 0<a_a-t\leq a_p \\ \varphi(a_a-t)\exp\left[-C_{pa}\int_{a_a-t}^{a_a}A\left(t-a_a+\sigma\right)d\sigma\right] \\ \quad \text{for } a_p<a_a-t \end{cases}.
\end{aligned}
$$

Relationships between the ctLPA model parameters and the LPA parameters are shown in Table 19.1.

An analysis of the FDE model in the special case $c_{el}=0$ appears in [9]. In unpublished work, we attempted to parameterize the FDE version of the ctLPA model with data from the route-to-chaos experiment. We used direct measurements to estimate some parameters, and obtained the rest by conditional least squares by splining three consecutive (l,p,a) data triples at a time in order to obtain the continuous delay segments needed to produce one-step predictions and residuals. With these parameter estimates, the ctLPA model could not predict the bifurcation to chaos that was predicted by the LPA model and which was observed in the data. It was not clear to us whether the ctLPA model's failure was due to a fundamental deficiency in the continuous-time model as applied to a system with strong discrete-time biological mechanisms, or whether it was due to our lack of sufficiently dense data collection, or to other reasons. We include this approach because we feel it warrants further thought and experimental effort.

19.3 CONCLUSIONS

In this chapter, we formulated two McKendrick PDE models for the flour beetle system; one with an egg stage and one without. These models have not been connected to data with the same rigor that the discrete-time LPA model has been. This chapter is intended to suggest the correct formulation of such continuous-time analogs and give a flavor of the continuous-time approach. We also aim to motivate further investigation by readers.

Several questions come to mind. The ctELPA and ctLPA models are dynamically different from the LPA model. Can a PDE model describe and predict *Tribolium* population data with anywhere near the accuracy that the LPA model has done?

And how is the LPA model related to its PDE analog? High-dimensional Leslie models have long been viewed as discretizations of their McKenrick PDE analogs. One example from the *Tribolium* literature used a 300×300 Leslie matrix model with a daily time step

[3]. Indeed, by shrinking the discrete time step to zero and simultaneously increasing the number of age classes without bound, one obtains in the limit the McKendrick PDE model, as rigorously demonstrated by Uribe [16]. But what about low-dimensional Leslie models with large time step such as the LPA model? The time step is much too large for the LPA model to be considered a discretization of a PDE. Could low-dimensional Leslie models be connected to PDEs in other ways? In [8] we showed how an LPA-type model can be a "stroboscopic snapshot"[2] of an impulsive McKendrick PDE, that is, a PDE in which some vital rates are pulsed periodically in time. Such a formulation was not straightforward, and it required careful and illuminating considerations of census timing relative to the timing of pulsing.

More fundamentally, which point of view (discrete or continuous time) is the more realistic, and which is the simplification? If we think the biological system is fundamentally continuous in time, we can ask whether a Leslie model can serve as an approximation of the flow. If we think the biological system is primarily discrete in time, we might ask whether the dynamics of the Leslie model can be *extended* to a continuous-time flow, and would that be helpful? Although traditionally dynamicists have come from the first point of view, we now realize that some biological systems are driven largely by more discrete mechanisms, and many of us take the second point of view for systems such as the *Tribolium* populations addressed in this book. Nevertheless, we feel there is room here for further inquiry.

Bibliography

[1] Costantino RF, Desharnais RA (1991) Population Dynamics and the *Tribolium* Model: Genetics and Demography. New York, Springer-Verlag. pp.24–36.

[2] Cushing JM (1998) An Introduction to Structured Population Dynamics. Conference Series in Applied Mathematics 71. Philadelphia, SIAM.

[3] Desharnais RA, Liu L (1987) Stable demographic limit cycles in laboratory populations of *Tribolium castaneum*. Journal of Animal Ecology 56:885–906. https://doi.org/10.2307/4955

[4] Gurtin ME, MacCamy RC (1974) Nonlinear age-dependent population dynamics. Archive for Rational Mechanics and Analysis 54:281–300. https://doi.org/10.1007/BF00250793

[5] Hastings A (1987) Cycles in cannibalistic egg-larval interactions. Journal of Mathematical Biology 24:651–666. https://doi.org/10.1007/BF00275508

[6] Hastings A, Costantino RF (1987) Cannibalistic egg-larva interactions in *Tribolium*: An explanation for the oscillations in population numbers. American Naturalist 130:36–52. https://doi.org/10.1086/284696

[7] Hastings A, Costantino RF (1991) Oscillations in population numbers: Age-dependent cannibalism. Journal of Animal Ecology 60:471–482. https://doi.org/10.2307/5292

[8] Henson SM (1998) Leslie matrix models as "stroboscopic snapshots" of McKendrick PDE models. Journal of Mathematical Biology 37:309–328. https://doi.org/10.1007/s002850050131

[9] Henson SM (1999) A continuous, age-structured insect population model. Journal of Mathematical Biology 39:217–243. https://doi.org/10.1007/s002850050169

[2] For periodic flows, the "stroboscopic snapshot" would be a Poincaré map.

[10] Hoppensteadt F (1975) Mathematical Theories of Populations: Demographics,Genetics and Epidemics. Regional Conference Series in Applied Mathematics 20. Philadelphia, Society for Industrial and Applied Mathematics.

[11] Hoppensteadt F (1982) Mathematical Methods of Population Biology. Cambridge, Cambridge University Press.

[12] Iannelli M (1994) Mathematical Theory of Age-Structured Population Dynamics. Applied Mathematics Monographs 7. Pisa, Giardini Editori e Stampatori.

[13] Kuang Y (1993) Delay Differential Equations with Applications in Population Dynamics. San Diego, Academic Press, Inc.

[14] Robertson SL, Henson SM, Robertson T, Cushing JM (2018) A matter of maturity: To delay or not to delay? Continuous-time compartmental models of structured populations in the literature 2000–2016. Natural Resource Modeling 31:e12160. https://doi.org/10.1111/nrm.12160

[15] Smith HL (1993) Reduction of structured population models to threshold-type delay equations and functional differential equations. A case study. Mathematical Biosciences 113:1–24. https://doi.org/10.1016/0025-5564(93)90006-V

[16] Uribe G (1993) On the Relationship Between Continuous and Discrete Models for Size Structured Population Dynamics. PhD Dissertation, Interdisciplinary Program in Applied Mathematics, University of Arizona, Tucson.

[17] Webb GF (1985) Theory of Nonlinear Age-Dependent Population Dynamics. New York, Marcel Dekker, Inc.

CHAPTER 20

Life Stages: Interactions and Spatial Patterns

One mechanism by which similar species can avoid competitive exclusion is through spatial separation. Similarly, spatial segregation of life-cycle stages within a single species is a common occurrence which presumably enhances survival. Segregation can arise from stage-specific resource niches, but it also can arise from intra-specific interactions. In cannibalistic species, for example, spatial segregation of life stages in heterogeneous environments can provide refugia for prey stages, physically separating them from predator stages [5, 6, 8, 10, 11, 16, 22]. Indeed, many cannibalistic species exhibit age- or stage-dependent habitat selection [5, 6, 16]. The utilization of available refuges in the presence of cannibals has been demonstrated experimentally in the benthic isopod *Saduria entomon* [8], the Socorro isopod *Thermosphaeroma thermophilum* [22], and the amphipod *Gammarus pulex* [11].

Spatial partitioning of the habitat by life stages has been observed in flour beetles of the genus *Tribolium*. Ghent [3] showed experimentally that, for both the species *T. castaneum* and *T. confusum*, adult- and larval-stage densities peaked at different depths in deep cylindrical vials, and that for *T. castaneum*, pupation tended to occur at the flour surface, where adults were more common. The vertical profile of the habitat quality was heterogeneous, due to compression of the flour medium under gravity. In flat and shallow homogeneous environments, spatial patterns have not been observed in *T. castaneum* and *T. confusum*.[1] Thus, the vertical life stage distributions seen by Ghent may have more to do with feeding, metamorphosis, and reproduction than with refugia.

In contrast to *T. castaneum* and *T. confusum*, pupae and adults of the giant flour beetle *T. brevicornis* spatially segregate in homogeneous habitats on the surface of large containers of various shapes [17, 20]. Given that adults cannibalize pupae, this suggests that life-stage interactions alone may be sufficient for the formation of these spatial patterns in *T. brevicornis*.

Can spatial segregation of life-cycle stages be caused solely by density-dependent, inter-stage interactions, without stage-specific resource niches? In this chapter, we explore this question using spatial dispersal versions of the LPA model. In Section 20.1 we use integrodifference equations to account for the dispersal, throughout a continuous spatial domain, of a structured population whose dynamics are described by a discrete-time model [9]. We apply this method in Section 20.2 to address the spatial separation of larvae and adults in *T. castaneum* observed by Ghent [3]. We do this using the LPA model, which, as shown in previous examples, is an accurate model of *T. castaneum* dynamics. There are major

[1] This statement is based on personal observations made from maintaining stock cultures of beetles in large flat containers with a shallow layer of flour.

DOI: 10.1201/9781003467120-20

biological differences between *T. castaneum* and *T. brevicornis* that require modification of the LPA model in order to account for the population dynamics of *T. brevicornis*. In Section 20.3, we use integrodifference equation versions of this modified LPA model to address the spatial patterns observed *T. brevicornis*. This research is from the dissertation of S. Robertson [17] and is also reported in [18, 19, 20].

20.1 INTEGRODIFFERENCE EQUATION MODELS FOR SPATIAL DISPERSAL

The population dynamic models considered in previous chapters ignore any spatial dimensions in the species habitat and hence any dispersal patterns the population might form. One method used to incorporate dispersal, throughout a continuous spatial habitat, into a discrete-time population model is by means of integrodifference equations built on a dispersal kernel, as introduced Kot and Schaffer [9] (also see Weinberger [23]). For example, consider a difference equation

$$p_{t+1} = f(p_t)$$

of the type considered in Chapter 3 for the dynamics of an unstructured population p_t. For a population continuously distributed in a linear habitat denoted by $a \leq x \leq b$, an integrodifference equation (IDE) model that incorporates both population dynamics and dispersal is

$$p_{t+1}(x) = \int_a^b K(x, y) f(p_t(y))\, dy \tag{20.1}$$

where $p_t(x)$ denotes the population density at location x. In this model, it is assumed the population dynamics occur in two steps. A sedentary interval occurs during which reproduction takes place and the population at location x produces $f(p_t(x))$ individuals at this location. This is followed by a dispersal period described by the probability $K(x, y)\, dy$ that an individual in the interval $[y, y + dy]$ will move to location x.[2] In [9], Kot and Schaffer give an equilibrium existence and stability analysis of the integrodifference equation (20.1) and show, among other things, that it is common for there to exist multiple stable equilibrium solutions, i.e., multiple possible stable, steady-state spatial distributions, and therefore, the one that is realized will be initial condition dependent.

In what follows, we apply this modeling methodology to flour beetle populations by including density-dependent dispersal. Toward that end we assume that the dispersal kernel K depends on population density. Given the assumption that the dispersal period in the model follows the population growth interval, this means that the kernel K is dependent on $f(p_t(x))$, and we write $K(x, y, f(p_t(x)))$. The integro-difference equation (IDE) model equation thus takes the form

$$p_{t+1}(x) = \int_a^b K(x, y, f(p_t(x)))\, f(p_t(y))\, dy.$$

Our working hypothesis is that spatial patterns observed in flour beetle cultures that separate life-cycle stages are caused by interactions among the life-cycle stages, specifically, by cannibalism and pupation suppression. This requires a structured population model, in place of a single equation (20.1), for total population density.

As a warm-up, consider the lowest dimensional case for a class-structured model, namely the two-dimensional system

$$\begin{aligned} J_{t+1} &= f(J_t, A_t) \\ A_{t+1} &= g(J_t, A_t), \end{aligned} \tag{20.2}$$

[2]This assumption is made for convenience. In reality, the adult and larval flour beetles stages move constantly throughout the medium.

where we have denoted the two class densities by J and A (for juvenile and adult, although more generally the classes could arise from any other categories of individuals). An IDE version of this model is

$$J_{t+1}(x) = \int_a^b K_J(x,y) f\left(J_t(y), A_t(y)\right) dy$$
$$A_{t+1}(x) = \int_a^b K_A(x,y) g\left(J_t(y), A_t(y)\right) dy,$$

with (density-free) dispersal kernels $K_J(x,y)$ and $K_A(x,y)$. For density-dependent dispersal, these kernels become functions of the two terms $f\left(J_t(x), A_t(x)\right)$ and $g\left(J_t(x), A_t(x)\right)$ in addition to x and y, yielding the notationally formidable equations

$$J_{t+1}(x) = \int_a^b K_J(x, y, f\left(J_t(x), A_t(x)\right), g\left(J_t(x), A_t(x)\right)) f\left(J_t(y), A_t(y)\right) dy$$
$$A_{t+1}(x) = \int_a^b K_A(x, y, f\left(J_t(x), A_t(x)\right), g\left(J_t(x), A_t(x)\right)) g\left(J_t(y), A_t(y)\right) dy.$$

Equilibria are solutions

$$\left[J^*(x), A^*(x)\right] = \begin{pmatrix} J^*(x) \\ A^*(x) \end{pmatrix}$$

of the integral equations

$$\begin{aligned} J(x) &= \textstyle\int_a^b K_J(x, y, f\left(J(x), A(x)\right), g\left(J(x), A(x)\right)) f\left(J(y), A(y)\right) dy \\ A(x) &= \textstyle\int_a^b K_A(x, y, f\left(J(x), A(x)\right), g\left(J(x), A(x)\right)) g\left(J(y), A(y)\right) dy. \end{aligned} \tag{20.3}$$

We are interested in populations closed to immigration so that

$$\begin{pmatrix} J^*(x) \\ A^*(x) \end{pmatrix} = \begin{pmatrix} 0 \\ 0 \end{pmatrix}$$

is an equilibrium, that is,

$$f(0,0) = g(0,0) = 0.$$

A non-extinction (or positive) equilibrium $\left[J^*(x), A^*(x)\right]$ is an equilibrium with positive components $J^*(x) > 0$ and $A^*(x) > 0$. The integral equations (20.3) are generally challenging to analyze, even with specified modeling components. Relative to our hypothesis we are interested in the existence and stability of positive equilibria and their properties as functions of the spatial variable x. There are general theorems, for IDE equations of arbitrary dimension, that establish the existence and stability of positive equilibria that arise when the extinction equilibrium destabilizes; see [1, 19]. We will not delve into this technical mathematical analysis here, but instead will use numerical simulations of models with components designed to address the problem at hand, namely the spatial segregation of life-cycle stages.

Before building IDE models for species of *Tribolium* in Sections 20.2 and 20.3, we illustrate how density-dependent dispersal can be a mechanism that results in spatial separation of juvenile and adult stages in a model of the two-dimensional type (20.3).

Consider the model components

$$f(J,A) = b\frac{1}{1+A}A$$
$$g(J,A) = (1-\mu_J)J + (1-\mu_A)A$$

in equations (20.2). Here adult reproduction is (adult) density dependent with density-free fertility $b > 0$, and juvenile and adult mortality rates are density-free with $0 \leq \mu_J, \mu_A < 1$. For the dispersal kernels, take

$$K_J = \left(\frac{1}{2}\sin x\right) e^{-g(J(x),A(x))D_J}$$
$$K_A = \left(\frac{1}{2}\sin x\right) e^{-f(J(x),A(x))D_A}.$$

This produces the IDE model equations

$$\begin{aligned} J_{t+1}(x) &= \int_0^{\pi} \left(\tfrac{1}{2}\sin x\right) e^{-g(J_t(x),A_t(x))D_J} b \tfrac{1}{1+A_t(y)} A_t(y)\, dy \\ A_{t+1}(x) &= \int_0^{\pi} \left(\tfrac{1}{2}\sin x\right) e^{-f(J(x),A(x))D_A} \\ &\quad \times \left((1-\mu_J) J_t(y) + (1-\mu_A) A_t(y)\right) dy. \end{aligned} \tag{20.4}$$

In this example, it is assumed that each stage avoids the other. That is, the presence of adults at a given spatial location decreases the probability of a juvenile moving there, and vice versa. The density-free dispersal kernels $\frac{1}{2}\sin x$ imply that the boundaries of the spatial interval $[0, \pi]$ are hostile and the most favored location is at $\pi/2$. The coefficients $D_J, D_A \geq 0$ can be thought of as *dispersal sensitivities*. If $D_J = D_A = 0$, there is no density-dependent dispersal. If a dispersal sensitivity coefficient is positive, then the probability of remaining in place decreases and dispersal is promoted.

The theory developed in [19] shows that the extinction equilibrium $[J^*(x), A^*(x)] = [0, 0]$ is (locally asymptotically) stable if the reproduction number $R_0 = b(1-\mu_J)/\mu_A < 1$. The destabilization of this equilibrium for $R_0 > 1$ results in the creation of positive equilibria (by means of a transcritical bifurcation). For R_0 near 1, the positive equilibria are approximately equal to positive multiples of

$$\begin{pmatrix} J(x) \\ A(x) \end{pmatrix} = \begin{pmatrix} \sin x \\ \frac{1-\mu_J}{\mu_A}\sin x \end{pmatrix}.$$

For these stable spatial distributions, both juvenile and adult stage densities are maximal at the center point $x = \pi/2$, and are zero at the hostile boundary points $x = 0, \pi$. For larger values of R_0, however, spatial segregation of juveniles and adults can occur, provided density-dependent dispersal is present (that is, at least one dispersal sensitivity coefficient is positive). This is demonstrated by the numerically computed examples shown in Figure 20.1.

Additionally, this example can generate spatially-segregated equilibria which oscillate periodically, forming temporal as well as spatial juvenile and adult niches [17, 18, 19]. In the following sections, we construct higher-dimensional IDE models based on the LPA model for *Tribolium*. For these models, we will find similar phenomena, as well as new features such as the existence of multiple attractors.

20.2 A SPATIAL LPA MODEL

In this section, we apply the modeling methodology in Section 20.1 to an experimental system in which spatial segregation of life-cycle stage patterns was observed to occur in species of *Tribolium*, namely Ghent's experiments reported in [3]. To do this, we base the spatial model on the LPA model

$$\begin{aligned} L_{t+1} &= bA_t e^{-c_{el}L_t - c_{ea}A_t} \\ P_{t+1} &= (1-\mu_l) L_t \\ A_{t+1} &= P_t e^{-c_{pa}A_t} + (1-\mu_a) A_t \end{aligned}$$

and utilize dispersal kernels appropriate for the experimental setup.

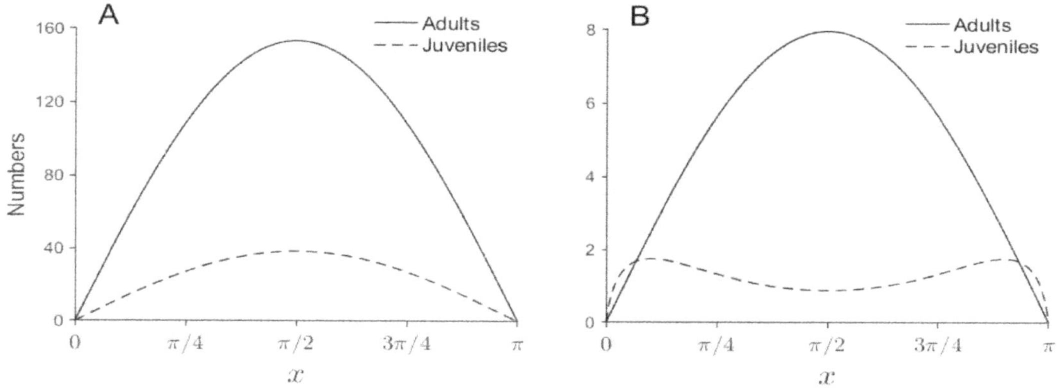

Figure 20.1 The equilibrium attractor of the juvenile-adult IDE model (20.4) with $b = 25$, $\mu_L = \mu_A = 0.2$ (hence $R_0 = 100$). A. $D_J = D_A = 0$. B. $D_J = 0.5$, $D_A = 0$.

Ghent studied the depth distribution of the life-cycle stages of the cannibalistic species *T. castaneum* and *T. confusum* in vials of flour [3]. For each species, he placed 200 adults on the surface of a cylindrical vial filled with 8 g of flour to a depth of 41-42 mm. The vial was divided horizontally into five rings, each being 10mm in height. After incubating the cultures for 26 days, Ghent counted the number of individuals in each stage in each of the five rings. (26 days is enough time to produce eggs, larvae and pupae, but not new adults.) Ghent found adults of both species to be most abundant near the surface of the flour, in the first ring at the surface, while larval densities peaked in the ring immediately below.

Our goal in this section is to construct an extension of the LPA model in order to determine if the spatial patterns observed by Ghent can be explained by density-dependent inter-stage interactions. There is substantial evidence that density plays a role in the dispersal of the mobile stages of both *T. castaneum* and *T. confusum* [12, 13, 14, 15, 24, 25].

The spatial patterns in a cylindrical vial of flour can be looked at as one-dimensional in space, where the spatial dimension is depth. Due to gravity, there is a gradient of increasing flour density from the surface to bottom of the bottle. Individuals can live on the surface but cannot pass above it. We model this environment as a one-dimensional domain with mixed boundary conditions, where the bottom of the bottle is a hostile boundary and the surface is a no-flux boundary. We make the following assumptions regarding the dispersal behavior of the two mobile life-stages, namely larvae and adults (pupae are immobile and have zero probability of dispersing). Adults prefer areas with larval densities that are lower than their current location, and larvae prefer areas with adult densities that are lower than their current location. Thus, we assume that the fraction of larvae that move from one location to another depends on the difference in adult densities, after population dynamics occur, between the two locations. Adult dispersal depends similarly on density of larvae. Furthermore, we assume both larvae and adults prefer flour near the surface over the more dense flour near the bottom. We choose our spatial units so that the spatial domain interval $0 \leq x \leq \pi/2$, representing depth in a cylindrical vial with 0 corresponding to the surface and $\pi/2$ corresponding to the bottom of the vial. We use the term $\cos^m x$ to incorporate density-independent spatial preference, where larger values of the integer m increase the undesirable region at the bottom of the vial. These assumptions, and after some mathematical simplifications, lead to the follow IDE LPA model (for details see [17, 18]):

$$L_{t+1}(x) = \int_0^{\pi/2} \frac{1}{c_1} \exp\left\{-D_{LA}\left[e^{-c_{pa}A_t(x)}P_t(x) + (1-\mu_A)A_t(x)\right]\right\} \times (\cos^m x)\left[bA_t(y)e^{-c_{el}L_t(y)-c_{ea}A_t(y)}\right]dy$$

$$P_{t+1}(x) = (1-\mu_l)L_t(x) \tag{20.5}$$

$$A_{t+1}(x) = \int_0^{\pi/2} \frac{1}{c_2} \exp\left\{-D_{AL}\left(bA_t(x)e^{-c_{ea}A_t(x)-c_{el}L_t(x)}\right)\right\} \times (\cos^m x)\left[e^{-c_{pa}A_t(y)} + (1-\mu_a)A_t(y)\right]dy$$

where

$$c_1 = \int_0^{\pi/2} \exp\left\{-D_{LA}\left[e^{-c_{pa}A_t(x)}P_t(x) + (1-\mu_A)A_t(x)\right]\right\}\cos^m x dx$$

$$c_2 = \int_0^{\pi/2} \exp\left\{-D_{AL}\left[bA_t(x)e^{-c_{ea}A_t(x)} - c_{el}L_t(x)\right]\right\}\cos^m x dx.$$

(Recall the notation $\exp(z) = e^z$.) We choose initial conditions $L_0(x), P_0(x), A_0(x)$ that are continuous functions on the domain $0 \le x \le \pi/2$. Appropriate for the hostile and no flux conditions on the boundaries $x = \pi/2$ and $x = 0$, respectively, we require of these initial conditions that $L_0(\pi/2) = P_0(\pi/2) = A_0(\pi/2) = 0$ and $dL_0(0)/dx = dP_0(0)/dx = dA_0(0)/dx = 0$. It is a feature of the model equations (20.5) that $L_t(x), P_t(x), A_t(x)$ are continuously differentiable functions that satisfy the hostile and no flux boundary conditions, i.e., $L_t(\pi/2) = P_t(\pi/2) = A_t(\pi/2) = 0$ and $dL_t(0)/dx = dP_t(0)/dx = dA_t(0)/dx = 0$, for all t.

To illustrate that the IDE LPA model (20.5) can corroborate the observation that spatial segregation between larvae and adults can occur, with maximal adult density near the boundary $x = 0$ and maximal larvae density below this boundary, the results of some selected numerical simulations are shown in Figure 20.2 in which model parameters for the population dynamics were chosen so as to fall within the maximum likelihood 95% confidence intervals of the estimates obtained in Chapter 5 and initial conditions all of which are positive constant multiples of $\cos x$ were used. In Figure 20.2A we see that in the absence of dispersal at temporal equilibrium all life cycle stages are maximal at the most favorable location at $x = 0$. However, in Figure 20.2B when density-dependent dispersal is in play, we see that the temporal equilibrium displays life cycle stage segregation, with maximal adult density at location $x = 0$, but maximal larvae and pupal densities occurring near the center of the spatial habitat.[3] For large values of the dispersal sensitivities D_{LA} and/or D_{AL}, numerical simulations show temporally fluctuating patches as well, although these have not been observed in *T. castaneum* and *T. confusum*. For more see [17] and [18].

20.3 A SPATIAL SLPA MODEL

Spatial patterns observed in laboratory cultures of the giant flour beetle *T. brevicornis* (Leconte) are different from those observed in species *T. castaneum* and *T. confusum* that inspired the modeling effort in Section 20.2. The adults and pupae of *T. brevicornis* often occupy different locations in a homogeneous habitat, namely on the surface of media containers of various shapes. These life cycle stages cluster in separate groupings rather that disperse uniformly over the surface of the media and do this in domains of many different shapes and sizes including rectangular boxes and cylindrical vials. Adults have been

[3]In Ghent's experiments, *T. castaneum* pupae were most abundant at the surface of the flour.

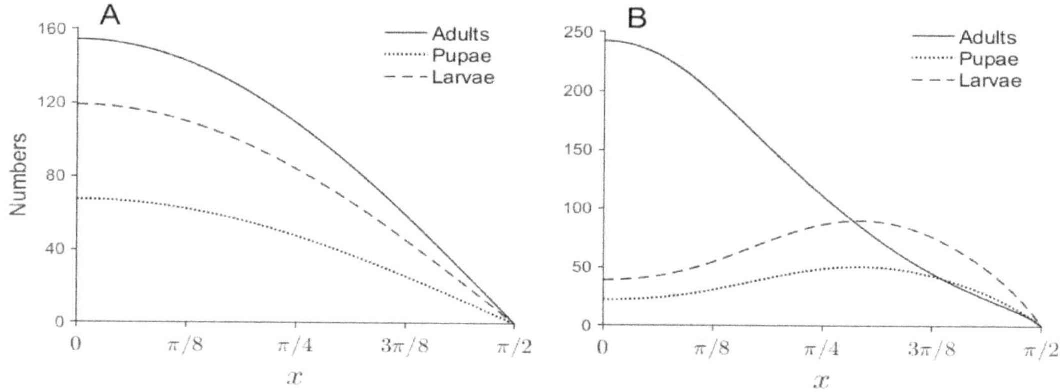

Figure 20.2 Temporal equilibria for all life cycle stages obtained by simulations of the IDE LPA model (20.5) with population parameter values $b = 10$, $c_{el} = 0.01$, $c_{ea} = 0.0175$, $\mu_l = 0.435$, $c_{pa} = 0.016$, $\mu_a = 0.075$. In the dispersal kernal, $m = 1$. A. $D_{LA} = D_{AL} = 0$. B. $D_{LA} = D_{AL} = 0.01$.

observed aggregating along the boundaries of the domains as well as in the interior; the specific location varies greatly among cultures, even among containers of the same size and shape [20]. Such spatial patterns do not occur in the *T. castaneum* or *T. confusum* nor have been reported in any of the other 25 species in the genus *Tribolium*. Among the species in the genus, there are currently five known types of interactions among the life stages: Larvae eat eggs, adults eat eggs, adults eat pupae, adults inhibit larval metamorphosis, and larvae inhibit larval metamorphosis. These interactions appear in different species in different combinations. They do not all appear in any one species. In cultures with two or more species, these interactions form the basis for interspecific competition in this genus. The time spent in each life stage also varies among species.

In this section, we focus on the species *T. brevicornis*, which has the same four life cycle stages, namely egg, larva, pupa, and adult, and share the same stage transitions and interactions as *T. castaneum* (and *T. confusum*). However, there are several major biological differences between the species that are important for our study here [7, 21]. First, *T. brevicornis* adults inhibit larval metamorphosis [7], which does not occur in *T. castaneum*. *T. brevicornis* larvae may remain in the larval stage indefinitely until local adult densities lower and they can pupate. Second, the innate length of the larval stage for *T. brevicornis* (in the absence of adults) is four weeks, two weeks longer than *T. castaneum*. Larvae are noticeably larger and more mobile in the latter two week period than the former. A third notable biological difference between these species is the absence of pupal cannibalism by adults in *T. brevicornis*. Pupal cannibalism is a mechanism that controls adult recruitment in *T. castaneum*; *T. brevicornis* has the alternate control mechanism of larval metamorphosis inhibition.

In order to model the spatial dynamics observed in *T. brevicornis*, we first need a non-spatial model to describe the population dynamics of the species. The LPA model was developed as a model for the population dynamics of *T. castaneum*. Given the biological differences between the two species described in the preceding paragraph, we need to modify the LPA in order to account for these differences.

To account for the longer larval stage of *T. brevicornis*, we split the L stage of the LPA model into two new stages and denote them by S and L. Then S_t represents the number of younger or "small" larvae at time t and L_t now represents the number of "large" larvae

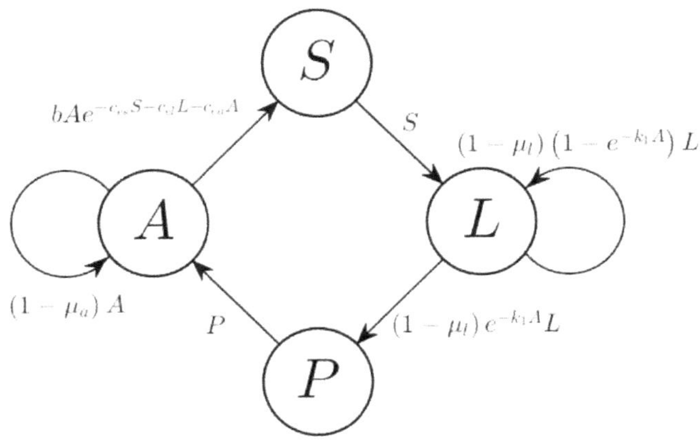

Figure 20.3 The flow diagram for the SLPA model (20.6).

at time t, i.e., the third week and fourth week old larvae, as well as older larvae who have failed to pupate due to inhibition. The time step of the model remains two weeks. We model larval inhibition with a Ricker (i.e., an exponential) type nonlinearity. This is appropriate given the assumption that inhibition is a result of random tactile encounters of larvae with adults at a rate k_i and the fraction of larvae inhibited increases with the density of adults. This is the same modeling methodology used to describe cannibalism in Chapter 3.

T. brevicornis eggs are subject to cannibalism by small larvae as well as large larvae. Since large larvae have been observed to be more voracious eaters than small larvae, we allow each larval stage its own cannibalism rate [4]. Thus, c_{es} is the cannibalism coefficient of eggs by small larvae, and c_{el} is the cannibalism coefficient of eggs by large larvae. Since adults do not eat pupae the coefficient c_{pa} for this term, which appears in the LPA model, equals zero [7]. Pupal mortality is zero.

The SLPA model, whose flow diagram appears in Figure 20.3, is given by the following equations:

$$\begin{aligned} S_{t+1} &= bA_t e^{-c_{es}S_t - c_{el}L_t - c_{ea}A_t} \\ L_{t+1} &= S_t + (1-\mu_l)\left(1 - e^{-k_i A_t}\right) L_t \\ P_{t+1} &= (1-\mu_l)e^{-k_i A_t} L_t \\ A_{t+1} &= P_t + (1-\mu_a)A_t. \end{aligned} \tag{20.6}$$

The parameters in the SLPA model equations (20.6) were estimated in [17], using *T. brevicornis* data collected by R. F. Costantino and R. A. Desharnais, by means maximum likelihood estimation based on a demographic stochastic version of the model (see Chapter 5 and [2]). The results are shown in Table 20.1.

With these parameter estimates the model predicts an equilibrium

$$\begin{pmatrix} S^* \\ L^* \\ P^* \\ A^* \end{pmatrix} = \begin{pmatrix} 12.21 \\ 71.77 \\ 2.59 \\ 163.84 \end{pmatrix}.$$

Numerical simulations show that as the inhibition parameter k_i increases, holding all other parameters remaining fixed, the number of large larvae in the equilibrium stage vector increases until eventually it is the dominant stage $L^* > A^*$.

TABLE 20.1 The maximum likelihood parameter estimates for the SLPA model [20]. Here μ_a is calculated from data on numbers of dead adults.

Parameter	Estimate
b	11.4096
c_{es}	0.0135
c_{el}	0.0169
c_{ea}	0.0223
μ_l	0.1339
k_i	0.0194
μ_a	0.0158

Our goal is to construct an IDE version of the SLPA model, in the same way we did for the LPA model above for *T. castaneum*, by taking into account the dispersal and stage interaction characteristics of *T brevicornis*. Not all life-stages of *T. brevicornis* disperse. Pupae are sedentary, of course. Since younger larvae in their first two weeks are smaller and slower than older larvae, we make the simplifying assumption that larvae in the S class also do not disperse. The remaining two stages, L and A, are dispersers in our model. We assume adults always disperse and the fraction of large larvae that disperse depends on the local density of adults.

Although the spatial patterns we observed in *T. brevicornis* were on two-dimensional surfaces, we take advantage of an approximate cross-sectional symmetry in some patterns to use one spatial dimension in our model. This is the case for the experimental cultures used to corroborate the model predictions, as described below. In these experiments, there were no heterogeneities in the surface habitat so that we assume no explicit spatial dependence of movement. Instead, we assume that adult beetles tend to prefer locations with lower pupal densities than that of their starting location. (This is biologically reasonable since pupae are more likely to pupate and enter the sexually mature adult class with a lower incidence of tactile contact with adults.) We choose the interval $0 \leq x \leq 1$ as the spatial habitat. We incorporate density-dependent dispersal into the adult kernel by an exponentially decreasing function of pupal density, recalling that pupal density is determined by the density of larvae and adults at the previous time step:

$$\frac{1}{c}\exp\left(-D_{AP}\left((1-\mu_l)\,e^{-k_i A_t(x)}L_t(x)\right)\right)$$

where

$$c = \int_0^1 \exp\left(-D_{AP}\left((1-\mu_l)\,e^{-k_i A_t(x)}L_t(x)\right)\right)dx.$$

Large larvae do not avoid small larvae or pupae, but they do avoid adults. With high adult densities, large larvae are inhibited and are unable to pupate; consequently, we assume that local adult density affects the fraction of large larvae dispersing at any time and location. We model the fraction of large larvae dispersing at each time step by an increasing function of local adult density, making the simplifying assumption that dispersing large larvae then redistribute uniformly over the entire habitat. These assumptions lead to the kernel for large larvae dispersal that is the probability of dispersing:

$$1 - e^{-D_{LA}(P_t(x)+(1-\mu_a)A_t(x))}.$$

Using these dispersal kernels and the interval $0 \leq x \leq 1$ as the spatial domain, we obtain the following IDE SLPA model:

$$S_{t+1}(x) = bA_t(x)\, e^{-c_{es}S_t(x) - c_{el}L_t(x) - c_{ea}A_t(x)}$$

$$L_{t+1}(x) = \int_0^M \tfrac{1}{M}\left[1 - e^{-D_{LA}(P_t(y)+(1-\mu_a)A_t(y))}\right] \times \left[S_t(y) + (1-\mu_l)\left(1 - e^{-k_i A_t(y)}\right)L_t(y)\right] dy$$
$$+ e^{-D_{LA}(P_t(x)+(1-\mu_a)A_t(x))}\left[S_t(x) + (1-\mu_l)\left(1 - e^{-k_i A_t(x)}\right)L_t(x)\right]$$

$$P_{t+1}(x) = (1-\mu_l)\, e^{-k_i A_t(x)} L_t(x)$$

$$A_{t+1}(x) = \int_0^M \tfrac{1}{c} \exp\left(-D_{AP}\left((1-\mu_l)\, e^{-k_i A_t(x)} L_t(x)\right)\right) \times \left[P_t(y) + (1-\mu_A)A_t(y)\right] dy.$$

Note that this model preserves no-flux boundary conditions at both endpoints

$$\left.\frac{dS_t(x)}{dx}\right|_{x=0} = \left.\frac{dL_t(x)}{dx}\right|_{x=0} = \left.\frac{dP_t(x)}{dx}\right|_{x=0} = \left.\frac{dA_t(x)}{dx}\right|_{x=0} = 0$$
$$\left.\frac{dS_t(x)}{dx}\right|_{x=1} = \left.\frac{dL_t(x)}{dx}\right|_{x=1} = \left.\frac{dP_t(x)}{dx}\right|_{x=1} = \left.\frac{dA_t(x)}{dx}\right|_{x=1} = 0$$

for all $t > 0$.

To simulate this model, we must choose initial conditions

$$\begin{pmatrix} S_0(x) \\ L_0(x) \\ P_0(x) \\ A_0(x) \end{pmatrix}.$$

Note that an initial condition with a uniform spatial distribution will remain a uniform spatial distribution for all time since no preferences for different spatial locations are built into the model. (The fraction of individuals leaving or settling at a given location depends only on the population density at that location.) There are many non-uniform initial distributions one could consider, but we will restrict our attention to initial conditions of the form

$$\begin{pmatrix} 0 \\ L_0(x) \\ 0 \\ A_0(x) \end{pmatrix}$$

which can be thought of as representing a biological invasion of a new environment, given that the only possible invaders are the dispersing stages L and A. Specifically, we assume $L_0(x)$ and $A_0(x)$ have uniform distributions on a subinterval of the domain $0 \leq x \leq 1$:

$$L_0(x) = \begin{cases} C_L & \text{on } 0 \leq x \leq m \\ 0 & \text{on } m < x \leq 1 \end{cases} \tag{20.7}$$
$$A_0(x) = \begin{cases} C_A & \text{on } 0 \leq x \leq m \\ 0 & \text{on } m < x \leq 1 \end{cases}$$

TABLE 20.2 Two different equilibrium attractors of the IDE SLPA model with parameter values from Table 20.1 and $D_{AP} = 1$, $D_{LA} = 0.25$, $m = 0.25$.

Initial condition (20.7)	Equilibrium $[S^*(x), L^*(x), P^*(x), A^*(x)]$
$C_L = 100$, $C_A = 0$	Uniform $[12.21, 71.77, 2.59, 163.8]$ on $0 \leq x \leq 1$
$C_L = 0$, $C_A = 100$	Patchy $\begin{cases} [0.015, 13.93, 12.06, 0.0016] \text{ on } 0 \leq x \leq 0.25 \\ [6.58, 13.92, 0.08, 259.4] \text{ on } 0.25 < x \leq 1 \end{cases}$

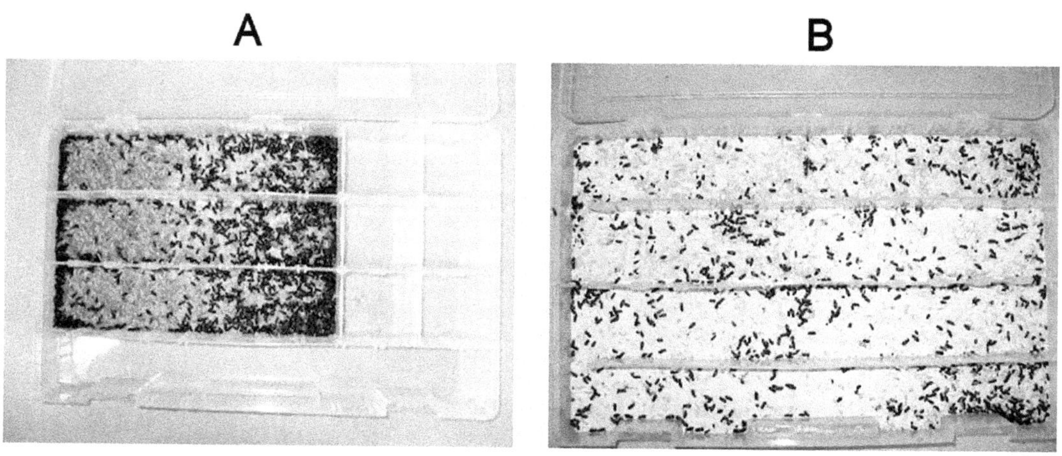

Figure 20.4 A. Three replicate cultures of *T. brevicornis* in linear horizontal habitats show segregation of adults and pupae, with a pupal nest formed on the left side of each row. B. Four replicate cultures show uniformly distributed stages, with no pupal nest.

for some constant $0 < m < 1$ and positive constants C_L and C_A. These kinds of initial conditions are utilized in the experimental cultures described below.

Extensive numerical simulations show that for these initial conditions and the maximum likelihood SLPA model parameter estimates in Table 20.1, the spatial SLPA model admits multiple attractors. These attractors include a spatially uniform, equilibrium distribution across the interval $0 \leq x \leq 1$ and a "patchy" equilibrium distribution on the subintervals $0 \leq x \leq m$ and $m \leq x \leq 1$. Which attractor is approached depends on initial conditions. An example is shown in Table 20.2 with $m = 0.25$. Note in this example how, in the patchy equilibrium, the adults dominate in the right patch $0.25 < x \leq 1$ while in the left patch $0 \leq x \leq 0.25$ they are virtually absent and the immature stages predominate, forming a pre-adult or pupal "nest."The two attractors (Table 20.2) are not the only possible attractors for those same parameter values, but they are the most common for the set of initial conditions we investigated (a uniform distribution of dispersing stages on a subinterval of the domain) and they are also the two attractors seen in our laboratory experiments. See Figure 20.4. Simulations show other attractors for other initial conditions, including a temporally recurrent pupal nest, but these were not observed in our experiments. For more on the pattern formations predicted by the IDE SLPA model and the experiments performed to corroborate the patterns, see [17, 20].

20.4 CONCLUSIONS

In preceding chapters we showed that the LPA model, together with many of its variants, is capable of an accurate accounting of a large variety of dynamic patterns observed in experimental cultures of flour beetles. These mechanistic models provide not only descriptions of observed dynamic patterns, but predictions of patterns that were subsequently validated by controlled and replicated experiments. In this chapter, we looked at other variants of the LPA model that are designed to investigate observed spatial patterns in flour beetle cultures. These variants are in the form of integrodifference equations (IDE) derived using dispersal kernels to describe spatial movement of some life cycle stages. The goal is to investigate the hypothesis that interclass interactions and density dependent dispersal can account for observed spatial segregation of life stages. For *T. castaneum*, the IDE LPA model (20.5) showed that the tendency of adults and larvae to avoid one another results in an equilibrium spatial distribution with adults near the top and larvae below the surface of their habitat vials, as observed by Ghent [3]. The significant differences between the biology of *T. castaneum* and *T. brevicornis* led to the derivation of the SLPA model (20.6) for *T. brevicornis* whose integrodifference version IDE SLPA predicted the formation of pupal nests as observed in the laboratory cultures and as observed in experiments designed to corroborate the model predictions. Extensive simulations of both spatial models show the existence of multiple attractors and therefore the models predict initial condition dependent outcomes. For example, in addition to predicted uniform and patchy (life cycle stage segregated) equilibrium distributions, as observed in experiments, the IDE SLPA model predicts temporally fluctuating patchy distributions. The latter have yet to be observed in *T. brevicornis* cultures. A rigorous mathematical treatment of equilibrium existence and stability for IDE models is given in [1].

Bibliography

[1] Alzoubi MYM (2010) Stability and bifurcation in a system of integro-difference equations model. Applied Mathematical Sciences 4:3175–3188. https://www.m-hikari.com/ams/ams-2010/ams-61-64-2010/index.html

[2] Dennis B, Desharnais RA, Cushing JM, Costantino RF (1995). Nonlinear demographic dynamics: Mathematical models, statistical methods, and biological experiments. Ecological Monographs 65:261–281. https://doi.org/10.2307/2937060

[3] Ghent AW (1966) Studies of behavior of the *Tribolium* flour beetles. II. Distributions in depth of *T. castaneum* and *T. confusum* in fractionable shell vials. Ecology 47:355–367. https://doi.org/10.2307/1932976

[4] Hastings A, Costantino RF (1991) Oscillations in population numbers: Age-dependent cannibalism. Journal of Animal Ecology 60:471–482. https://doi.org/10.2307/5292

[5] Hill C (1988) Life cycle and spatial distribution of the amphipod *Pallasea quadrispinosa* in a lake in northern Sweden. Ecography 11:298–304. https://doi.org/10.1111/j.1600-0587.1988.tb00813.x

[6] Hunte W, Myers RA (1984) Phototaxis and cannibalism in gammaridean amphipods. Marine Biology 81:75–79. https://doi.org/10.1007/BF00397628

[7] Jillson DA, Costantino RF (1980) Growth, distribution, and competition of *Tribolium castaneum* and *Tribolium brevicornis* in fine-grained habitats. American Naturalist 116:206–219. https://doi.org/10.1086/283623

[8] Jormalainen V, Shuster SM (1997) Microhabitat segregation and cannibalism in an endangered freshwater isopod *Thermosphaeroma thermophilum*. Oecologia 111:271–279. https://doi.org/10.1007/s004420050235

[9] Kot M, Schaffer W (1989) Discrete-time growth-dispersal models. Mathematical Biosciences 80:109–136. https://doi.org/10.1016/0025-5564(86)90069-6

[10] Leonardsson K (1991) Effects of cannibalism and alternative prey on population dynamics of *Saduria entomon* (Isopoda). Ecology 72:1273–1285. https://doi.org/10.2307/1941101

[11] McGrath KE, Peeters ETHM, Beijer JAJ, Scheffer M (2007) Habitat-mediated cannibalism and microhabitat restriction in the stream invertebrate *Gammarus pulex*. Hydrobiologia 589:155–164. https://doi.org/10.1007/s10750-007-0731-5

[12] Naylor AF (1959) An experimental analysis of dispersal in the flour beetle, *Tribolium confusum*. Ecology 40:453–465. https://doi.org/10.2307/1929763

[13] Naylor AF (1961) Dispersal in the red flour beetle *Tribolium castaneum* (Tenebrionidae). Ecology 42:231–237. https://doi.org/10.2307/1932074

[14] Naylor AF (1965) Dispersal responses of female flour beetles, *Tribolium confusum*, to presence of larvae. Ecology 46:341–343. https://doi.org/10.2307/1936339

[15] Prus T (1966) Emigrational ability and surface numbers of adult beetles in 12 strains of *Tribolium confusum* duval and *T. castaneum* Herbst (Coleoptera, Tenebrionidae). Ekologia Polska Seria A 14:547–588.

[16] Ribes M, Coma R, Zabala M, Gili J (1996) Small scale spatial heterogeneity and seasonal variation in a population of a cave-dwelling Mediterranean mysid. Journal of Plankton Research 18:659–671. https://doi.org/10.1093/plankt/18.5.659

[17] Robertson SL (2009) Spatial patterns in stage-structured populations with density dependent dispersal. PhD Dissertation, Interdisciplinary Program in Applied Mathematics, University of Arizona, Tucson.

[18] Robertson SL, Cushing JM (2011) Spatial segregation in stage-structured populations with an application to *Tribolium*. Journal of Biological Dynamics 5:398–409. https://doi.org/10.1080/17513758.2010.503283

[19] Robertson SL, Cushing JM (2011) A bifurcation analysis of stage-structured density dependent integrodifference equations. Journal of Mathematical Analysis and Applications 388:490–499. https://doi.org/10.1016/j.jmaa.2011.09.064

[20] Robertson SL, Cushing JM, Costantino RF (2012) Life stages: Interactions and spatial patterns. Bulletin of Mathematical Biology 74:491–508. https://doi.org/10.1007/s11538-011-9705-x

[21] Sokoloff A, Albers JC, Cavataio P, Mankau SK, McKibben S, Mills S, Perkins R, Roberts LD (1980). Observations on populations of *Tribolium brevicornis* (Le conte) (Coleoptera, Tenebrionidae). I. Laboratory observations of domesticated strains. Researches on Population Ecology 22:1–12. https://doi.org/10.1007/BF02513532

[22] Sparrevik E, Leonardsson K (1995) Effects of large *Saduria entomon* (isopoda) on spatial distribution of their small *S. entomon* and *Monoporeia affinis* (amphipoda) prey. Oecologia 101:177–184. https://doi.org/10.1007/BF00317281

[23] Weinberger HF (1978) Asymptotic behavior of a model of population genetics. In Chadam J (ed.), Nonlinear Partial Differential Equations and Applications. Lecture Notes in Mathematics 648. Springer.

[24] Ziegler JR (1976) Evolution of the migration response: Emigration by *Tribolium* and the influence of age. Evolution 30:579–592. https://doi.org/10.2307/2407581

[25] Zyromska-Rudzka H (1966) Abundance and emigrations of *Tribolium* in a laboratory model. Ekologia Polska Serial A 14:491–518.

CHAPTER 21

In Summary

The studies contained in this book initially were motivated by Robert May's suggestion that some of the complexity seen in the dynamics of biological populations might be attributed to simple deterministic laws [5]. To address this assertion, we first needed to investigate a more fundamental question: Can "simple" (low-dimensional) deterministic mathematical models both describe and predict the dynamics of real biological populations with sufficient quantitative accuracy to allow an inquiry into complex dynamics? In dealing with this question, we were confronted with a number of significant issues concerning modeling methodology. These include model derivation and construction, model parameterization (calibration) and model validation of prediction capabilities. Another especially important issue for us was noise: the inevitable presence of stochastic effects that cause deviations from deterministic model predictions.

The second central feature of our studies was the implementation of controlled and replicated experiments which had two broad goals. One was to obtain time-series data which we could use to parameterize and validate our dynamic models. The other was to determine whether predictions made by the validated model could be observed in our experimental populations. These predictions ranged from simple to complex dynamic patterns of many sorts. The interplay between model and data involved a two way feedback: Occasionally an experimentally observed pattern, not initially foreseen in the model, stimulated mathematical analysis to discover the pattern in the model and, conversely, patterns predicted by the model—ones that might even be unexpected from a biological point of view—were seen to occur in data from carefully designed experiments.

In a nutshell, at a general level our studies dealt with the question: Can simple mathematical models for the dynamics of a biological population "work"? Once we believed that the answer to this question was "yes," we felt confident to conduct further studies into questions like May's hypothesis and many other nonlinear dynamical questions as described in the chapters of this book. The fact that many other dynamic phenomena were explained and predicted in subsequent projects further confirmed to us that simple models can work.

Another important general aspect of our studies involved arriving at new insights into the dynamics of biological populations. The studies in the previous chapters are not really about flour beetles. The beetles simply serve as a highly useful animal model for addressing a variety of questions and issues—ones that are important in the general study of population dynamics and ecology.

The next section contains a review of the main highlights from the flour beetle project and our investigations into nonlinear population dynamics.

DOI: 10.1201/9781003467120-21

21.1 HIGHLIGHTS

1. Model assessment. As we pointed out in the Preface, the projects described in this book were about more than just experiments dealing with the population dynamics of flour beetles, but from a broader scientific point of view were also about interfacing population data with mathematical models (deterministic and stochastic). In that regard, it is important to distinguish between parameterizing (calibrating) a model and validating a model. As J. D. Aber puts it [1], there should always be "some attempt to compare model predictions against independent data sets: data not used in any way in the derivation of the model's parameters ... there are very few aspects of ecology for which no validation data exist." If new data not used in a parameterization procedure are not readily available (or obtainable by further observation or experiments) for testing the prediction capabilities of the parameterized model, then another procedure is to use only some of the available data for parameter estimation and then study how well the resulting parameterized model fits (that is, "predicts") the remaining data.

Prediction is, of course, at the heart of the scientific method, and the ultimate goal is to obtain confidence not only in the model's ability to fit data, but also in the model's ability to predict data. To accomplish this goal, E. D. Ford, in the quote reproduced in the Preface [3], lays out three criteria for model assessment: model parameterization, model prediction, and the prediction of new, sometimes unexpected, results (validated by controlled and replicated experiments). We think the reader will find that these goals were accomplished in the many projects described in this book by using the LPA model (and its many variants) and the flour beetle laboratory system. The project involving chaotic dynamics is the highlight example, but many other dynamic phenomena were studied as well, once the basic validation and prediction capability of the LPA model was firmly established.

2. Importance of "mechanistic" models. At the core of the successes in our projects using the deterministic LPA model (and its variants) is that the model's design is based on dominant mechanisms known to drive the dynamics of the species used in the experiments. The interactions stemming from inter-stage cannibalism in flour beetles are reflected in the number and type of state variables appearing in the model and in the placement and type of the nonlinearities utilized. An advantage of mechanistic models (over phenomenological and nonparametric fitting models) is that they are, in general, lower dimensional (hence more mathematically tractable) and provide quantitative predictions when shown to be accurate.

Even though the laboratory beetle system might be considered a simple biological system (and it certainly is when compared to most natural ecosystems[1]), it is considerably more complicated than the LPA model suggests. Left out of the model are many important biological factors that affect fertility, mortality, and growth, including sex, the egg stage, chronological age, and many other genetic, physiological, and behavioral traits of the beetles. Certainly one could derive more accurate models by incorporating more such factors. However, to do so would come at the expense of higher dimensionality and an increase in parameters, costs that should be introduced only if warranted by failures of the LPA model and if the availability of data is sufficient to support (i.e., parameterize and validate) such models.

3. Stochastic versions of models are necessary. Another important aspect of the mechanistic modeling approach taken in this book is the inclusion of stochasticity. The connection of the deterministic model to population data is accomplished by modeling the expected deviations of data from model predictions in terms of probabilistic assumptions concerning environmental and/or demographic sources of noise. However, the stochastic

[1]While likely its natural habitat was under the bark of trees, the flour beetle has lived and evolved in stored grain containers used in agricultural societies for millennia. So, perhaps, the laboratory cultures in our experiments were not so far removed from its "natural" habitat as one might at first think.

versions of the LPA model are not used only to accomplish the statistical tasks needed for model parameterization and prediction validation. They also provide predictions of the beetle population dynamics and how they are influenced by both stochastic and deterministic factors. Like deterministic models, stochastic models can serve as exploratory tools for studies of temporal and state space patterns expected to be observed in data. They provide explanations of detailed patterns appearing in data that result from the intricate interplay of deterministic forces and noise, patterns that are not explained by either deterministic or stochastic forces alone. Stochastic models also provide, as do their deterministic counterparts, testable predictions that can be confronted with data.

4. Low-dimensional models can "work." One conclusion that we can draw from the studies in this book is that the deterministic LPA model provides a remarkably accurate description of flour beetle population dynamics under a variety of laboratory conditions. For example, the model made accurate *a priori* predictions of the long-term dynamics of beetle populations when various demographic parameters were changed by direct manipulation, as demonstrated in Chapters 6 and 7.

Equally impressive is the ability of the LPA model to account for population dynamics in the bifurcation experiments described in Chapters 6 and 7 when population numbers were not near an attractor, either because they were not initially near the attractor at the start of the experiment or because stochastic disturbances occurred during the experiment that moved them away from the attractor. Census data taken from these experimental treatments often showed state space patterns and temporal episodes unlike those of the predicted attractor and yet the LPA model offered the means to explain these events in terms of, for example, "saddle flybys" (Chapters 9, 12, and 14), i.e., dynamic motion predicted to occur in the vicinity of *unstable* deterministic entities. Saddle flybys have proved to be very common in flour beetle population data. The ability of the LPA model to account in this way for unusual episodes of quiescence in some data time series has provided a previously unavailable explanation for these puzzling temporal patterns.

Another particularly striking example of the explanatory power of the LPA model comes from the chaos treatment in Chapter 7 where intermittent episodes of a complicated 11-cycle pattern were observed (by plotting data in the three dimensional LPA state space) and were subsequently explained by the existence of an unstable, saddle 11-cycle that lies on the chaotic attractor (Chapter 9). Another prominent cyclic pattern in state space observed in the experimental data was a distinctive six-cycle that, although not related to a six-cycle on the chaotic attractor, is explained by the existence of a six-cycle in the lattice LPA model (Chapters 8 and 9).

The bifurcation and route-to-chaos studies in Chapters 6 through 9 are not the only success stories for the LPA model. In a variety of other circumstances the model, or a modification of the model, not only provided new explanations for patterns observed in data, but made predictions—sometimes unexpected and non-intuitive predictions—that were borne out by further experiments designed to validate them.

For example, the existence of multiple attractors accounted for certain kinds of patterns seen in data and explained differences among replicate populations in terms of stochastic "bouncing" between attractors (or, more accurately, their basins of attractions) or between different phases of cycle attractors (Chapters 13 and 14). In a study described in Chapter 14, we used a periodically-forced version of the LPA model to explain resonance dynamics historically observed in periodic habitats. Surprisingly, the model also predicted multiple attractors—not only resonant cycles but also attenuant cycles—along with associated phenomena such as basins of attraction, saddles, and stable and unstable manifolds, all of which we documented with controlled and replicated laboratory experiments. The startling feature of those experiments was not only the unexpected existence of two attractors but, from a biological point of view, the non-intuitive nature of the attenuant attractor.

We derived spatial versions of the LPA model in Chapter 20 and used them to account for observed heterogeneities in the spatial distributions of the L, P, and A stages that are observed in cultures of flour beetles. We conducted experiments to validate predicted spatial heterogeneities, as well as the model prediction of multiple attractors (and the resulting initial condition-dependent outcomes).

All of these studies illustrate and support the (often uncanny) accuracy of the LPA model as a mathematical description of the dynamics of flour beetle cultures and the model's ability to generate explanations and accurate predictions that can be documented by experimental evidence.

Of course, the laboratory system we used in these studies—while close to the "natural" habitat of *Tribolium*—is certainly far less complex than communities of interacting species found in nature. Ecological systems in nature are complicated dynamical systems, and a significant problem is to determine to what extent, if any, their dynamics can be described and predicted by low-dimensional models. Certainly there is little to no chance of doing this, if mathematical models cannot accurately describe and predict the dynamics of simpler biological systems. The flour beetle system and the studies in this book provide a cornerstone example of a successful modeling effort that both describes and predicts a wide range of complex dynamics of a real biological population.

5. Nonlinearity and bifurcations. One of the features of nonlinearity is that responses are not proportional to disturbances. Indeed, responses might be very unexpected and non-intuitive. Not only can magnitudes of quantities change in unusual ways, but so can the properties of and patterns in their temporal dynamics.

For example, the dynamic changes predicted by the bifurcation diagrams appearing in Chapters 6 and 7 are not intuitive (mathematically or biologically). That a population would change from an equilibrium dynamic to an oscillatory crash-and-boom dynamic and then return to an equilibrium dynamic is not an obvious or expected outcome of a steady increase in adult mortality. The complicated dynamics and route to chaos predicted by the bifurcation diagram in Chapter 7 is even less intuitive. Yet, our experimental populations of flour beetles indeed follow this bifurcation scenario when their demographic parameters are appropriately manipulated.

One can single out a treatment lying along this route to chaos where chaotic attractors are predicted to reside and study the replicates from that experimental treatment as examples of chaotic populations, as done in Chapters 7, 8, and 9. However, a stronger message is contained in the whole bifurcation diagram. The route-to-chaos experiment clearly demonstrates that real biological populations can display the range of dynamics—various bifurcations and complexities—predicted by simple deterministic rules embodied in low-dimensional model. This is the heart of May's hypothesis [5].

6. Confidence intervals. Parameter estimates come with confidence intervals. Model dynamics should be investigated for parameter values ranging throughout these intervals. If the dynamics of a well-validated model are robust (i.e., are qualitatively unchanged) throughout the confidence intervals, then one has a strong support for reaching the conclusion that the biological population also has these same dynamics. If, for example, the model possesses a stable equilibrium for all parameter values lying in the confidence intervals, then one concludes that the deterministic component of the population's dynamics is an equilibrium. The same can be said about other types of attractors, including chaotic attractors.

However, it is possible for bifurcations to occur within confidence intervals and as a result it is possible that several qualitatively different types of dynamics occur for parameter values ranging throughout these intervals. Indeed, when dealing with complex dynamics, such as chaos, it is likely that many bifurcations occur within confidence intervals. In such a situation it may be inappropriate to assign a specific dynamic property to the modeled biological population according to the predicted dynamic at selected point estimates for parameters. Instead, one must in some way assess the various influences of the different attractors that are distributed throughout the confidence intervals. For example, one needs to decide what it means to assert that a population is "chaotic" if, within the confidence intervals for the estimated parameters, there is a complicated array of chaotic attractors and periodic locking windows containing stable cycles. Moreover, there can be additional complications in assigning a particular attractor type to a population—such as the complications that arise, for example, from transients, stochasticity, and lattice effects.

7. Lattice effects. An interesting insight, with potentially widespread implications for the study of populations dynamics in general, resulted from a study of the chaos treatment in Chapters 8 and 9. Observed patterns in the experimental data were found to be caused by the fact that census numbers are necessarily discrete integers. Animal numbers or densities come from a finite set of discrete numbers—a demographic "lattice"—and being so constrained must ultimately display, in the absence of noise, equilibrium or periodic dynamics. It is another tribute to the LPA model that it provided, by its integerization, an accurate explanation of these originally unanticipated patterns.

In a real sense, population time series data cannot—because they lie on a lattice—display quasiperiodic or chaotic dynamics, even if they are purely deterministic. In Chapters 8 and 9 studies of some deterministic and stochastic lattice models revealed the relationship between the dynamics on a lattice and those on the underlying continuous state space.

First of all, it was found that transient lattice dynamics (i.e., the dynamics prior to becoming locked into a cycle) tend to resemble the continuous state space dynamics. In this way, the dynamics of a stochastic system on a lattice—being a system continually perturbed into transient behavior—tends to resemble the underlying continuous state space dynamics. This, then, is one way the continuous state space attractor exerts its influence on a lattice system. Note how, in this case, stochasticity is an aid—rather than a hindrance—to discovering the characteristics of the underlying dynamics.

Second, as one might expect, the finer the demographic lattice, the more the lattice dynamics will resemble the continuous state space dynamics. One way (but not the only way) in which lattices can be refined is by considering the population density in larger habitat sizes, at least for populations whose abundance scales approximately with habitat size. It follows that habitat size can be a significant factor in determining population dynamics. In particular, the potential for complex dynamics is lessened and obscured in smaller habitat sizes. Or, put another way, a study that determines a particular dynamic for a population in one habitat size might be invalid in a larger habitat size because the simplified dynamics on the coarser lattice of the smaller habitat hide the underlying tendency for complexity—a complexity that is revealed, perhaps unexpectedly, in a larger habitat. A kind of bifurcation occurs in the dynamics as a function of habitat size.

It should be noted that these lattice effects are not necessarily the result of extremely small population numbers. Nor are they manifested only on very coarse lattices. For example, the lattice in the chaos treatment in Chapters 8 and 9 contains over a million points.

Finally, as discussed in Chapter 10, for populations dominated by demographic, as opposed to environmental, stochasticity, the coefficient of variation decreases to zero with increasing population abundance. It follows, for such populations, that as habitat size increases the stationary distribution of the stochastic lattice model converges to the deterministic attractor of the underlying continuous state space dynamics. In particular, if the underlying continuous state space dynamic of a population is chaotic, the population will more clearly exhibit the properties of this chaos in larger habitats.

8. Mixes of deterministic/stochastic, continuum/lattice, and attractor/transient dynamics. A conclusion that emerges from our studies, particularly Parts II and III, is that an adequate explanation of ecological times series data is unlikely to be found by analyses that rely solely on deterministic model attractors. Instead, what one is likely to see in time series data is a mix of different patterns predicted by a deterministic model that occur—in whole or in part—during random temporal episodes. These patterns can include attractors (perhaps more than one), but also transients, unstable invariant sets and their stable manifolds, and lattice cycles (perhaps more than one). Stochasticity, besides simply "blurring out" deterministic attractors, can repeatedly cause trajectories to visit regions in state space away from attractors and thereby bring populations perpetually under the influence of deterministic dynamics other than those of an attractor.

If an adequate explanation of observed dynamic patterns found in time series data from a single population under highly controlled conditions in which stochasticity is minimized—as in our laboratory studies—cannot be obtained by means of deterministic model attractors alone, then certainly this fact is even more likely to be true for ecological data obtained from field observations or experiments. It is widely recognized that insufficient length of time series data is a major problem in the study of chaos in ecological data [7]. In addition, if ecological data are typically a blend of deterministic properties and stochasticity as described above, then one has added difficulties. The problem is not only the length of the data set, but the length of time spent on or near the attractor (or any other deterministic entity of interest).

9. Chaotic dynamics in ecological systems. The issues described above apply particularly to the question of identifying chaotic dynamics in ecological time series data; namely, the data may be "contaminated" not only by noise but by other deterministic dynamics (transient and even unstable). Moreover, a converse holds. Non-chaotic dynamics may be mistakenly identified as chaotic for the same reason, i.e., that stochasticity allows visits to regions of state space away—even far away—from attractors. A time series of data might often visit regions in state space where trajectories are separating, for example near an unstable equilibrium or cycle, producing a kind of sensitivity to initial conditions (as measured, for example, by the stochastic Lyapunov exponent SLE) that is not due to deterministic chaos. This phenomenon occurs, for example, in the route-to-chaos experiment reported in Chapter 7. For the treatment with $c_{pa} = 0.50$ the deterministic LPA model possesses a stable cycle (and no chaotic attractor) throughout the confidence intervals of the estimated parameters, even though the SLE is positive throughout those intervals. In these cases, the stochastic Lyapunov exponent is uninformative and misleading with regard to the detection of chaos, especially when one considers that a noisy stable equilibrium or cycle is a leading alternative hypothesis to chaotic dynamics. Moreover, it is often forgotten that a focus on Lyapunov exponents overlooks the fact that sensitivity to initial conditions, which is what Lyapunov exponents measure, is only a necessary and is not a sufficient condition for deterministic chaos.

Efforts to find unequivocal evidence of chaos in ecosystems generally have focused on the analysis of available data sets. While different approaches have been taken in order to undertake this analysis, we have seen that there are many difficulties associated with this task. Another approach to the question of chaos in ecology (indeed, complex nonlinear dynamics in general) is to study the response of a system to perturbations. Depending on the nature of the perturbations, an ecosystem in or near a region of complex nonlinear dynamics such as chaos is likely to respond quite differently from one in or near a robust state of stable equilibrium (or period cycling)—and to do so often in unusual, surprising and unintuitive ways, even to small perturbations. (See for example, Chapter 11.) In many ways, this is perhaps a more interesting and important question to ask anyway, rather than whether or not a specific system is, or is not, chaotic in its present state and configuration.

Therefore, in our studies we did not focus on problems associated with the detection of chaos in available data sets. Nor have we addressed the often debated and important questions concerning the role of chaos in "natural" populations, such as the frequency or rarity of its occurrence or whether it is selected for or against. Instead our studies relate to different questions: What roles do complex dynamics, such as chaos, play in the response of ecosystems to disturbances? Given that population interactions are nonlinear, what are the possible consequences of changes in environmental and demographic parameters? What unexpected and unintended consequences due to nonlinearity might occur? Given the continual disturbances that affect the physical and biological environments of many, if not most, biological species (especially when the activities of humans are considered), these are important questions to ask about natural ecosystems. Experiments using laboratory systems are a natural way to begin a rigorous study of these questions. However, rigorous study demands a close connection between the biological system and the mathematics of nonlinear theory, that is, between data and models.

For an in-depth look at the problem of detecting chaos in ecological data see [6].

10. Interacting species. Having a well-validated model for a single species, as documented in Parts I–III, we have an opportunity to investigate multi-species interactions of, in this case, beetles from the genus *Tribolium*. As discussed in Chapter 15, flour beetle experiments played a central role last century in the formulation of the classic competitive exclusion principle. A coupled system of two LPA models for two species of flour beetles is based on the dominant interference mechanisms of egg and pupae cannibalism within and between species. Exploring available historical data using the resulting competitive LPA model (the cLPA model), we found that not only could anomalous, inconclusive, and indeterminate results from the historic investigations of Thomas Park and collaborators be explained, but that new phenomena in competition theory arise. Classic competition theory, based as it is on Lotka-Volterra dynamics, is a theory focused entirely on equilibrium dynamics. As seen in Chapter 15, the cLPA model, on the other had, has a rich abundance of attractors and, as a result, there are more competitive scenarios than the classic four equilibrium scenarios of Lotka-Volterra theory. Within the range of model parameters that we found for the LPA model and the experimental cultures in our laboratory, there can be coexistence periodic cycles, aperiodic oscillations arising from invariant loops in state space, and chaotic attractors. This contrast to Lotka-Volterra theory can be attributed to the fact that Lotka-Volterra theory lacks nuance with regard to the mechanisms involved in the inter-species interactions. Notably, we found that there are robust instances (within parameter estimates for the flour beetle species used) in which the classic competitive exclusion principle is violated. Specifically, we found that *increased* intensities of inter-species competition can result in coexistence (not competitive exclusion, as the classic competitive principle would have it).

11. Ecological dynamics in the frequency domain. The vast majority of ecological analyses of population data are conducted in the "time domain," that is, population numbers are examined graphically and/or statistically as a function of time. This approach was used in Parts II–IV. In Chapters 16 and 17, we conducted analyses of flour beetle time series in the "frequency domain." This involved the use of spectral analysis applied to stationary population time series that fluctuate around a stable attractor. We showed that these methods are useful tools that can reveal insights not apparent using traditional techniques.

In Chapter 16, we applied spectral analyses to data from beetle populations fluctuating around nonlinear attractors. In the case of a deterministic cycle, power spectra revealed the cycle periodicity. For populations influenced by deterministic chaotic attractor, spectral analyses detected the relative influence of lattice three-cycles and six-cycles, as well as episodic two-cycles. These spectral results complement and extend the conclusions reached using the lag metric comparison (LMC) in Chapter 9. For populations fluctuating around a stable equilibrium, we used the LPA model to make experimentally testable predictions in the frequency domain. We used linear filter theory to derive predicted power spectra from the stochastic LPA model, and, using parameter estimates, we compared them directly to spectral analyses of the data. We were able to infer, for example, the time scales of the stochastic fluctuations of beetle numbers for the different life stages. Linear filter theory applied to the LPA model also allowed us to predict the effects of "colored noise" on the dynamics of beetle numbers and evaluate those predictions with a designed population experiment.

In Chapter 17, we extended the LPA model to metapopulations—populations that are separated geographically but are linked by environmental fluctuations and dispersal. We used spectral methods to examine ecological synchrony in the frequency domain. We applied linear filter theory to the metapopulation LPA model and obtained predictions for how dispersal and the color of environmental noise affect the time scales of population correlations. We conducted a *Tribolium* experiment to test these predictions. Spectral analyses of the census data supported the theory. Our results show that changes in the time scale of the environmental covariance can induce changes in both the total synchrony and the time scale structure of synchrony and that population stability and dispersal affect how these changes occur.

12. Natural selection and population dynamics. Changes in the genetic structure of populations can affect their dynamics. In Chapter 18, we presented a genetic version of the LPA model. Genotypes for a single genetic locus differ with respect to their life history parameters. These differences cause changes in allele frequency which alter the nonlinear dynamics of the populations. We used the gLPA model to predict how natural selection can result in "evolutionary rescue" from population extinction and compared these predictions to experimental data. We also reframed the classic concept from population genetics of the "cost of natural selection" using an ecological genetic approach, derived predictions from the gLPA model, and tested those predictions using experimental data. This provides another demonstration of how the LPA model and the *Tribolium* laboratory system can be used to investigate rigorously concepts in population biology.

21.2 CONCLUDING REMARKS

We have seen in this book that it is possible to describe and, importantly, to predict a large variety of dynamic patterns observed in a biological population by means of a low-dimensional mathematical model. This was done for the most part (but not exclusively) by documenting the consequences that environmental disturbances and demographic manipulations have on the dynamics of flour beetle populations in carefully designed, controlled, and replicated experiments. As P. Kareiva writes [4], "... simply documenting direct demographic effects of environmental change or environmental stress has little predictive power. Only with a solid understanding of underlying dynamics can ecologists ever hope to anticipate the consequences of environmental perturbations." Our studies described in this book show how, as Kareiva continues to say, "the marriage of nonlinear models and experiments can help to accomplish this task."

Furthermore, as a result of our modeling and experiment efforts, we have seen that a real biological population can indeed follow a so-called model-predicted route to chaos, the possibility of which had so intrigued May in his seminal papers. While documenting this fact, we gained new insights into the complex dynamic patterns that can arise from the mix of nonlinear, deterministic, and stochastic elements inevitably found in ecological systems [2]. However, despite the many complications that arise from this mix of elements, one should not lose sight of the fact that it was a simple, low-dimensional model whose predictions were fulfilled, and upon whose foundation were based accurate explanations for complex dynamic patterns observed in the data. The *Tribolium* project provides an example that indeed corroborates May's hypothesis that complex dynamics in ecological data can be the result of simple rules.

We hope the studies in this book not only provide insights into nonlinear population dynamics, but also provide a modest step toward the "hardening" of ecological science—a step toward raising its explanatory and predictive power beyond purely theoretical speculation and a satisfaction with qualitative accuracy, reasonable "guesstimates," and verbal metaphors. It is true that the laboratory system used in our studies is a relatively simple biological system, and that the low-dimensional LPA model is a simple mathematical model of even that system. Nonetheless, we can find motivation and inspiration for the study of such systems from another quote taken from May's seminal 1976 paper [5] :

> Not only in research, but also in the everyday world of politics and economics, we would all be better off if more people realized that simple nonlinear systems do not necessarily possess simple dynamic properties.

Bibliography

[1] Aber JD (1997) Why don't we believe the models? Bulletin of the Ecological Society of America 78:232-233. http://www.jstor.org/stable/20168170

[2] Bjrnstad ON, Grenfell BT (2001) Noisy clockwork: Time series analysis of population fluctuations in animals. Science 293:638–643. https://doi.org/10.1126/science.1062226

[3] Ford ED (2000) Scientific Methods for Ecological Research. Cambridge, Cambridge University Press. pp. 390-391.

[4] Kareiva P (1995) Predicting and producing chaos. Nature 375:189–190. https://doi.org/10.1038/375189a0.

[5] May RM (1976) Simple mathematical models with very complicated dynamics. Nature 261:459–467. https://doi.org/10.1038/261459a0

[6] Munch SB, Rogers TL, Johnson BJ, Bhat U, Tsai C-H (2022) Rethinking the prevalence and relevance of chaos in ecology. Annual Review of Ecology, Evolution, and Systematics 53:227–249. https://doi.org/10.1146/annurev-ecolsys-111320-052920

[7] Perry JN, Smith RH, Woiwod IP, Morse DR (2000) Chaos in Real Data: The Analysis of Nonlinear Dynamics from Short Ecological Time Series. Population and Community Biology Series. Volume 27. Dordrecht, Kluwer Academic Publishers.

Index

For Product Safety Concerns and Information please contact our EU
representative GPSR@taylorandfrancis.com
Taylor & Francis Verlag GmbH, Kaufingerstraße 24, 80331 München, Germany

www.ingramcontent.com/pod-product-compliance
Ingram Content Group UK Ltd.
Pitfield, Milton Keynes, MK11 3LW, UK
UKHW052051230926
475702UK00025B/1096
9781032736594